Annals of Scientific Society for Assembly, Handling and Industrial Robotics 2023

Steffen Ihlenfeldt · Thorsten Schüppstuhl ·
Kirsten Tracht
Editors

Annals of Scientific Society for Assembly, Handling and Industrial Robotics 2023

 Springer

Editors
Steffen Ihlenfeldt
Fraunhofer Institute for Machine Tools
and Forming Technology (IWU)
Chemnitz, Sachsen, Germany

Thorsten Schüppstuhl
Institute of Aircraft Production Technology
(IFPT)
Hamburg University of Technology
Hamburg, Germany

Kirsten Tracht
Bremen Institute for Mechanical Engineering
(BIME)
University of Bremen
Bremen, Germany

ISBN 978-3-031-74009-1   ISBN 978-3-031-74010-7   (eBook)
https://doi.org/10.1007/978-3-031-74010-7

Wissenschaftliche Gesellschaft für Montage, Handhabung und Industrierobotik (MHI e. V.) Geschäftsstelle Patrick
Rückert

This Springer imprint is published by the registered company Springer Nature Switzerland AG
The registered company address is: Gewerbestrasse 11, 6330 Cham, Switzerland

If disposing of this product, please recycle the paper.

Contents

Latest Challenges in the Development of Scalable Assembly Systems for Fuel Cell Stacks

Fabian Klaus⬤, Lennard Margies⬤, and Rainer Müller

Abstract

The assembly of fuel cell stacks presents some unresolved engineering challenges. The aim of this work is to provide a comprehensive solution concept that addresses the identified issues. A product analysis was conducted, from which the assembly process chain could be derived. Subsequently, the problems encountered during assembly were structured and presented. Finally, a solution concept was developed that addresses both the overarching problem and the process-specific difficulties. The analysis shows the necessity of a cost degression of the stacks through full automation of the assembly systems. However, this represents a significant economic risk, which is why a scalable assembly in terms of automation level is proposed. The problem of process complexity and a generally low level of product knowledge is addressed by the introduction of cognitive and physical assistance systems in the low automation stages. To increase the speed of knowledge building in terms of product, process and equipment, algorithms for self-optimizing processes or for outputting action recommendations are used. The control complexity arises from high speed and accuracy requirements of the stacking process, which is addressed in a novel approach using continuous drive technology.

Keywords

Assembly automation • Fuel cell • Assistance systems • Process optimization • Scalability • Cam mechanisms

F. Klaus (✉) · R. Müller
Chair of Assembly Systems, Saarland University, 66121 Saarbrücken, Germany
e-mail: f.klaus@zema.de

L. Margies
Center for Mechatronics and Automation gGmbH (ZeMA), 66121 Saarbrücken, Germany

S. Ihlenfeldt et al. (eds.), *Annals of Scientific Society for Assembly, Handling and Industrial Robotics 2023*, https://doi.org/10.1007/978-3-031-74010-7_1

1   Introduction

Notably, due to an increasing societal demand for emission reduction and support from government and economy, the fuel cell is gaining ground as a substitute for conventional energy conversion technologies, especially in the transport and energy sector. Now it is essential to align and adjust production capacities from the automotive, equipment and supplier industry to the new requirements in order to achieve the climate targets set by the federal government and remain economically competitive. For this purpose, the rapid development of competencies, processes and technologies in the field of fuel cell production systems is crucial to achieve a global competitive advantage [1].

With the planning, use and urgently required scaling of the respective production systems, however, come certain challenges for which engineering solutions must be found as soon as possible. This contribution therefore aims to provide an overview of both generic and process-specific issues and then to offer a holistic solution concept that fully addresses them. The approach will be elaborated within the research project "H2SkaProMo" until 2024 and integrated into a prototypical assembly demonstrator in three automation degrees "manual", "semi-automated" and "fully automated". The focus is on the assembly of the fuel cell stacks downstream of manufacturing, since control of this process is generally decisive for the economic viability of production [2].

Most of the issues identified within this contribution and "H2SkaProMo" in general can be attributed to the assumed lack of familiarity of process participants with the product as well as the complexity of the fuel cell and its production systems.

2   State of the Art

The complexity and heterogeneity of the necessary assembly scopes will be elaborated below using an exemplary process chain for the assembly of fuel cell stacks. The design of the processes is dependent on the product, which is why the fuel cell type must be determined initially. Despite the abundance of fuel cell technologies currently available on the market, a concentration on the polymer electrolyte membrane fuel cell (PEMFC) for mobile and stationary applications as well as the solid oxide fuel cell (SOFC) mainly for stationary applications (e.g. building and charging technology) can be observed [3]. Mobile drive technology based on fossil fuels continues to be a global mass market, which is to be gradually supplemented or substituted by alternative technologies in order to meet the environmental goals of many countries. This brings PEMFCs and their production into the focus of research and industry. For this reason, the joint project "H2SkaProMo" deals with a variant of the PEMFC stack and its assembly. The knowledge gained can also partially be transferred to other technologies.

The exemplary stack assembly can be seen from its schematic drawing in Fig. 1.

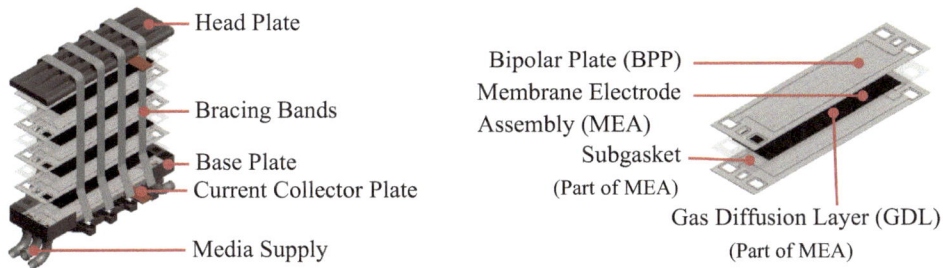

Fig. 1 Structure of a PEM fuel cell stack (simplified)

The left half of the figure shows the complete stack. The abstraction level of the component description has been chosen in such a way that it is possible to speak of a relatively generic structure. In the considered case, it consists of the sub-assemblies head section, the fuel cell stack itself, foot section, current collectors, media supply and bracing mechanism, which in this case is realized by bracing bands. The fuel cell assembly itself consists of 250 stacked individual bipolar plates and MEAs.

Process-wise, the repetitive, highly precise stacking of these components is one of the most complex and time-consuming assembly tasks. The remaining presented sub-assemblies can be scheduled as pre-assembly processes and must be partly carried out before and partly after stacking.

With the help of the bill of materials associated with the stack, a precedence graph can be derived from which a general assembly sequence for the stack can be constructed. This general assembly sequence will be introduced below.

The assembly process begins with the **pre-assembly of the head and foot sections**. In addition to joining operations such as the insertion and bolting of spring assemblies, this also involves measuring and leveling operations, which require a high level of manual effort. This is followed by high-precision **stacking of the bipolar plates and MEAs** 200–300 times on the head section. After completion, the base plate is placed on top and the stack is **compressed** under high forces. Under the applied pressing force, the stack must then be braced with the aid of the bracing bands. This is followed by various tests, including a **leak test** in the form of a flow and pressure drop test and a load test. Finally, the stack is **labeled** and **packaged**.

Once the assembly sequence has been set up, the more detailed planning of the necessary assembly systems can usually be initiated.

3   Challenges in the Assembly of PEM Fuel Cell Stacks

The challenges in planning the assembly of fuel cell stacks need to be mastered quickly and efficiently to achieve rapid and widespread placement of hydrogen systems on the market. Due to the variety of the issues involved, a subdivision of the problems to be solved is considered useful. In the following, these problems named by the expert consortium of "H2SkaProMo" are thus split into overarching and process-specific topics.

The overarching challenge is cost reduction. The cost per unit of fuel cells and other hydrogen technologies are currently still considered very high compared to conventional technology alternatives. This can be attributed to material costs on the one hand and production costs on the other. Even with prospectively decreasing material costs, the costs for production technology and processes remain. A strong lever for reducing the process costs incurred in production is the use of economies of scale, thus achieving low unit costs through a high throughput. A large part of the process costs is accounted for by assembly, so a full automation of the assembly would help to maximize the number of units and minimize unit costs. However, a closer look reveals further challenges:

1. According to the current state of the art, it can be assumed that most developments in the field of PEM fuel cells are not yet at their final stage. Instead, work is being done mostly with small series developments, which in the future will still undergo significant product adjustments [3]. Rigidly automated systems, by their very nature lacking adaptability, create a limitation in the possibilities of adjustment. Adaptability only exists within the initially planned system flexibility.
2. Market entry for less financially strong companies is hindered when undergoing a portfolio restructuring or expansion, as they would be exposed to a considerable economic risk in acquiring a fully automated fuel cell assembly line.

The process-specific problems are illustrated by the assembly sequence as follows.

- Pre-Assembly: Within the stack considered in H2SkaProMo there are a large number of components which must be assembled for the completion of the head and foot sub-assemblies. Due to their complexity, most of the assembly steps have to be carried out manually. This complexity combined with manual efforts is particularly problematic as PEM fuel cells are often relevant as a substitute or complement to established technologies (e.g. automotive drive technology), but the necessary product and process knowledge does not yet exist among the employees and must therefore be learned step by step. A search for solutions to reduce complexity and qualify employees is thus inevitable [4]. If also performance impaired employees are involved in assembly, the described problem is exacerbated.

- Stacking: When stacking the cell assemblies of BPP and MEA, the main problem arises from the positioning accuracy requirements in the range of tenths to a few hundredths of a millimeter, which are necessary to ensure tightness and functionality, in conjunction with the required stacking frequency. In the corporate landscape, current publications speak of targeted stacking frequencies of 0.5–2 Hz [5]. Additional constraints in the stacking process are caused by the sensitivity of the components. While contamination of the active field or the seals of the BPP can cause functional impairments that must be avoided, a controlled environment in terms of temperature and humidity is necessary for the MEA to avoid significant component changes and inutility [6]. Moreover, the MEA is a flexible component, which complicates the automated handling. In summary, the highest requirements apply to environmental conditions, accuracy, component handling and speed.
- Compressing: Compressing the stack poses special requirements for process control and is a major guarantee for the performance and quality of the end product [7]. The joined individual cells, for example, may have different curing dates of the sealant, which must be considered when defining the pressing force or pressing path, in order to avoid damaging the seals.
- Bracing: The stacks should be braced with bracing straps, which are joined to the head and base plate in a stress-free state while in the pressed condition. Upon subsequent release of the press, the applied pressing force is transferred to the bracing bands; this maintains the tension. A potential issue in this regard may be the parallel automatic handling and joining of multiple bracing bands on both sides of the stack.
- Testing: The primary issue with the necessary testing cycle, consisting of leak and load testing, is its temporal scope. The load testing and conditioning of a completed stack can take several hours, while the leak testing can be estimated to take 10-15 min. The preceding tests of insulation resistance and open-circuit voltage only take a few minutes. Should the test result of one of the tests fail to meet the required standards, a very time-consuming disassembly of the entire stack, in which each cell must be tested individually, must be conducted. Even a failure rate in the parts per thousand range, which leads to a total failure of a single cell, can therefore result in immense personnel and financial costs. After testing, the stack assembly is finished.

An overview of the challenges presented is shown in Fig. 2.

4 Definition and Adaptation of the Solution Space to the Considered Assembly System

To be able to establish a successful assembly concept, a holistic solution space is required, which addresses the problem statements adequately. Within the scope of the BMWi project "H2SkaProMo—Scalable Cyber-Physical Systems for the Assembly of Fuel Cell Stacks",

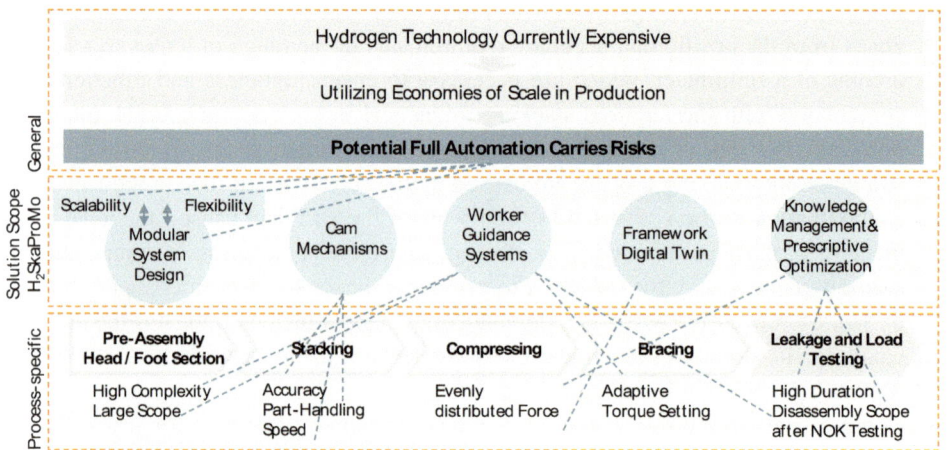

Fig. 2 Challenges in the assembly of fuel cells

this solution space is created by 15 actors within an interdisciplinary consortium of users, equipment suppliers and scientists. The current status and adaptation of the solutions to the processes considered in the project be seen in Fig. 3.

The figure shows the totality of overarching or process-specific solutions for the challenges presented colored yellow. The central element is the assembly sequence which can be scaled in the three stages manual, semi and fully automated due to the modular structure. The modularity also allows for a certain degree of flexibility, which also has a cross-process effect. The digital twin is represented as all-encompassing and includes all further solutions. The dual model for parameter optimization allows the adaption of

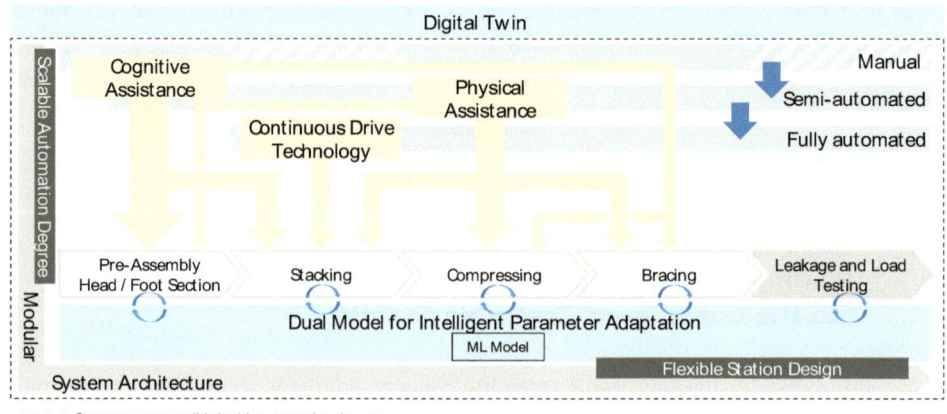

Fig. 3 Solution space adaption to H2SkaProMo fuel cell stack assembly

ongoing assembly processes based on knowledge acquired by all process participants and the findings of past assembly runs. Thus, the predictive and prescriptive model is fed by process data and in return feeds the process parameters with the findings. The process-specific solutions presented are integrated to varying degrees depending on the process, as visualized by the differently sized arrows in the figure.

The distribution of the individual building blocks of the solution construct on the assembly sequence will be discussed in the following.

4.1 Digital Twin

As an overarching and functional assurance framework, the digital twin serves as a repository for all necessary data on product, process, and equipment and thus enables the execution of services necessary for data collection and utilization through its connection to a graph-based database. All development approaches must therefore be compatible with the digital twin. The digital twin is currently in its test phase, in which the basic graph model is connected with a graphical user interface and fed with test data.

4.2 Modular System Design

As a further process-encompassing framework, the modular system structure is shown in Fig. 3, which enables the development of assembly systems to have a flexible station design and the ability to scale their automation level. In the event of future product adjustments, which are expected at the current stage of development, it is possible to continue using an existing assembly system for new product variations. This is possible through the exchange of individual process, feeding or transport modules in combination with a versatile workpiece carrier. Also, a significant increase in demand for H_2 technologies is expected, provided that they are cost-equivalent or cheaper than conventional technologies [8]. A then required cost-effective partial automation can also be represented by the simple exchange of modules of higher automation levels.

4.3 Cognitive and Physical Worker Assistance

One major challenge in stack assembly is providing adequate assistance for employees of varying abilities who are faced with the new product. This will be addressed with cognitive and physical assistance systems in the manual and semi-automated stages.

An assistance system generally is an arrangement of components with the aim of providing support for a specific task, controlled by an information technology unit that is sufficiently complex to enable information processing or appropriate adjustment of system behavior [9]. Further differentiated, perception and decision assistance systems are cognitive support and execution assistance systems are physical support [10].

While determining the need for physical assistance is based on ergonomic guidelines, determining cognitive assistance needs is significantly more complex. For this reason, a concept will be developed to enable an estimation of the need for each process, resulting in the technology-neutral selection of the required capability categories.

For this purpose, the assembly process is initially divided into sufficiently small pieces using MTM analysis. Subsequently, for each sub-process step, the impact of human error, cognitive load, likelihood of human error, and the need for assistance are examined and quantified. After that, the necessary system capabilities are selected. Here, a distinction is made between information input, output, feedback, retrieval, transmission and perception. With this concept, the first relevant processes have already been identified. In particular, an increased need for support was found in pre-assembly, which should be addressed by capabilities of information perception and output. For example, work area monitoring should be used to monitor handling operations and display subsequent tasks. In addition, employees should be warned in the event of mistakes. The specific technology solutions to be used are still being evaluated.

Ultimately, the "accompaniment" of the employees in cognitively demanding processes is expected to solve the problem presented and thus significantly shorten the learning time, reduce the error rate, and increase the motivation of the employees.

4.4   Continuous Drive Technology

In current developments in stack assembly, automated solutions utilizing robots are commonly employed to perform the complex task of precise stacking at maximum speed. The robot technology used varies from articulated robots, to SCARA, to delta robots (such as current developments in the research project "H2FastCell"). However, these technologies cannot be scaled to arbitrary speeds with sufficiently accurate repeatability, which can pose a significant problem in the case of 500-layer repetitive stacking (250-cell), particularly in the context of mass production. With an expected increase in the demand for stacking frequencies exceeding 1 Hz, robot technology will reach its limits. This is due to the mechanical construction which, although it typically provides high flexibility in path planning through the use of multiple axes to span a large handling range, still relies on intermittent drives for pick-and-place operations. The braking and acceleration processes necessary for gripping and releasing are thus the bottleneck for achieving high stacking frequencies.

A promising solution for this is the use of a continuously driven system that operates in a mechanically constrained mode. A cam drive in the necessary dimensions for stack assembly, in conjunction with a lightweight gripper at the output, ensures a strictly constant motion with achievable repeat accuracies of 20 μm. The maximum stacking frequency is dependent on various factors such as vibration, mass inertia of the components, the adhesion between gripper and component, and upstream issues such as the realization of a correspondingly fast-acting feed and alignment of the components.

Apart from the advantages in terms of accuracy and speed, there are also disadvantages, particularly due to the mechanical lockstep. In contrast to the flexibility in gripping points that can be achieved with a robot, this is hardly or not at all present with a cam drive. Since the components still need to be aligned with high precision, an alignment station must be added. An unregulated feed of the components, for example through mechanical stops, cannot ensure fulfillment of the accuracy requirements, especially due to the component tolerances in conjunction with the handling speeds in automated assembly. The component alignment must provide translational and rotational adjustment of BPP and MEA in the micrometer range. Using optical detection, the exact component position is captured and adjusted to the desired setpoint, allowing cam drive and gripper to precisely pick them up and place them on the stack.

The custom cam mechanism in conjunction with an optical inspection system and an intelligent stacking algorithm represents an innovation in fuel cell assembly that could be a game changer in terms of series production due to the achievable speeds.

4.5   Dual Model for Intelligent Adaptation of Assembly Process Parameters

As previously stated, the fuel cell is a complex product, and its mass production requires both extensive product knowledge and highly optimized processes. The building of knowledge is time-consuming and expensive when there is a high product complexity and a lack of product experience. Also, companies are often under constant competitive pressure during this learning process, and a rapid market launch is desired. Still, initial experiences can be gained in a prototype phase. These experiences are gained by process planners and involved blue-collar workers, which allows for appropriate measures to be taken when process errors occur, in order to avoid them in the future. This type of feedback loop is dependent on human action and therefore slow and unpredictable.

To address this, a hybrid software system is developed, which allows the collection of expert knowledge from process participants and then uses and continuously expands this knowledge for process optimization. This way, the assembly system can be enabled to become a self-regulating process expert and the market launch can be accelerated.

In order to achieve the goal, the presence and quality of existing knowledge are crucial. Therefore, it must be ensured that the knowledge input possibilities for blue-collar workers, planners, and experts is not only designed as simple as possible, but is also encouraged. If necessary, incentives should be created to support the acquisition. In order to ensure simplicity for the parties concerned, it should be possible to create information primarily on the basis of simple logics (IF–THEN) with as few textual input as possible. To ensure an easy assignment of information, the system structure is based on the assembly sequence and the related product, process, and equipment.

Based on the logics entered, a process model is developed that always reflects the most up-to-date knowledge of the process participants. The assembly is now repeated several times with the established parameters. At this point, the self-optimization as the second component of the system comes into play. If processes such as leak testing evaluate the product as "not OK", the system will vary individual parameters and gather information about cause-and-effect relationships, which are then incorporated into the feedback loop. In manual assembly, recommendations for changing the assembly parameters will be given to the process participants and the effectiveness of the changes will be monitored (Prescriptive Quality). In addition, if a sufficient amount of data is available, a statement can be made about the probability of an "OK" testing outcome (Predictive Quality). Futhermore, an increased probability of failure can be detected during assembly, for example in stacking, and a rejection for rework can be initiated.

5   Discussion

The solution concept presented here addresses very diverse challenges with an interlocking modular model. Each of these modules requires in-depth consideration and thus a high level of time and specific expertise from individual disciplines. In this case, all topics are handled within "H2SkaProMo", thus the modular clustering is still meaningful and leads to a comprehensive solution in the form of assembly system prototypes.

Due to the lack of familiarity with the product, further adjustments to the requirements and the concept are expected during the development as expertise increases. A comprehensive generality or transferability to similar products cannot be assumed.

Solutions such as cognitive assistance systems and the parameter optimization model have the greatest impact in the early stage of the product and assembly system life cycle. At this stage, the product, process and operating resources as well as their optimized parameters are often largely unknown. As knowledge penetration occurs and mass production is established, the achievable benefit of these systems slightly decreases. The isolated concept for acquiring and exploiting knowledge for accelerated optimization of the assembly and final product can be considered transferable to new products.

Regarding the scalability of automation, a strict increase in the automation degree of all modules in all three stages of the assembly system is not considered sensible. For example, semi- or fully automated pre-assembly is not economically sensible for the current product stage, as human dexterity, elaborate measuring and testing are required. In the medium term, an adaptation of the product is considered advantageous compared to its automation in the current state. Furthermore, manual pre-assembly and automated stacking both have approximately the same cycle time. Therefore, there is currently no immediate need for automation in terms of time. Another example are transport modules such as electrically driven belt conveyors, which could also be necessary in manual assembly due to the high component weights and only represent a marginal additional financial burden compared to manual conveyor belts. The examples show that the feasibility of each module in different automation stages must be carefully examined.

The modular system structure implies that entire stations are highly flexible in their physical arrangement. For pre-assembly, testing, rework and manual stacking, a flexible relocation is feasible with little effort, in contrast to automated stacking. The cam drive combined with the handling masses under high acceleration as well as the correspondingly large geometries and travel paths require a solid basic structure that can withstand the occurrence of high inertias and effectively dissipate vibrations. At the same time, maximum precision is required for free stacking without stops. The solid stacking station thus must be firmly anchored in the hall floor and cannot change its position without high effort, also due to the interfaces to component alignment.

6   Summary and Conclusion

Within the present work, current engineering problems in fuel cell assembly were illuminated. As was shown, these can be summarized mainly under the headings of production costs and product complexity. Based on the identified general and process-specific problems, a holistic solution model was developed that addresses all sub-problems in a modular way. This allows for the lack of knowledge about novel products to be addressed using a dual model for intelligent adaptation of assembly parameters, supported by cognitive assistance systems and a digital twin as a data framework. The market penetration of assembly systems with low economic risk is supported by scaling the level of automation and a modular design of the overall system. The main process-specific problem of the need for high precision and extremely fast stacking of components can be solved using continuous drive technology in situations where robots no longer produce sufficient results. The individual solutions presented are highly specific to the challenges of fuel cell assembly, but the majority of the model is also adaptable to completely different technologies and can therefore be considered generic.

The future research needs are derived from the solution building blocks presented. These as well as the validation of the overall concept will be addressed in H2SkaProMo in collaboration between industry and research. In summary, it is expected that the obstacles currently hindering market penetration of fuel cells from a production perspective can be eliminated in the near future.

Acknowledgements This work is supported by the German Federal Ministry for Economic Affairs and Climate Action (BMWK) with funds provided through the project "H2SkaProMo" (03EN5014I) on the basis of a decision by the German Bundestag.

References

1. BMWi: Die Nationale Wasserstoffstrategie. Die Bundesregierung, Berlin (2020)
2. Mueller, R.: Montagegerecht. In: Bender, B., Gericke, K. (eds.) Pahl/Beitz Konstruktionslehre. 9th edn. Springer, Bochum (2021)
3. Fuehren, D., Graw, M., Kroell, L.: Wertschöpfungskette Brennstoffzelle – Metastudie. N. GmbH, Berlin (2022)
4. Nationale Plattform Zukunft der Mobilität, AG 4 "Sicherung des Mobilitäts- und Produktionsstandortes, Batteriezellproduktion, Rohstoffe und Recycling, Bildung und Qualifizierung": AG 4 – Zwischenbericht – Positionspapier Brennstoffzelle (2021)
5. ThyssenKrupp. https://www.thyssenkrupp-system-engineering.com/de/automobilindustrie/e-motor-montage/brennstoffzelle. Accessed 26 Jan. 2023
6. Wienk-Borgert, B.: Einfluss der klimatischen Fertigungsumgebung auf die Mechanik und Rissstrukturierung der elektrodenbeschichteten Membran einer PEM-Brennstoffzelle. F. J. GmbH, Jülich (2019)
7. Kampker, A., Ayvaz, P., Schön, C., Reims, P., Krieger, G.: Produktion von Brennstoffzellen-Systemen. Aachen (2020)
8. AVL List GmbH, ZSW: Systemvergleich zwischen Wasserstoffverbrennungsmotor und Brennstoffzelle im schweren Nutzfahrzeug. Stuttgart (2021)
9. Beetz, S.: Beitrag zur Methode der Arbeitsplatz-integrierten Assistenz am Beispiel Formmesstechnik. Shaker-Verlag GmbH, Erlangen-Nürnberg (2006)
10. Bengler, K., Lock, C., Teubner, S., Reinhart, G.: Der Mensch in der Produktion von Morgen – Grundlegende Konzepte und Modelle. Carl Hanser Verlag, München (2017)

Experimental Investigation of the Influence of Different Nozzle Exit Geometries on the Depositing of Strands in Fused Layer Modeling

Michael Krampe⬧ and Bernd Kuhlenkötter⬧

Abstract

The aim of this paper is to investigate the influence of different exit geometries of the nozzle in the FLM process. In order to determine the effects of nozzle exit shapes, an experimental investigation was carried out. Individual strands were deposited with different nozzle exit shapes, whose shape was subsequently determined by laser scanning and microscopic images of the strand cross-sections. In addition, the orientation in relation to the deposition direction and five different layer heights were investigated. Circular, triangular, square, and hexagonal nozzle exit shapes with cross-sectional areas equivalent to a circle of 1, 0.8 and 0.6 mm diameter were prepared for these tests. Strands were printed with these different geometries and then scanned with a laser scanner and examined in cross-section with an optical microscope. Afterwards, the strand shapes were compared with each other using the created profiles and strand cross-sections. The results show that different geometries of the nozzle exit lead to different shapes of strand cross sections. It can be observed that the deviation from the cross section of the circular nozzle decreases with an increase in the number of corners of the nozzle geometry. Furthermore, the extent of the deviation increases with increasing layer height. In addition to the outlet shape, a different orientation of the same nozzle leads to a deviating strand cross-section. Especially with the triangular nozzle exit geometry, this made a big difference in the shape of the deposited strand.

M. Krampe (✉) · B. Kuhlenkötter
Chair of Production Systems, Ruhr-Universität Bochum, Universitätsstraße 150, 44801 Bochum, Germany
e-mail: krampe@lps.ruhr-uni-bochum.de

S. Ihlenfeldt et al. (eds.), *Annals of Scientific Society for Assembly, Handling and Industrial Robotics 2023*, https://doi.org/10.1007/978-3-031-74010-7_2

Keywords

Additive manufacturing • Fused layer modeling • Nozzle exit geometry • Strand deposit

1   Introduction

At present, additive manufacturing is becoming increasingly relevant in the industry, hightech technologies and even environmental sustainability [1, 2]. Probably the most widespread additive manufacturing process is fused layer modeling (FLM), which is particularly suitable to produce prototypes and small series, as the financial entry hurdle is low. In the FLM process, a wire-shaped thermoplastic is conveyed via an extruder into a melting system, where it is heated to a pasty state and then extruded through a nozzle. Via the nozzle, the extruded material is applied in defined lanes to a building platform or the previously created layer, where it melts the already deposited strands of material and forms a permanent bond after cooling [3]. In this process, the nozzles of an FLM system have a circular outlet opening of constant size as standard. If the desired part requires a high level of detail, a nozzle with a small diameter, for example between 0.1 and 0.4 mm, can be used at the expense of production speed. The use of nozzles with a larger diameter, up to 2 mm and even beyond, can shorten the build time, but also lowers the possible level of detail. The operator of the FLM system can thus decide by the choice of nozzle diameter whether the focus should be more on speed or on the level of detail. One possible solution is provided by modern multi-nozzle systems, which allow several different nozzles in one FLM system. This technology is used, for example, to switch between construction material and support material in the process. A benefit for different nozzle diameters has not yet been established. Another solution could be a nozzle whose diameter can be variable adjusted, meaning it can be enlarged or reduced. Thus, it would be possible to manufacture with a small diameter at the points in the component that require a high level of detail, and with a larger diameter at other points. Several different approaches already exist for this. Some of these approaches originate from research work, others have been described directly in patent specifications. Brooks et al. [4] describes a concept for a two-stage nozzle, which would allow switching between two different nozzle diameters. However, approaches of a continuously adjustable nozzle are of more interest for this work. In the Chinese patent by Ge et al. [5], a nozzle is presented which has a fine moving mechanism to allow the nozzle opening to be continuously adjusted. In this case, the opening of the nozzle consists of six components, which result in a hexagonal opening area. The nozzle diameter is therefore not circular. Other approaches also produce cross-sectional areas of the nozzle diameter that are not circular. A patent from Desktop Metal [1] has a triangular geometry and Papon et al. [6] have developed a nozzle for printing concrete that has a square opening geometry. While the FLM process usually uses a circular nozzle opening geometry, a mechanically infinitely variable opening diameter seems to require a non-circular geometry. This seems to be mainly due to the low form factor of the necessary mechanics. This paper therefore aims to investigate the influence of a non-circular opening geometry on the deposition of a plastic strand in the FLM process. The usual FLM parameters will be used and only the opening geometry will be changed. For this purpose, nozzles with different exit geometries are manufactured with cross-sectional areas equivalent to a circle of 1, 0.8 and 0.6 mm diameter. The aim is to deposit individual strands, which are then measured and examined under a microscope. In addition to the shape of the nozzle, the direction of movement of the nozzle is also compared. Thus, the difference is examined whether the nozzle moves with the corner or the edge of the exit geometry in the direction of printing. The investigations are necessary because it is unclear whether and how the different exit geometries can influence the deposited strand and thus also subsequent strands, layers or entire components. For example, the influences can arise from different friction ratios within the angles and radii of the different geometries. In this paper, however, the focus is initially on individual strands.

2   Related Work

Despite the numerous approaches with non-circular nozzle geometries, the influence on the FLM process is not or only slightly dealt with in the literature, while for example the influences of different raster orientations or printing speeds have already been investigated [2]. A few papers that consider different exit geometries are investigations by Papon et al. [7–9].

First, the authors investigated in a numerical simulation, among other things, how a rectangular and star-shaped outlet geometry (with the same cross-sectional area as a circular nozzle with a diameter of 0.4 mm) affects the pressure loss and the distribution of the melt outlet velocity [9]. Compared to the circular cross-section, the pressure drop was slightly increased for the star-shaped and rectangular orifice. The velocity distribution in the radial direction showed a decrease from the centre toward the outer edge for all shapes. The star-shape resulted in a more uniform decrease. The authors hypothesized that the geometry-related changes may lead to a deviating strand cross-section, which affects the size of the air pockets and thus the mechanical properties of the FLM component.

In following experimental tests, Papon et al. compared the mechanical properties of FLM components produced using circular and rectangular outlet geometries [7]. The rectangular nozzle significantly reduced the air pockets between the strands, which led to an increase in the mechanical strength of the component.

In the latest publication by Papon et al. [8], a numerical simulation was performed to investigate how different exit geometries affect the shape of the extruded strand. The predictions of the simulation were then compared with experimental results. The three exit geometries from the previous studies (circle, rectangle and star) were investigated.

The cross sections were adjusted so that all shapes had the same diameter of 0.4 mm. In addition, the edges of the rectangular and star cross-sections were rounded to avoid possible instabilities due to stress concentrations and material jamming in advance. Extrusion followed by strand deposition on a moving platform was simulated. When the strand is deposited on the platform, there is no contact between the nozzle and the material, so that the material is not deformed by squeezing.

However, even the described work does not show an ordinary process workflow using a nozzle with a modified exit geometry, since ordinary layer heights and the squeezing of the layer onto the build platform were not performed. This research gap will be addressed with the experiments explained in the following.

3   Experimental Setting

The experiments are carried out in the Learning and Research Factory of the Chair of Production Systems. The test bench for robot-assisted additive manufacturing, which has an ABB IRB 2400-16 industrial robot with an extrusion head can be seen in Fig. 1. The robot was used because it had already been converted to an FLM system from previous projects. In addition, the future research plans require the robot because of its flexibility and payloads.

The extrusion head consists of a Bulldog XL driven by a NEMA 17 stepper motor. The stepper motor controller is connected to the IRC5 controller of the robot and can thus be addressed directly via the program code in RAPID. The hotend consists of standard FLM components.

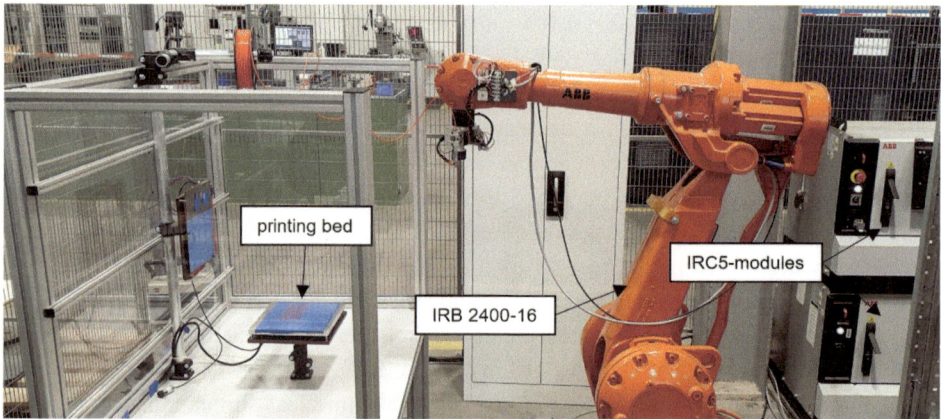

Fig. 1 The robotic cell used for the experiments

3.1   Nozzle

Since there are no commercial nozzles with non-circular exits, a total of twelve different extrusion nozzles with circular, square, triangular and hexagonal exit geometries are manufactured in the chair's in-house production facility. Nozzles in three different sizes are produced for each of the four exit shapes. The dimensions were adjusted so that the cross-sectional area of the respective nozzle exit, corresponds to a circle with a diameter of 0.6, 0.8 and 1.0 mm. The exit geometries of the nozzles are generated by milling. A milling cutter with a diameter of 0.3 mm is used for this purpose, resulting in the corners of the milled shapes having a radius r of 0.15 mm. Like most commercially available nozzles, the nozzles are made of brass.

3.2   Laser Scanner

To investigate the influence of the exit geometry, profile data of the deposited individual strands are generated by laser scans and compared with each other. A laser scanner from the manufacturer Micro-Epsilon of the type ScanCONTROL 2600-25, which is attached to the extrusion unit via a holder, is used to record the profile data. Data is transferred to the computer via an Ethernet interface. The laser scanner is attached to the robot of the test plant via a self-made bracket, which is screwed to the conveyor unit (Bulldog XL) of the extrusion unit.

3.3   Test Execustion

The focus of the investigation is the influence of the exit geometry. For this purpose, as already described, 12 different nozzles are manufactured, four geometries in three sizes each. In addition, the orientation of the nozzle to the printing direction is included, resulting in two variants, "corner" and "edge", for the triangular, square and hexagonal nozzles respectively. Finally, different layer heights are also investigated, since it is unclear how much they can amplify the influence of the exit geometry.

Figure 2 shows the planned arrangement of the strand lines of a test series on the construction platform. In each of the test series, 50 individual strands are produced, with the layer height h being increased in the x-direction after every 10 lanes. The subsequent area for measuring the deposited strands is marked by the vertical lines, which generate a strand length of 100 mm.

The start and end of the deposited paths are not measured, as these usually exhibit inaccuracies. In addition, a so-called "purge line" is deposited, which serves to prepare the nozzle for the process and thus increase process stability. PLA (polyactic acid) filament is used as the material. Table 1 shows the experimental parameters for the 1 mm nozzle.

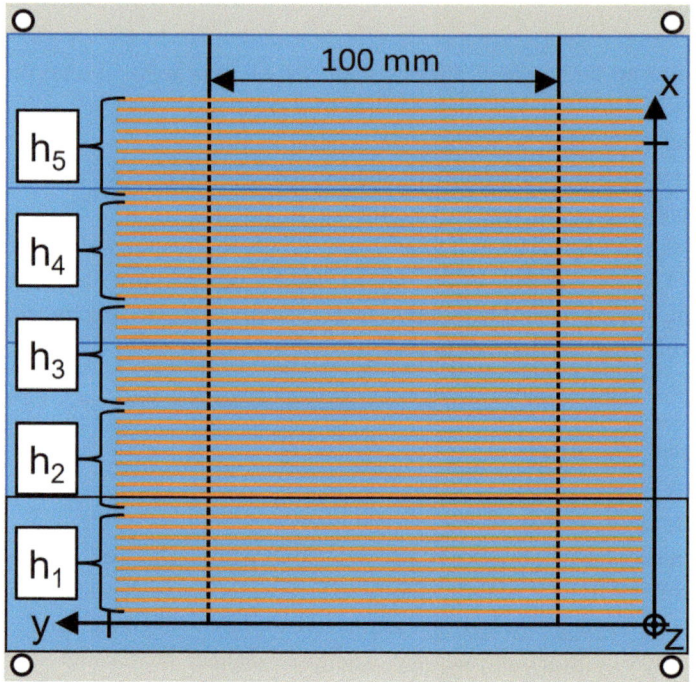

Fig. 2 The arrangement of the deposited lanes on the print bed

Table 1 Process parameters for the exit size 1 mm

Factor	Unit	Stages				
Geometry of the exit	–	Circle, Rectangle, Triangle, Hexagon				
Orientation of the nozzle	–	Corner, edge				
Nozzle diameter (eq.)	Mm	1				
Temperature heating bed	°C	60				
Temperature nozzle	°C	220				
Printing speed	mm/s	20				
Extruder speed	mm/min	44.5	64.86	83.96	101.8	118.37
Layer thickness	Mm	0.25	0.375	0.5	0.625	0.75

In the next step of the test procedure, the profile curves of the deposited strands are recorded by measurements with the laser scanner, immediately after strand deposition. The scanCONTROL 3D-View software package is used to record and export the measurement data. Since undercuts cannot be detected and measured with the laser line sensor, additional strand cross sections are cut out and examined and photographed under a digital microscope at a 200x magnification.

4   Results

In order to evaluate the influence of the nozzle exit geometry, the averaged curves of the strand profiles from the tests are presented. In addition, the microscopic images of the strand cross sections are shown, thus extending the observations regarding the strand profiles. Figure 3 shows the results of the profile measurements sorted by exit geometries, direction, and layer height for the 1 mm nozzles. At this point, it should be mentioned that no comparable experiments and results have been found in the literature so far. The focus should therefore first be on providing an overview of how the different geometries can have an effect. In later experiments and corresponding publications, it will certainly be possible to go into more detail in many cases.

A closer look at the strand profiles reveals that the shape of the deposited strands is influenced by the outlet geometry of the nozzle and, moreover, its orientation direction. Considering the profile of the circular nozzle as a reference, a varying degree of deviation can be seen for the exit geometries investigated. In general, it can be observed that the divergence of the strand profile decreases with an increasing number of corners of the exit shape. This can be seen by comparing the individual unusual geometric shapes with the standard round exit geometry. For example, the profiles of the triangular exit geometry show strong changes, in that they have curves with almost right angles in the "corner ahead" variant, while in the "edge ahead" variant they have a very strong peak in the middle. With the hexagonal nozzles, on the other hand, there are only very slight deviations in shape between the variants and in comparison, with the round nozzle. Again, it can be observed that the influence of orientation decreases as the number of corners increases. While the square nozzles differ in the orientation directions by a visible curvature of the upper side of the profile, the profiles of the various orientations of the hexagonal nozzle differ much less. Due to the fact that the approximation of a circular surface with regular polygons improves by increasing the number of corners, the described deviations of the strand profiles are comprehensible. While the strand profiles of the circular, square and hexagonal nozzles hardly differ at a low ratio of layer height to diameter, occurring effects or changes of the profiles intensify with increasing layer height. An explanation is provided by considering the general strand shape. With constant strand width and simultaneously increasing layer height, the actual shape of the extruded strand is less compressed due to less squeezing with the nozzle. Thus, the shape of the freely extruded strand is

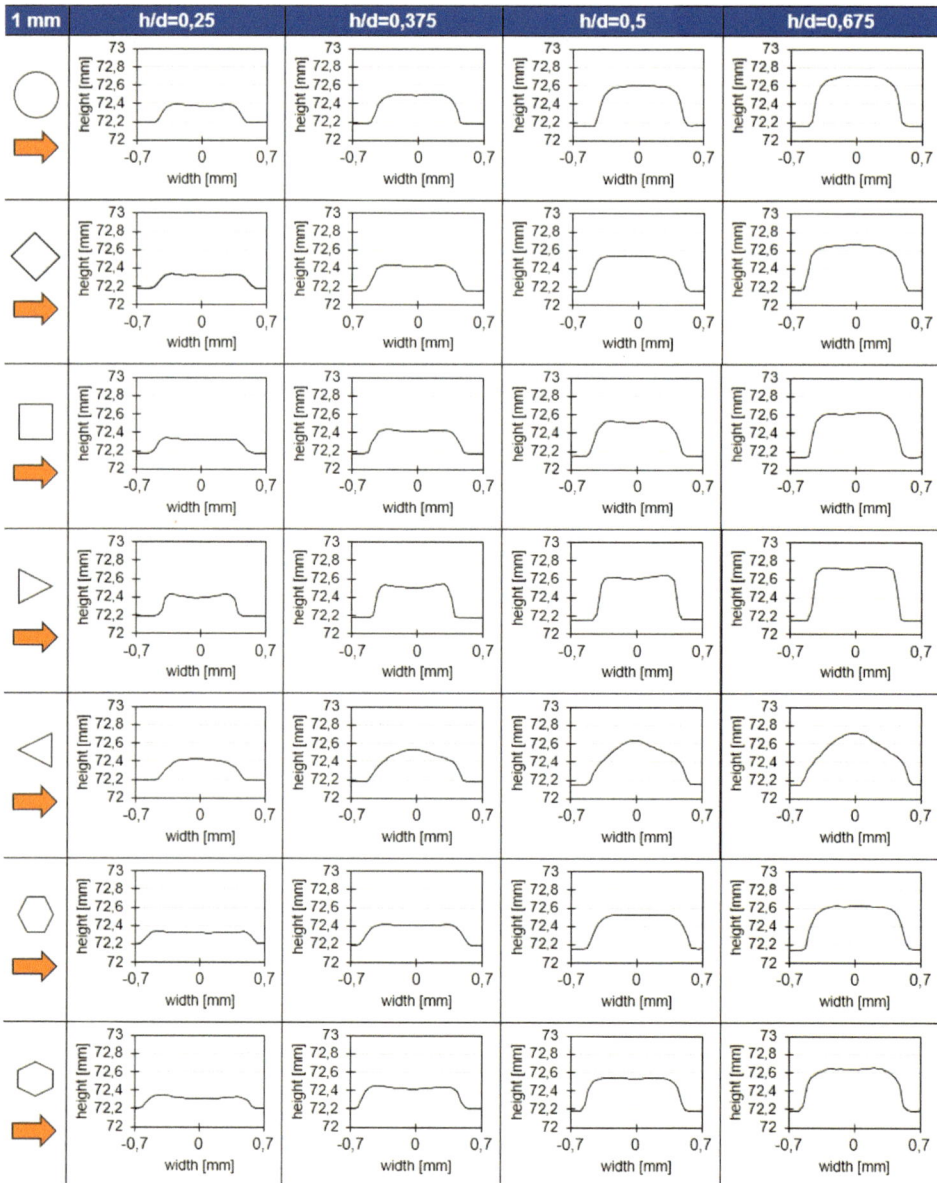

Fig. 3 The resulting profile cross section after laser scanning with the exit size 1 mm

largely retained. If the profiles of the different nozzle diameters are compared, the effects that occur decrease as the size decreases. Particularly for low layer heights and the smallest diameter, only slight or hardly perceptible changes in the profiles can be detected. This was to be expected, since in advance smaller experiments were carried out to isolate possible influencing factors on the appearance of the strand. As a result, only the shape of the exit geometry remained as the strongest influencing factor in the final experiments. The results thus show that the effects of the choice of geometry are scalable and depend on the level of detail of the printing process. It should also be noted that the nozzle outlets have production-related corner radii which have the same radius irrespective of the diameter. This reduces the expression of the shape for smaller diameters, which can have an additional effect on strand forming.

To obtain additional information about the strand geometry, cross sections were made for one strand of each layer height using a digital microscope. The results for the 1 mm nozzles can be seen in the Fig. 4.

For the square nozzle with the orientation direction "corner", a difference appears which is not obvious from the extrusion profiles. For all nozzle diameters, the shape of the cross sections approximates a rounded trapezoid, where the bottom side is narrower than the top side. The shape decreases with decreasing layer height. The changes in the top side of the strand can be observed analogously to the profile data. If the strands still possess a concave curvature for the lowest layer height, a flat to slightly concave curvature appears at $h/d = 0.375$ and $h/d = 0.5$, which increases with increasing layer height. For the "edge" orientation of the square nozzle, the cross-section is also trapezoidal, but the widest point is at the contact surface with the build platform and decreases toward the top. The side surfaces are much steeper than in the "corner" orientation, making the shape almost rectangular. These observations are comparable to those of Papon et al. [8], who uses numerical simulations and experimental tests to produce a similar, trapezoidal strand cross-section with a rectangular nozzle.

For the triangular nozzle with the orientation "corner", there is a clear difference from the profile resulting from the laser measurement. Due to the shading, the actual shape of the cross-section is not accurately recognized. For high layer heights, the strand exhibits a rounded triangular shape with the tip pointing downward. While the contact area to the building platform increases as the layer height decreases, the concave curvature of the top of the strand increases at the same time. The cross sections of the triangular nozzle with the orientation "edge" show a great similarity to the measured profiles. However, an effect is also hidden here by the shading of the strand. Here, the lower edges of the triangular strand are slightly raised upwards, reducing the contact area with the build platform.

About the hexagonal nozzles, it can be observed that the shapes of the different orientations differ only slightly from each other. Here it can be observed that the cross-section of the orientation "corner", resembles the cross-section of the square nozzle with the orientation "edge". In both cases, the strand is slightly wider at the bottom than at the top, but the contact area with the build platform is reduced in the hexagonal nozzle. The

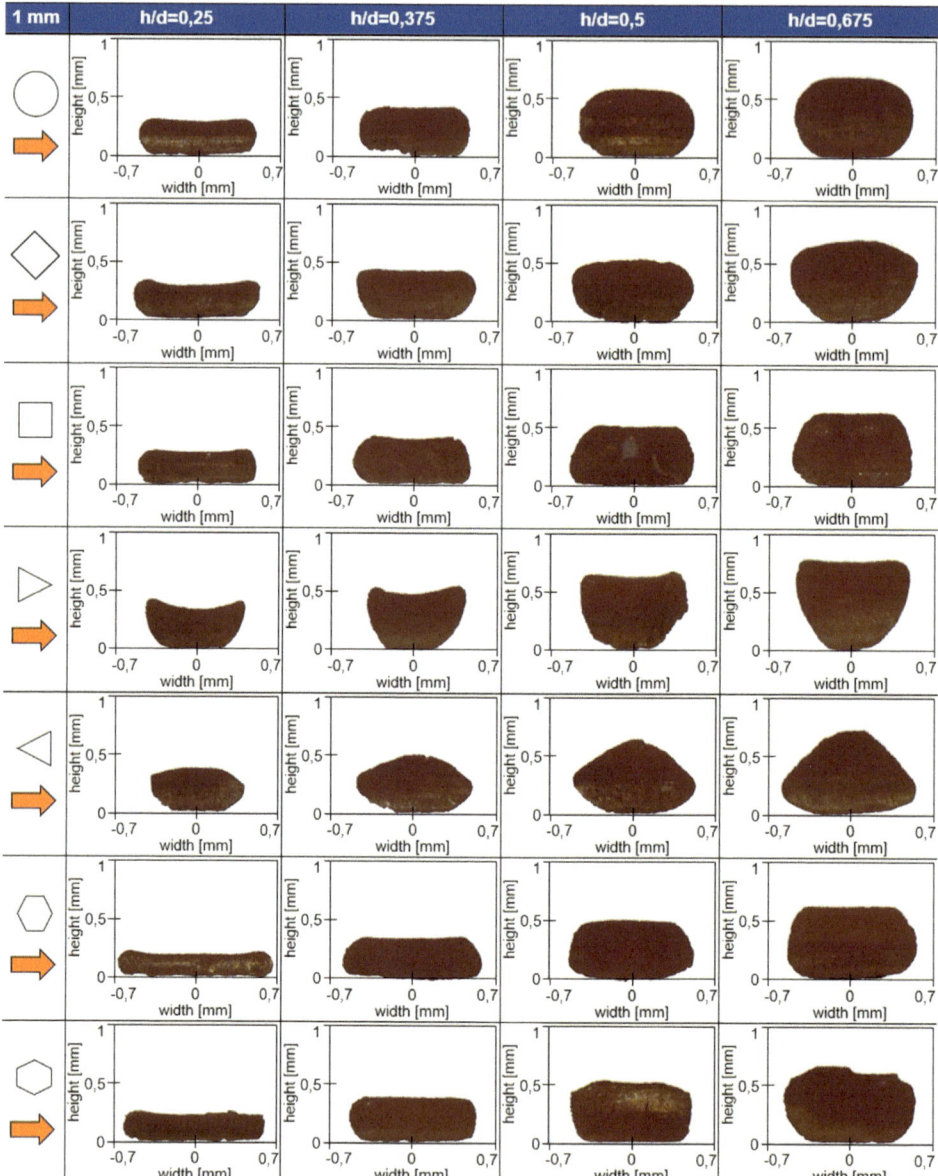

Fig. 4 The microscopy images of the cross-sectional area with the exit size 1 mm

strand shape of the "edge" orientation, on the other hand, is comparable to that of the circular nozzle. However, the upper corners of the strand geometry are slightly angled, like a hexagon.

5　　Conclusion

In conclusion, the results show that a change in the nozzle outlet shape leads to a change in the strand cross-section. With the increasing number of corners of the exit geometry, a decreasing difference to the strand of the circular nozzles can be seen. Furthermore, the expression of the change in shape increases with increasing layer height while the strand width remains constant. About the investigated variables, the changes in strand shape can be observed for all diameters. However, the amplitude of the deformation, and therefore the visibility, decreases with increasing diameter. For the circular nozzles, the strand cross-section has a rectangular shape with semicircular sides. However, the cross section becomes slightly elliptical with increasing layer height. For the square nozzles, two different strand shapes result depending on the orientation direction. If the corner of the nozzle is oriented in the direction of deposition, the basic shape of the strand cross-section is trapezoidal with rounded corners, the narrow side being on the underside of the strand. Moreover, from a ratio of the layer height to the equivalent diameter of the nozzle of 0.675, a convex curvature of the upper side of the strand can be observed, which increases with increasing layer height. If the square nozzle is oriented with one edge facing the deposition direction, the cross-section is also trapezoidal, but with the narrow side facing upward. In contrast to the other orientation, the top of the strand shows a flat to slightly concave curvature. The strands with the greatest deviation in relation to the circular nozzle are produced with the triangular exit. Again, the strand shape differentiates depending on the orientation. If a corner is oriented in the deposition direction, the strand cross-section up to a layer height of 50% of the equivalent diameter has a crescent-like shape with the tips directed upwards. For greater layer heights, the shape becomes a bulging triangle, with concave top. If the nozzle is oriented with one edge in the deposition direction, the strand will have a triangular shape with rounded corners and the tip pointing upwards. The bottom corners bend slightly inward, reducing the contact area with the build platform. The strand shape for the hexagonal nozzles shows the least deviation from the circular nozzle. Nevertheless, minor differences between the orientation directions can also be seen here. If the corner is oriented parallel to the direction of deposition, the strand is slightly wider in the lower region and thus has a weakly defined trapezoidal shape. For the smallest diameter of the test, the change in shape is hardly noticeable. If the nozzle is rotated with one edge to the deposition direction, the cross-section is like the strand shape of the circular nozzle. However, the strand is in the upper area, slightly beveled, instead of a constant rounding. If the strand shapes of the different exit shapes differ as described, the cross sections of the circular, square and hexagonal nozzle show a strongly similar shape for the smallest layer heights investigated. Especially for the low layer thicknesses, a concave curvature of the surface of the strands was shown, which is no longer visible with increasing layer thickness. That the results of the hexagonal exit geometry are very close to the standard round exit geometry is advantageous, since planned concepts for a size-variable nozzle are based on hexagonal mechanics and could therefore provide promising results.

The investigation results allow initial assessment of exit geometries' influence in the FLM process but are limited to qualitative descriptions of the general cross-sectional shape. Further studies on the impact on components and properties, including strand connection and layer analysis, are recommended.

References

1. Javaid, M., Haleem, A., Singh, R.P., et al.: Role of additive manufacturing applications towards environmental sustainability. Adv. Indus. Eng. Polym. Res. **4**, 312–322 (2021). https://doi.org/10.1016/j.aiepr.2021.07.005
2. Khorasani, M., Ghasemi, A., Rolfe, B., et al.: Additive manufacturing a powerful tool for the aerospace industry. RPJ **28**, 87–100 (2022). https://doi.org/10.1108/RPJ-01-2021-0009
3. Gebhardt, A., Hötter, J.-S.: Additive manufacturing. In: 3D Printing for Prototyping and Manufacturing. Hanser Publications; Hanser Publishers, Cincinnati, OH, Munich, Cincinnati (2016)
4. Brooks, H., Rennie, A., Abram, T. et al.: Variable fused deposition modelling—concept design and tool path generation (2011)
5. Ge, M., Lu, J.: Nozzle bore adjustable 3D print. CN208774065U (CN208774065U)
6. Papon, M.E.A., Haque, A., Sharif, M.A.R.: Effect of nozzle geometry on melt flow simulation and structural property of thermoplastic nanocomposites in fused deposition modeling. In: American Society for Composites 2017. DEStech Publications, Inc., Lancaster, PA (2017)
7. Khosravani, M.R., Berto, F., Ayatollahi, M.R., et al.: Characterization of 3D-printed PLA parts with different raster orientations and printing speeds. Sci. Rep. **12**, 1016 (2022). https://doi.org/10.1038/s41598-022-05005-4
8. Papon EA, Haque, A.: Tensile properties, void contents, dispersion and fracture behaviour of 3D printed carbon nanofiber reinforced composites. J. Reinf. Plast. Compos. **37**, 381–395 (2018). https://doi.org/10.1177/0731684417750477
9. Papon, E.A., Haque, A., Sharif, M.A.R.: Numerical study for the improvement of bead spreading architecture with modified nozzle geometries in additive manufacturing of polymers. RPJ **27**, 518–529 (2021). https://doi.org/10.1108/RPJ-05-2019-0142
10. Myerberg, J.S.: Fused filament fabrication extrusion nozzle with concentric rings. US Patent Application Publication US 2017/0173879 A1(US 2017/0173879 A1)
11. Xu, J., Ding, L.: Volume—forming 3D concrete printing using a variable—diameter square nozzle. In: Creative Construction Conference 2018—Proceedings. Budapest University of Technology and Economics, pp. 104–113 (2018)

Exploratory Pilot Study Investigating Effects of Exoskeletons on Movement Patterns

Steffen Jansing, Elena Vakhnina, Vanessa Weßkamp, and Jochen Deuse

Abstract

The ergonomic configuration of processes is becoming increasingly important, especially considering the changing demographics and increasing shortage of skilled workers. Exoskeletons are widely discussed as a means of protecting employees from overstraining at the level of personal protective measures. The field of industrial exoskeletons research is still relatively new and has many unanswered questions. For example, there have not yet been sufficient studies on the influence of exoskeletons on the movements of employees. This publication discusses the effects of exoskeletons in manual processes. For this purpose, exemplary physical activities are carried out in a pilot study by a subject collective, whereby the tasks are executed with and without an exoskeleton. During the execution, a motion capturing system is used to record the movement data. Different back-supporting exoskeletons are taken into account in the study. The evaluation is based on the joint angles of the participants while performing tasks with and without exoskeletons. It is shown that the use of exoskeletons has a significant effect on the movement patterns, with a distinction made between rigid and soft support structures.

S. Jansing (✉) · E. Vakhnina · V. Weßkamp · J. Deuse
Institute of Production Systems, TU Dortmund University, Leonhard-Euler-Straße 5, 44227 Dortmund, Germany
e-mail: steffen.jansing@tu-dortmund.de

J. Deuse
Centre for Advanced Manufacturing, University of Technology Sydney, 15 Broadway, Ultimo, NSW 2007, Australia

© The Author(s) 2025
S. Ihlenfeldt et al. (eds.), *Annals of Scientific Society for Assembly, Handling and Industrial Robotics 2023*, https://doi.org/10.1007/978-3-031-74010-7_3

Keywords

Exoskeletons • Movement patterns • Motion capturing

1 Introduction

Musculoskeletal disorders (MSDs) are the most frequent cause of physical functional limitations, chronic pain, and loss of quality of life worldwide, regardless of age group [1]. MSDs are a leading cause of lost work days and high healthcare costs in Germany [2, 3]. In particular, men engaged in manual tasks in production and logistics have a high risk of work incapacity [4]. Against this background and in view of demographic change, ergonomic work system design is becoming increasingly important as a preventive tool for the long-term protection of the ability to work [5, 6]. Conventional methods often reach their limits in this regard. Not all activities, especially not those with low production runs, manual tasks that change location, and large components, can be technically or economically sensibly automated [7, 8].

Exoskeletons (Exos) are considered to offer a high economic potential to reduce employee absenteeism through improved ergonomics [9] and to increase productivity [10]. Early studies indicate that work can be performed for longer periods without discomfort [11] and the reduction in effective handling weight can shorten execution times [12].

The present study examines the impact of industrial exoskeletons on human movement behavior during various typical industrial tasks, for which the use of Exos is often considered to have potential. Movements during activities with and without support, as well as with rigid and soft support structures, are compared. The conducted pilot study is understood as a preliminary experiment, which will precede a detailed investigation.

2 State of the Art

Movement patterns are defined in many different ways. In the context of logistics tasks, they can be defined as a standardized and consistent sequence of coordinated muscle contractions at one or more joints during work tasks, which is unique to each person [13, 14]. Especially in the context of manual production and logistics, movement patterns with high stress levels occur. Repetitive movements, especially those under load combined with ergonomically unfavorable bending and twisting movements, tire the body and promote physical injuries. An example of activities with a high repetition rate, forced postures and frequent lifting and lowering movements is order picking, where pain and musculoskeletal disorders often occur as a result of one-sided stress.

The effect of the use of Exos has already been investigated in studies. However, these studies mainly focused on the areas of performance [15] and biomechanical parameters [16]. Thus, studies have demonstrated a positive effect of Exos on physiological demands [11, 16, 17]. Previous studies have investigated restrictions in the freedom of movement caused by Exos via subjective queries [6, 15] or have used a combination of recorded movement data and EMG measurements to assess changes in physical strain [16]. Despite the relevance of possible movement restrictions due to the additionally handled weight as well as the support structure, the movement behavior of the employees is hardly investigated separately in current research [18].

Industrial Exos represent a personal protective measure for the employees and are understood as an external support structure that mechanically supports certain body segments [6]. Depending on the activity, Exos can provide targeted support for different body regions. Load reduction in the areas of the hands and arms, for example, is useful for overhead activities. In contrast, support of the trunk is advantageous for load handling and static posture work [19]. Furthermore, Exos can be divided into active and passive systems [20]. In contrast to active systems, passive exos have no external energy sources and use only stored energy (e.g. springs or gravity) that is charged by the user [21]. Passive systems can be further divided into rigid Exos with dimensionally stable structures for force transmission, and soft Exos with textile elements that transmit forces [22].

3   Concept and Procedure

The study presented below serves to investigate the influence of Exos on movement patterns. Therefore the investigation is based on five representative processes with typical industrial loads.

3.1   Selection of Activities and Definition of Hypotheses

Studies (e.g. [6, 23]) report that production and logistics workers often suffer from MSDs due to the physical work involved. Despite the automation of several work processes, manual tasks characterized by forced postures and repetitive load handling make up a high proportion in both manufacturing and logistics [24]. Therefore, manual subtasks, which on the one hand cause high stress for production and logistics employees, and on the other hand occur frequently in production and work processes, were selected for the pilot study. Therefore, especially the lifting and carrying loads is strongly represented; typical work tasks include loading and unloading as well as order picking in production and assembly [25]. These activities have a highly effect on the lumbar spine. The more often a load has to be lifted and the heavier it weighs, the higher the risk of complaints in the lower back area [26]. In addition, other factors such as the lifting technique, the posture adopted, the type of load handling (one or two-handed), and the restriction of the range of motion can be identified as relevant. Specifically, forced postures can also lead to negative consequences such as muscle tension and pain in body segments.

On this basis, five activities are defined as representative test scenarios. The execution serves to identify potential influences of Exos on the movement patterns of the participants. Following the definition of movement patterns, the joint angles assumed by the participants during the execution of the tasks serve as evaluation parameters for hypothesis testing. The variables used for hypothesis testing are the flexion of the knee and hip joints, the flexion and axial rotation of the vertebral joints L5S1, L4L3, L1T12 and T9T8, the flexion of the ergonomic angles from upper arm to upper body (T8-Upper Arm) and the flexion of the ergonomic angles from upper body to pelvis (Pelvis-T8).

Lifting loads. The participants stand upright at a floor marker and bend forward to a mass. They pick it up at a height of 50 cm and bring it to 110 cm (final hand height). They turn their upper body 90° to the right and place the mass on a roller conveyor at the same height. The mass is then pushed out of the working area with the left hand and the participants return to their starting position. Whereas the specification for the process included the sequence of movements, each participant was free using a freely selected but always identical lifting technique. The experimental setup for this process is exemplified in the following Fig. 1.

Carrying loads. The participants lift the mass in the same way as in process one. After lifting, they turn 90° to the right. They walk unhindered a straight distance of 6 m, turn around and return to the starting point. After a 90° turn to the right, the participants set the mass down at the starting position.

Fig. 1 Experimental setup for lifting loads with floor marking for the starting position

5-Min Waist-Height lift task. The participants have the task of lifting a mass to waist height as often as possible within a period of five minutes, whereby the participants can take breaks independently. Lifting and setting down is performed in the same way as task one and two from a pallet; there is no torso rotation.

Loading a mesh box. The participants pick up a mass at waist level, turn 180° to the left and place the mass in an open mesh box. Due to the spatial restriction of the mesh box, the upper body is slightly bent forward and the mass is extended in front of the body. After placing the mass at a height of 60 cm (final hand height), the participants straighten up and turn back to the starting position.

Sorting small parts in forced position. The participants bend their upper body over a work surface at knee height for a period of five minutes and perform a sorting task with small parts. The flexion is not interrupted during the test.

3.2   Technical Systems

Two passive Exos are considered in this study, as shown in Fig. 2. On the one hand, the Exo Paexo Back (PB) from Ottobock SE & Co. KGaA as representative for a system with rigid support structure and the Exo LiftSuit (LS) by Auxivo AG as representative for textile, soft Exo. Both Exos have been sold commercially since 2020 and mainly support the user's back and trunk. In order to prove an influence it is necessary to determine the joint angles during the execution of the activities. For this purpose, motion data acquisition is performed using the Xsens-MVN-Awinda system from the Dutch supplier Xsens Technologies B. V., which records the participant's body movements with 17 inertial sensors at a sampling rate of 240 Hz. The data is translated into a biomechanical model of 23 body segments connected by joints [27], while taking into account anthropometric variables such as the participant's height.

In addition to the recording of objective parameters, subjective parameters are recorded by questioning the participants. The perceived difficulty of the task, local and general discomfort and perceived movement restrictions as well as a general assessment of the support systems used are queried as parameters. All parameters are assessed by the participants using the Visual Analog Scale (VAS), which is used to evaluate subjective parameters, e.g. in medical studies to classify the sensation of pain [30].

3.3   Procedure

The study took place from May 2 to May 18, 2022, at the training center of the Institute of Production Systems at TU Dortmund University, with one to two participants take part in the data acquisition per day. The duration of the test is a maximum of three hours.

Fig. 2 Illustration of PB (left) and LS (right), view of the back (acc. to [28, 29])

Beginning with the questioning about anthropometric and personal data, the examination of the exclusion criteria takes place. This is followed by the introduction to the study procedure. Each participant runs through the defined scenarios both without Exo and with both support systems, randomizing and balancing the order of support used to eliminate fatigue and potential interference effects.

After the application of the motion capturing system, the application of the corresponding Exo takes place. This is followed by the execution of the examination scenarios, whereby the activities are run through in the following order: Five times lifting loads, once sorting small parts in a forced posture, once loading a lattice box, three times carrying loads and once lifting for time. After each scenario has been carried out, the subjective parameters are queried by a questionnaire. After a complete series of measurements with an Exo, a query is made for the general evaluation. Between the series, there is a recovery break of five minutes plus the time required to change the Exo (approx. 15 min).

4   Results and Discussion

The described experimental procedure is run through a test collective in randomized order to exclude fatigue and possible interference effects. The study population consists of twelve participants with an average age of 27.9 years, half of whom are women

(29.8 years) and half men (26.0 years). Exclusion criteria are defined as functional limitations of the extremities and general physical complaints. Due to the limited adaptation possibilities of the technical systems, the body height is defined as 1.60–1.90 m, with an average body height of 1.75 m. The majority of the participants are located in the student environment and do not perform manual activities on a regular basis.

4.1   Objective Data

The basis for the objective analysis of the movement patterns is provided by the recorded movement data, for which an initial plausibility check is performed using video recordings made in parallel with the measurements serve. Subsequently, in a first step, the maxima and minima of all participants and test executions for the described activity are determined.

In order to reduce the target variables to be considered, an examination for correlation takes place. Thus, the examination of the vertebral joints shows that the vertebral joints L5S1, L4L3, L1T12 as well as T9T8 have a strong positive correlation ($r > 0.975$) for all activities in the presented study, which is why it can be limited to a consideration of the vertebral joint L5S1 as a representative variable for the hypothesis testing of the vertebral joints. A comparable study for the hip joints shows that the angles for the left and right hip joint only show a strong positive correlation ($r > 0.81$) for the activities of lifting loads, sorting and lifting for time. For the other activities, as well as for the angles of the knee joints, no relationship can be determined. With regard to the ergonomic joint angles, only a weak correlation can be found in each case. An exception is the flexion and extension of the upper arms in relation to the T8 vertebra. For activities with strong arm flexion (carrying loads, loading a lattice box as well as lifting for time), T8-Upper Arm right can be determined as a representative.

Following the correlation test, multiple regression models are set up to take into account the general conditions of Exos, activity and gender. Including the interactions of the three conditions as well as a normally distributed error variable, a regression model is obtained that is suitable for testing the established hypotheses in a post hoc model test. In this significance test, the mean values of the joint angles for the comparison of the Exo configurations are compared pairwise and examined for statistical differences. The significance level is defined as $\alpha = 5\%$, with an adjustment using the p-values.

In a consideration of the first activity, the adjusted exceedance probabilities listed in Table 1 provide statements about the hypotheses that were made with reference to the defined significance level.

The results show that the PB in particular has an influence on the movement patterns. With this Exo, an influence on the flexion of the back (L5S1) and the hip can be demonstrated for all participants, whereby a significant influence can also be demonstrated in the flexion of the knees and shoulder-arm for female participants.

Table 1 Results for max. joint flexion in "lifting loads" task with adjusted p-values

		p*-value						
		Knee joint		L5S1	Hip joint		T8-Upper arm	
		Left	Right		Left	Right	Left	Right
W/O versus PB	M	1.000	0.644	<0.001	0.021	0.006	1.000	1.000
	F	0.004	0.001	<0.001	<0.001	<0.001	<0.001	<0.001
W/O versus LS	M	1.000	0.338	0.522	1.000	1.000	1.000	0.134
	F	0.016	0.040	1.000	0.001	0.394	0.097	0.033

A significant influence on the movement patterns can also be identified for the activity of carrying loads, especially for the PB. While the LS is only significantly expressed in the extension of the hip joints for female participants ($p* < 0.001$), a gender-independent influence of the PB on the flexion of the hip ($p* < 0.001$) as well as for female participants for the flexion of the left knee joint ($p* = 0.016$) can be shown.

When examining the activity "loading a mesh box", it can be seen that when lowering the load, significantly less flexion of the vertebral joints (represented by L5S1) ($p* = 0.004$) can be identified by wearing the LS for female participants. In contrast, the effect of the PB is particularly evident in the axial rotation to the right in male participants ($p* < 0.001$) and in the flexion of the hip in female participants ($p* < 0.001$).

For the other activities, no influence on the movement patterns can be statistically demonstrated for the LS. For the PB, the effects are limited to male participants and relate to flexion of the upper body to the pelvis (angle pelvis-T8; $p* < 0.001$) for "sorting small parts" and to flexion of the L5S1 ($p* < 0.03$) for "5-Min Lift Task".

4.2   Subjective Data

The participants were asked about the discomfort they felt during the performance of each activity. In addition, the participants were asked to compare the two Exos used with regard to perceived movement restrictions after all activities have been performed. The questions were answered using the VAS. Due to the small number of test persons, only a comparison of the frequency of statements was made, but not their severity.

With regard to the perceived discomfort, for example, a significant reduction in the area of the back is shown for the activity "lifting for a period of time" through the use of Exo. Here, for example, 8 out of 12 participants testify independently of the system that the perceived discomfort in the area of the back is significantly reduced. Similarly, at least seven participants testify to support in the lower back for the activity "carrying", regardless of the specific system. For the activity "Sorting small parts in a forced posture", however, a more differentiated picture emerges. Here, nine of 12 participants rated the LS

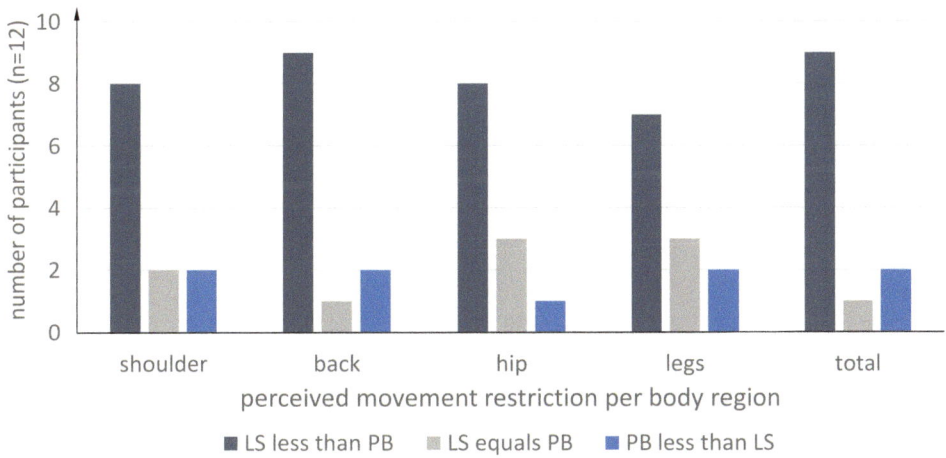

Fig. 3 Comparative representation of the participant survey on movement restrictions during the task "loading a mesh box"

as uncomfortable in the chest area. Likewise, 11 of the 12 participants rated the LS as more uncomfortable in the lower back.

If the results of the survey on movement restrictions are considered, it is noticeable that for all activities and all body regions, the movement restrictions exclusively with the PB were rated as higher by the participants than with the LS. An example is shown in Fig. 3 for the activity "Loading a mesh box".

4.3   Discussion

The analysis of the objective parameters confirms in principle the assumption that the use of Exos has an effect on the execution of movements in the test persons. It could be identified that the influence can be observed more strongly in dynamic load handling. For example, for compliance with static forced posture, only a small effect can be demonstrated.

In addition, it can be determined that in the comparison of both systems investigated as representatives for passive Exos with rigid and soft structure, differences in the influence on movement can be seen. Especially the PB with rigid structure shows a significant influence on the movement execution compared to the LS with soft structure.

In the context of the analysis of the subjective parameters, the basic suitability of Exos for reducing the perceived discomfort can be established. In particular, the systems have a significant effect in the area of the back. Regardless of the activity, however, the restriction of movement is perceived as higher exclusively by wearing the PB with rigid support structure.

5   Conclusion and Outlook

The findings obtained should be classified in that it is a study of a small sample of test persons and activities. However, it was possible to provide evidence that an influence of the supporting structure on the movement patterns of the participants could be demonstrated in both the objective and subjective parameters. The influence of Exos with rigid support structure could be identified as higher compared to systems with soft structure. At the same time, however, rigid structures were associated with less discomfort in movement performance than soft systems.

When interpreting the results presented, however, it should be noted that only the maximum flexion of the joint angles was used as a relevant evaluation criterion. The temporal course of the joints during the execution of the movement was not considered. In addition, it must be taken into account that the Xsens system calculates the movements in the lumbar region in particular based on only a few markers. This may affect the validity and accuracy of the target variables, which may be a problem in the case of small differences and adjustments of the movement patterns. Similarly, the results represent only a short-term investigation of processes; therefore, no statement can be made about long-term effects due to wearing the Exos on movement patterns.

Therefore, further investigations are needed to identify additional critical influencing variables and to increase the data base. Furthermore, the assessment of the ergonomic effects of the changed movement patterns should be considered in more detail in further studies, since the exclusive identification of a changed movement does not allow any statement about the ergonomic consequences.

Acknowledgements This work was part of the IGF-projects with the grant numbers 21605 and 22827 N, supported by the Federal Ministry for Economic Affairs and Climate Actions (BMWK) on the basis of a decision by the German Bundestag.

References

1. Fuchs, J., et al.: Prävalenz ausgewählter muskuloskelettaler Erkrankungen: Ergebnisse der Studie zur Gesundheit Erwachsener in Deutschland (DEGS1). Bundesgesundheitsblatt Gesundheitsforschung Gesundheitsschutz **56**, 678–686 (2013)
2. Knieps, F., et al. (eds.): Krise - Wandel - Aufbruch: Zahlen, Daten, Fakten. BKK Gesundheitsreport, vol. 2021. Medizinisch Wissenschaftliche Verlagsgesellschaft, Berlin (2021)
3. BMAS/BAuA: Sicherheit und Gesundheit bei der Arbeit - Berichtsjahr 2017. Bundesministerium für Arbeit und Soziales (BMAS) (2018)
4. Liebers, F., et al.: Alters- und berufsgruppenabhängige Unterschiede in der Arbeitsunfähigkeit durch häufige Muskel-Skelett-Erkrankungen. Rückenschmerzen und Gonarthrose. Bundesgesundheitsblatt Gesundheitsforschung Gesundheitsschutz **56**, 367–380 (2013)

5. Schneider, J., et al.: Arbeitswelten der Logistik im Wandel: Automatisierungstechnik und Ergonomieunterstützung für eine innovative Arbeitsplatzgestaltung in der Intralogistik. In: Hermeier (ed) Arbeitswelten der Zukunft, vol. 21, Wiesbaden, pp. 51–66 (2019)
6. Bednorz, N., et al.: Ergonomieunterstützung in der Logistik: Industrieller Einsatz von Exoskeletten an Palettier-und Kommissionierarbeitsplätzen zur körperlichen Entlastung von Mitarbeitern. GfA-Frühjahrskongress (2019)
7. Groos, S., et al.: Belastungs- und Beanspruchungsanalyse beim Einsatz eines passiven Exoskelettes zur Bein- und Rückenunterstützung während simulierter Montagetätigkeiten. Z Arb Wiss **54**, 212 (2019)
8. Schick, R.: Einsatz von Exoskeletten in der Arbeitswelt. Zbl Arbeitsmed **68**, 266–269 (2018)
9. Hoffmann, N., et al.: Towards a framework for evaluating exoskeletons. In: Wulfsberg, J.P., et al. (eds.) Proceedings of the 9th Congress of the German Academic Association for Production Technology (WGP), Hamburg, pp. 441–450. Springer. Heidelberg (2020)
10. Lotter, B.: Wirtschaftlichkeit und Ergonomie – kein Gegensatz. In: Lotter, B., Deuse, J., Lotter, E. (eds.) Die Primäre Produktion, pp. 125–141. Springer Vieweg, Berlin (2016)
11. Bosch, T., et al.: The effects of a passive exoskeleton on muscle activity, discomfort and endurance time in forward bending work. Appl. Ergon. **54**, 212–217 (2016)
12. Hempen, S., et al.: Ergonomie in logistischen Prozessen. In: Landau, K. (ed.) Produktivität im Betrieb: Tagungsband der GfA-Herbstkonferenz. Ergonomia, Stuttgart (2009)
13. Walter de Gruyter & Co.: Pschyrembel - Klinisches Wörterbuch, 268 edn., Berlin (2020)
14. Schmidt, A.: Movement pattern recognition in basketball free-throw shooting. Hum. Mov. Sci. **31**, 360–382 (2012)
15. Baltrusch, S.J., et al.: The effect of a passive trunk exoskeleton on functional performance in healthy individuals. Appl. Ergon. **72**, 94–106 (2018)
16. Glitsch, U., et al.: Biomechanische Beurteilung der Wirksamkeit von rumpfunterstützenden Exoskeletten für den industriellen Einsatz. Z Arb Wiss **21**, 456 (2019)
17. Theurel, J., Desbrosses, K.: Occupational exoskeletons: overview of their benefits and limitations in preventing work-related musculoskeletal disorders. IISE Trans. Occup. Ergon. Human Factors **7**, 264–280 (2019)
18. DGUV: Einsatz von Exoskeletten an gewerblichen Arbeitsplätzen. Fachbereich Aktuell (2019)
19. de Looze, M.P., et al.: Exoskeletons for industrial application and their potential effects on physical work load. Ergonomics **59**, 671–681 (2016)
20. Bostelman, R., et al.: Cross-industry standard test method developments: from manufacturing to wearable robots. Front. Inf. Technol. Electron. Eng. **18**, 1447–1457 (2017)
21. Lee, H., et al.: The technical trend of the exoskeleton robot system for human power assistance. Iint. J. Precis. Eng. Man **13**, 1491–1497 (2012)
22. Mayer, T.A., et al.: Einfluss eines aktiven Hand-Exoskeletts auf die Muskelaktivität im Unterarm bei industriellen Montage-Greifarten. ASU **2020**, 503–512 (2020)
23. Andersen, L.L., et al.: Association between occupational lifting and day-to-day change in low-back pain intensity based on company records and text messages. Scand. J. Work Environ. Health **43**, 68–74 (2017)
24. Bäuerle, I.: Konzept zur biomechanischen Bewertung von Exoskeletten in Feldversuchen mittels inertialer Messtechnik. GfA-Frühjahrskongress B.8.1 (2020)
25. DGUV (ed.): Mensch und Arbeitsplatz – Physische Belastungen: DGUV Information 208-053, Berlin (2019)
26. Serafin, P., et al.: Manuelles Heben, Halten und Tragen: Gefährdungsbeurteilung mit der Leitmerkmal- methode. Baua: Praxis. Baua, Dortmund (2022)
27. Schepers, M., et al.: Xsens MVN: consistent tracking of human motion using inertial sensing (2018)

28. Ottobock SE & Co. KGaA: Exoskelett Paexo Back (Medieninformationen): Spürbare Entlastung der Rückenmuskulatur beim Heben, Duderstadt (2022)
29. Auxivo, A.G.: Auxivo (Medieninformation): wearable support systems (2022)
30. Schomacher, J.: Gütekriterien der visuellen Analogskala zur Schmerzbewertung. Physioscience **4**, 125–133 (2008)

Robot-Based Assembly of Hydrogen Tube Fittings for Large-Scale Electrolyzers

Patrick Adler⊙, Daniel Syniawa⊙, Lukas Christ⊙,
and Bernd Kuhlenkötter⊙

Abstract

Hydrogen is one of the main pillars in the transition to renewable energy and can be used in particular for buffering and storing energy. Electrolyzers are needed to produce sustainable, green hydrogen. Today, these electrolyzers are mainly manufactured by hand. An electrolyzer plant consists of two main components, the stack in which the actual electrolysis takes place and the balance of plant that ensures the operation of the stack. Different electrolysis technologies have essential similarities in the balance of plant so that automation can achieve optimization potentials at this point. Automation technologies such as industrial robots are intended to bring this production to series maturity. For this reason, large-scale electrolysis plants are analyzed with regard to their design, the combination of different electrolysis technologies, and the connection technology used. A significant proportion of the necessary assembly steps are cable, hose, and tube connections, which automation technology has not yet been able to assemble in a process-safe and economical manner. In this paper, the process of tube connection using tube nuts is explained in more detail, and a design system for a robotic tool is presented. This end-effector is designed for industrial robots and is intended to make series production more economical. For this purpose, the necessary information processing, material flows, and energy transformations are investigated, and their interrelationships are presented. This publication aims to evaluate a suitable physical operating principle for bolting hydrogen tube nuts.

P. Adler (✉) · D. Syniawa · L. Christ · B. Kuhlenkötter
Ruhr-University Bochum, Chair of Production Systems (LPS), Industriestraße 38C,
44894 Bochum, Germany
e-mail: adler@lps.ruhr-uni-bochum.de
URL: https://www.lps.ruhr-uni-bochum.de

© The Author(s) 2025
S. Ihlenfeldt et al. (eds.), *Annals of Scientific Society for Assembly, Handling and Industrial Robotics 2023*, https://doi.org/10.1007/978-3-031-74010-7_4

Keywords

Large-scale electrolyzer • Hydrogen production • Production system • Robot-based assembly

1   Introduction: Automation of Electrolyzer Production

Hydrogen is one of the key elements to replace fossil fuels with eco-friendly energy sources. Potential applications for hydrogen are the long-term storage of energy [1]—especially from renewable sources—or the production of fossil-free steel [2] and synthetical fuels [3]. In addition, other industrial processes could also benefit from economically produced hydrogen by replacing highly greenhouse gas-emitting processes with hydrogen-powered ones. Industrial decarbonization through the increased use of hydrogen can only be truly successful if green hydrogen is used. Hydrogen is produced by a water electrolysis process, which takes place in a water electrolyzer. Electrical energy is used to separate water molecules into their constituent parts, hydrogen, and oxygen. If this process is powered by renewable energies without exception, the produced hydrogen is called green hydrogen and is entirely sustainable.

Because of the ecological benefits like the reduction of CO_2-emissions, countries worldwide are developing and pursuing strategies for increasing the industrial usage of green hydrogen [4–6]. For example, Germany forecasted a demand for 110 terawatt hour electrolysis capacity by 2030 [4], which is approximately equivalent to a production of three and a half million tons of hydrogen per year and below the currently installed capacity of solar plants or even onshore wind energy [7].

In order to provide the demand mentioned above, it is necessary to research intelligent solutions for the efficient production of water electrolyzers and related engineering processes. Primarily because of the fact that the production of water electrolyzers is currently based almost completely on manual processes.

This paper is organized as follows: Sects. 1.1 and 1.2 will provide an overview of electrolyzers as products as well as an introduction to German research projects in the hydrogen context. Further on, Sect. 2 describes which processes of water electrolyzer assembly are suitable for automation and why special consideration should be given to the assembly of tubes, hoses, and cables. Subsequently, Sect. 3 will summarize already-developed approaches for robot-based bolting processes. Before Sect. 5 draws conclusions about this contribution and gives an outlook on future research, Sect. 4 presents a first concept of an end-effector for the stationary in-factory robot-based assembly of hydrogen tube fittings and describes the preceding development process.

1.1   Brief Introduction to Electrolyzers, Stack, and Balance of Plant

Hydrogen is already being used in a wide range of industries, although it often does not yet come from renewable energy sources. This energy is converted in electrolyzers using electrical energy and chemical processes. In the course of this, water (pure or in an alkaline solution) is split, and hydrogen, oxygen, and process waste heat are generated. Locations with access to renewable energies, such as wind or water power and solar energy, are particularly suitable for the production of green hydrogen.

An electrolyzer plant consists of two main components: the stack, in which the process just described takes place, and the balance of plant (BoP), which is responsible for the supply and disposal of the stack. Usually, a stack consists of different layers which fulfill different functions. More detailed information on the structure of the stacks of different electrolysis technologies can be found in [8].

The BoP consists of all systems necessary for the electrolysis stack operation. These include water purification to ensure that no foreign substances reduce the efficiency of the process. Transformers and power electronics are designed to supply the electrolyzers with power from the available grid or a stand-alone system. For the resulting process gases, hydrogen, and oxygen, treatment is provided, in which drying of the gases and the separation of impurities are integrated. The resulting process heat can be used on a large scale for higher efficiency for other processes but has so far often been released into the environment. Depending on the size of the plant and the intended use, the BoP can be extended and specifically adapted.

1.2   Situation of German Hydrogen Research

The Federal Ministry of Education and Research of Germany supports three different research projects, called hydrogen flagship projects, in the context of hydrogen, with a total amount of up to 740 million euros. One of these projects is the H_2Giga-project, which focuses on developing solutions for the serial production of water electrolyzers on a gigawatt scale, regardless of the used electrolysis technology [9].

As part of a consortium of industrial partners and research institutions, the Chair of Production Systems (LPS) is working in the H_2Giga-project HyPLANT100. The main goal of HyPLANT100 is to build electrolysis capacity up to the gigawatt scale as economically as possible. The approach is to achieve high capacity by numbering-up and interconnecting existing electrolysis technologies with their BoP into one large plant. A conceptual configuration of a modular plant, based on containerized water electrolyzers and supplying units—like the water purification and transformers—is visualized in Fig. 1.

In the context described, the work of the LPS focuses on the development of intelligent data-driven planning processes and on researching application potentials of flexible,

Main Components:

■ H$_2$ drying and storage
■ Transformers
■ Water purification
■ Electrolysis units
■ AEM-Units
■ PEM-Units
■ Power Electronics
■ Cooling units
⚡ Renewable energy sources

Media Flows:

↑ H$_2$O
↑ Electrical energy
↑ Waste heat
↑ H$_2$

Fig. 1 Modular design of a large-scale electrolyzer based on different electrolysis technologies

economically viable, and technically feasible automation solutions in upstream assembly processes of modular electrolyzers.

2   Automation Potential in Water Electrolyzer Assembly

For the most comprehensive exploration of the automation capability in the field of water electrolyzer production, it is necessary to consider all water electrolysis technologies. Basically, a distinction can be made between the four most popular technologies of water electrolysis. These four are the alkaline, proton exchange membrane, anion exchange membrane, and high-temperature electrolysis. The various technologies have fundamental procedural and material differences [8]. However, a lot of these differences relate exclusively to the stack. The further used technologies of peripheral components are similar for the different types and mostly differ in dimensioning. Therefore, water electrolyzers, both in the cabinet and in the container scale, have a comparable component structure across all technologies (see Fig. 2) [10].

Because of the described structures, especially connecting elements such as tubes, wires, and hoses, are assembled in large numbers in the hydrogen industry independent of the process technologies used.

Solutions for robot-based wiring are already being researched at the LPS [11]. In the further course, this paper focuses on the robot-based assembly of hydrogen tube

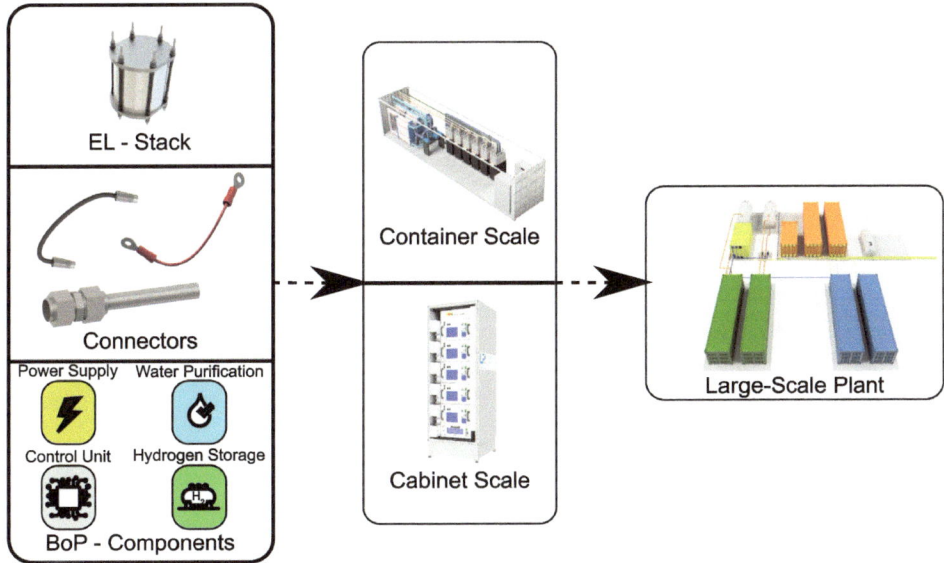

Fig. 2 Combination of the main components of one electrolysis unit

or double-ferrule fittings. These fittings are used to connect instruments and facilities of process plants. Technically tight connection for hydrogen-carrying tubes can only be achieved with the mentioned double-ferrule fitting, which minimizes the escape of the highly volatile gas. Thus, these fittings are used in almost all commercially offered electrolysis systems.

This paper aims to present a constructional solution of an end-effector for the robot-based assembly of hydrogen tube fittings. A significant challenge, but also an opportunity for automation based on industrial robots, lies in the repetitive assembly of an enormous number of double-ferrule fittings in compliance with all quality standards so that technical tightness is guaranteed. For example, an Enapter electrolyzer in container scale with a capacity of one megawatt has approximately 2,000 double-ferrule fittings [12]. The necessary assembly processes and bolting connections are multiplied by numbering up several of these one-megawatt units to the gigawatt scale. The use of industrial robots for the described bolting processes has the advantage of corresponding flexibility, which in turn enables their use for the assembly of diverse electrolyzer systems.

3   State of the Art: Robot-Based Bolting Processes

Many industrial processes require the fastening technology of bolting. Usually, manual technologies are used to fulfill this task with the use of common tools, such as wrenches. However, previous works demonstrate applications that perform automatic and robot-based bolting processes. These developments towards automation solutions are pursued in order to increase process efficiency.

An example is a presented robotic end-effector for the lateral screwing and unscrewing of bolts. It consists of three main parts: a novel screwing mechanism, a holding mechanism with force magnification as well as a self-locking drive, and a height-adjustable connection for the adjustment of the distance between the screwing mechanism and the holding mechanism. Due to this structure, it is possible to assemble different components and various component sizes [13].

An alternative concept for a robot-based gripper to screw bolts is presented by [14]. This solution includes a multifunctional three-finger gripper, which is used for grasping and bolting tasks. These two operation modes cannot be run simultaneously. The gripper is designed for screwing hexagon socket head cap bolts. The front of the gripper fingers is designed to fit the inside geometry of the screw heads of different sizes and to achieve a form-fit connection.

In addition to the described solutions for special robot end-effectors, there is a tool attachment developed for robotic two-finger parallel grippers as well. This entirely mechanical tool converts the translational gripping motion of the parallel gripper into a continuous rotation to realize bolting processes. It is based on a combined scissor-like element and double ratchet mechanism, which leads to the fact that no cable or additional drive is necessary [15].

In addition to the described developments of the gripper as the robot's end-effector, there is research on automatic bolting that goes beyond. In this context, the bolt tightening shaft for the autonomous bolt tightening operation is an example with considerable potential for industrial applications. It is used in a production module for automatically open and seal metal drums. The bolt tightening shaft locates the bolt hole and can screw bolts to the specified torque even when the bolt has a relative deviation of 5 mm from the bolt hole [16].

The robot-based automation system for steel beam assembly was presented by [17]. It consists of a bolting end-effector with two different bolting tools and a gantry-type robotic manipulator. In this case, the bolting process has two assembly steps to increase productivity. The first bolting tool of the end-effector is used in the first step to screw the bolt and the nut with high speed but low torque. The second bolting tool, in turn, finishes the tightening task with high torque but low speed as the second step. This automation solution helps to improve the productivity and safety of the steel beam assembly.

The robot-based bolting processes presented so far are stationary, but an even higher degree of automation for the automated bolting process can be achieved with mobile systems. The research of [18] includes an approach for an automatic bolting process with a mobile manipulator. The considered bolting task out of the steel beam assembly is described as uncomfortable, repetitive, non-ergonomic, and dangerous for employees. Based on the results of the investigations, the use of mobile robots for this bolting task is very potential. However, it is necessary to provide sufficient data with an adequate level of detail for the operation of collaborative mobile robots in construction.

Another solution for supporting employees in bolting tasks in hard-to-reach or dangerous locations is the unmanned aerial vehicle platform for aerial drilling and screwing. It enables performing drilling and screwing, which occurs in construction, decoration, and maintenance frequently, without the direct use of employees. This approach allows screwing very flexibly, omnidirectionally, and remotely [19].

4   End-Effector for Handling and Assembly of Tube Fittings

Robot-based assembly of hydrogen fittings is not possible with commercially available end-effectors. An end-effector is one or more tools that enable an industrial robot to perform a specific task and is usually defined by one or more tool center points (TCP). The present end-effector has to combine at least three independent functions, the rotation of the union nut, the clamping of the tube, and the integration of an optical measuring technology. These requirements result in the requirements list, which is the backbone of the performed product development.

The product development for the assembly of hydrogen tube fittings is therefore presented below. Special attention has been paid to the modular design to allow easy adaptability to other applications. For the present application, it is assumed that the following parts are provided individually. A bulkhead fitting is assembled, and a final torque fitting is mounted manually to evaluate the physical principle of operation.

4.1   Systematic Design Development for the Robotic End-Effector

The design decisions prioritize a compact overall design to minimize interference and improve accessibility to assembly areas while accommodating various electrolyzer systems. Subsequently, the decisive modules are designed, and preliminary designs are created, which in their composition, result in the entire product in the form of the final effector. The product documentation is prepared accordingly [20].

The fundamental design decisions are derived from the set of requirements for the development of the end effector, resulting in the geometry shown in Fig. 3. These include the combination of gripping and screwdriving. A minimized interference contour helps to avoid the risk of potential collisions and improves accessibility to complex or limited assembly areas. This is justified by the fact that a potential tool should be versatile enough to be used in connecting tubes for different electrolyzer systems with diverse geometric configurations.

The clamping system uses an electrically operated two-jaw gripper from FESTO, which is adapted to the application using additively manufactured gripper fingers. The optical measuring system is integrated through the medium of a visual image sensor concentric to the effective point (TCP) of the rotary gear system. It is implemented in the first prototype by a PiCam V2.1 since commercial image processing systems are not economical or suitable for the particular application.

Fig. 3 Overview of the prototype of robotic end-effector with gear, sensors, and gripper

The challenge in the assembly of a tube fitting is the application of rotational forces using torque to be generated. Several concepts have been developed to perform this function. One concept is to use a hexagon wrench and have the industrial robot perform the rotation, which results in a simple mechanical design but very complex path planning. Another concept is the use of a ratchet wrench, which is not pursued further due to the predetermined direction of rotation and the complicated integration of drive technology. Due to the combination of low interference contour, comparatively easy-to-realize robot movement, and good adaptability of the effective system, a toothed gear is used to implement the rotation function. The end-effector looks as shown in Fig. 3.

4.2 Practical Application of Rotational Motion Through Gear Train

For the implementation of the two main functions, clamping and screwing, a function separation approach is chosen. Although function integration is technically possible, it results in a much more complex mechanical structure as well as much more space-consuming. It is, therefore, not optimal for evaluating the physical operating principle. By separating the functions, the gripper already described can be used for clamping. For screwing, the previously described solution in the form of a gear is used to ensure access to the tube in a limited space, following the example of a brake pipe wrench.

Unlike manual tools, end-effectors for particular applications must mostly be equipped with sensors to verify the intended measurements. For this reason, the prototype uses an electric motor with control technology that can monitor torque and angle of rotation. In addition, optical measurement technology is integrated concentrically with the TCP. This measurement technology can identify the position of drill holes, tube ends, or hexagonal edges and transmit them to the control technology used.

The gear transmission consists of two larger and two smaller gears that are interlocked for form- and force-fit torque transmission. The drive is provided by a motor centrally located on the large input gear. It is transmitted via two satellite gears to the output gear, which has a hex recess to accommodate the union nut of the tube fitting. The two satellite gears ensure permanent meshing between the input and output gears, similar to a planetary gear. This provides a low backlash, form- and force-fit connection at all times. Plastic bearings are used because the prototype is designed with a minor possible interference contour. Forces and torques are chosen to be negligibly small since only the physical principle is to be tested.

The evaluation of the physical operating principle is ensured by an additively manufactured prototype, which enables the screwing of tube fittings. The end-effector shown in Fig. 3 is designed by iteratively improving the constructive design. Special attention is paid to making the design suitable for production so that it can be manufactured immediately using conventional machine tools—the prototype is produced to ensure the basic functionality of the process under consideration.

5   Conclusion and Next Steps

Innovative technical solutions are required to comply with the current climate challenges and the resulting changes in the energy industry. The development of solutions like the robot-based assembly of hydrogen tube fittings for large-scale electrolyzers contributes decisively to the automation of water electrolyzer production as part of the progressive automation and improvement of production processes in the renewable energy sector since more than 100,000 fittings need to be installed for a hydrogen plant with one gigawatt of power consumption.

This paper clarifies the need and the great potential of this automated screwing process in the context of water electrolyzer production. Furthermore, it shows a novel robot-based screwing approach and especially an innovative design solution, which already approved the accessibility with the robot as well as the physical principle of screwing in simulative and physical investigations. This multifunctional assembly tool contributes to the fulfillment of the H_2Giga-project HyPLANT100 and the associated research on the production of green hydrogen.

Further steps are to transfer this robot-based screwing solution to other applications. The automated hydrogen tube fitting assembly is a specific solution. Still, it offers a high potential for transferring it to other industries simultaneously due to the standardized external hexagon of the tube fittings used. It is possible to transfer this solution to all sectors in which tubes or tube fittings are mounted, for example, in process engineering, the chemical industry, or, more specifically, for procedural dashboards. Moreover, improving the automated tube fitting assembly as well as the identification of further automation potentials in the context of water electrolyzer production and the development of corresponding solutions is aimed.

Acknowledgements This work was supported by the Federal Ministry of Education and Research (BMBF) under the grant number 03HY114B within the research project H_2Giga—HyPLANT100.

References

1. Buttler, A., Spliethoff, H.: Current status of water electrolysis for energy storage, grid balancing, and sector coupling via power-to-gas and power-to-liquids: a review. Renew. Sustain. Energy Rev. **82**, 2440–2454 (2018). https://doi.org/10.1016/j.rser.2017.09.003

2. Vogl, V., Åhman, M., Nilsson, L.J.: Assessment of direct hydrogen reduction for fossil-free steelmaking. J. Clean. Prod. **203**, 736–745 (2018). https://doi.org/10.1016/j.jclepro.2018.08.279
3. Ueckerdt, F., Bauer, C., Dirnaichner, A., Everall, J., Sacchi, R., Luderer, G.: Potential and risks of hydrogen-based e-fuels in climate change mitigation. Nat. Clim. Chang. **11**(5), 384–393 (2021). https://doi.org/10.1038/s41558-021-01032-7
4. Federal Ministry for Economic Affairs and Energy (ed.): The National Hydrogen Strategy. Berlin (2020)
5. COAG Energy Council Hydrogen Working Group (ed.): Australia's National Hydrogen Strategy. Canberra, A.C.T. (2019)
6. U.S. Department of Energy (ed.): DOE National Clean Hydrogen Strategy and Roadmap. Washington, DC (2022)
7. Bundesnetzagentur (ed.): Kraftwerksliste (2022). https://www.bundesnetzagentur.de/DE/Fachthemen/ElektrizitaetundGas/Versorgungssicherheit/Erzeugungskapazitaeten/Kraftwerksliste/start.html. Accessed 29 Jan 2023
8. Shiva Kumar, S., Lim, H.: An overview of water electrolysis technologies for green hydrogen production. Energy Reports **8**, 13793–13813 (2022). https://doi.org/10.1016/j.egyr.2022.10.127
9. Federal Ministry of Education and Research: German Hydrogen Flagship Projects. https://www.wasserstoff-leitprojekte.de. Accessed 31 Jan 2023
10. Prior, J., Bartelt, M., Sinnemann, J., Kuhlenkötter, B.: Investigation of the automation capability of electrolyzers production. Procedia CIRP **107**, 718–723 (2022). https://doi.org/10.1016/j.procir.2022.05.051
11. Spies, S., Bartelt, M., Kuhlenkotter, B.: Wiring of control cabinets using a distributed control within a robot-based production cell. In: 2019 19th International Conference on Advanced Robotics (ICAR). Belo Horizonte, Brazil, 02.12.2019–06.12.2019, pp. 332–337. IEEE
12. Enapter S.r.l.: The AEM Multicore. Produce Megawatt-Scale Green Hydrogen. Simply. Rapidly. Anywhere. https://www.enapter.com/aem-multicore. Accessed 31 Jan 2023
13. Tao, R., Fan, J., Jing, F., Jun, H., Xing, S., Ma, Y., Tan, M.: A robotic end-effector for screwing and unscrewing bolts from the side. IEEE Robot. Autom. Lett. **7**(4), 9786–9793 (2022). https://doi.org/10.1109/LRA.2022.3192773
14. Pei, M., Xu, K., Ding, X., Jiang, S., Gao, X.: Design and analysis of continuous rotating multifunctional mechanical gripper. In: 2018 IEEE International Conference on Robotics and Biomimetics (ROBIO). Kuala Lumpur, Malaysia, 12.12.2018–15.12.2018, pp. 2007–2012. IEEE
15. Hu, Z., Wan, W., Koyama, K., Harada, K.: A mechanical screwing tool for parallel grippers—design, optimization, and manipulation policies. IEEE Trans. Robot. **38**(2), 1139–1159 (2022). https://doi.org/10.1109/TRO.2021.3091282
16. Liu, S., Ge, S.S., Tang, Z.: A modular designed bolt tightening shaft based on adaptive fuzzy backstepping control. Int. J. Control Autom. Syst. **14**(4), 924–938 (2016). https://doi.org/10.1007/s12555-015-0008-0
17. Chu, B., Jung, K., Ko, K.H., Hong, D.: Mechanism and analysis of a robotic bolting device for steel beam assembly. In: ICCAS 2010. 2010 International Conference on Control, Automation and Systems (ICCAS 2010). Gyeonggi-do, 27.10.2010–30.10.2010, pp. 2351–2356. IEEE
18. Brosque, C., Galbally, E., Khatib, O., Fischer, M.: Human-robot collaboration in construction: opportunities and challenges. In: 2020 International Congress on Human-Computer Interaction, Optimization and Robotic Applications (HORA). Ankara, Turkey, 26.06.2020–28.06.2020, pp. 1–8. IEEE
19. Ding, C., Lu, L., Wang, C., Ding, C.: Design, sensing, and control of a novel UAV platform for aerial drilling and screwing. IEEE Robot. Autom. Lett. **6**(2), 3176–3183 (2021). https://doi.org/10.1109/LRA.2021.3062305

20. VDI 2221, 2019-11: Design of Technical Products and Systems—Model of Product Design

Novel Concept for Micro-Assembly of Optical Devices Using a Planar Motor System

Niklas Terei, Lars Binnemann and Annika Raatz

Abstract

Research and Development in recent years has led to first commercially available, magnetically levitated planar motor systems. Considering the advantages of such systems, they perfectly fit as a substitute for multi-axis fine positioning stages in high-volume productions. To address the need for solutions for high-volume packaging of optical devices, we present a novel concept for a micro-assembly station using a planar motor. To qualify a novel, vacuum-operatable planar motor system for our design, we conducted experiments regarding positioning repeatability, positioning noise, and dynamics on a pre-series system.

Keywords

Maglev planar motor · Micro-assembly · Optical devices

N. Terei (✉) · L. Binnemann · A. Raatz
Leibniz University Hanover, Institute of Assembly Technology, An der Universität 2, 30823 Garbsen, Germany
e-mail: terei@match.uni-hannover.de
URL: https://www.match.uni-hannover.de/en/

Cluster of Excellence PhoenixD (Photonics, Optics, and Engineering - Innovation Across Disciplines), Hanover, Germany

S. Ihlenfeldt et al. (eds.), *Annals of Scientific Society for Assembly, Handling and Industrial Robotics 2023*, https://doi.org/10.1007/978-3-031-74010-7_5

1   Introduction

The maglev (magnetically levitated) planar motor has been well-developed in recent years and is showing great potential as a precision motion system. The electromagnetic thrust in maglev planar motors is generated by the interaction between the magnetic field of a permanent magnet array (mover) and the current in a stationary array of coils (stator). Thus, the mover can be controlled in six degrees of freedom (DoF). Due to its principle of operation, the mover is entirely passive, from which the many advantages of the system can be derived. To begin with, the stator tiles can be arranged and rearranged in user-tailored machine layouts, providing production flexibility. Each mover can be assigned to an individual path through the production facility to meet customer-specific demands and realize smart manufacturing. Due to the passivity of the movers, minimal cleaning and maintenance effort is required, and they are operatable in demanding environments such as cleanrooms, vacuum environments, and in hygienic applications. The exceptional dynamics, the motion capabilities in 6-DoF, and the high positioning precision make the system not only suitable as a transport system but also as a motion unit within individual manufacturing processes [1]. The main limitations of the system concern the limited travel range in the direction perpendicular to the stator plane, and in the rotational axis that restrain the technology to specific types of tasks. Moreover, due to the magnetic levitation, modeling and controlling of the system is complex, and the motor's stiffness is limited. Given the advantages and disadvantages of maglev motors, in past works, the system was recommended for use in lithography production processes as a single-stage fine-positioning stage [2, 3]. Apart from this application, we see particular potential for the usage of the maglev planar motor in silicon photonics production. There, silicon is used as an optical medium to realize highly integrated optical devices on a microchip level. The great benefits of silicon photonics result from the manufacturing processes using well-developed and scalable CMOS production technologies, thus providing low-cost mass manufacturing capability [4]. However, the packaging of these so-called photonic integrated circuits (PICs) is significantly more challenging and currently orders of magnitude more expensive than electronic packaging. This is due to the compelling necessity of micron-level alignment of the optical components (e.g., laser diodes, glass fibers) that are to be assembled to the PIC [5]. In order to make optical products economically viable in more applications, an overall cost reduction for the manufacturing of optical devices is required. Additionally, the production volume of photonic devices is smaller compared to electronic devices, so more flexible assembly machines are desired [6]. Hence, in this paper, we present a design for an assembly station using a maglev motor that has the potential to be cheaper, more flexible, smaller in size, and more dynamic than conventional optics assembly machines.

Before we go deeper into the concept of the assembly station, we first point out current technological challenges regarding maglev motors and what challenges have been focused on in the past. Further, we provide an overview of the few commercially available systems that have emerged from decades of research on maglev motors. Based on these systems, we present a novel design for an assembly station for optical devices. In order to qualify the system for application in optical assembly processes and our design, we conducted experiments regarding positioning accuracy and dynamic behaviour on a pre-series system.

2   Maglev Motors-Research and Commercial Systems

In the following, we give a short overview about the technological challenges and the research regarding maglev planar motors (Sect. 2.1). Further, we present the few commercially available systems and their industrial applications (Sect. 2.2).

2.1   State of Research

The first mention of a planar motor, which is magnetically levitated and does not use aerostatic bearings, was by Kim in his doctoral thesis from 1997 [1] and in his earlier related works. The high-precision planar motor by Kim consists of a levitated mover with a Halbach array and stationary stator coils and thus can be controlled in all 6-DoF. In a Halbach array, the magnetized elements are arranged vertically and horizontally to their magnetic axis, which results in an amplification of 40% of the magnetic field on one side of the array while cancelling the field to near zero on the other side [7]. Kim specified his motor with travel ranges in x, y, and z of 200, 200, and 0.15 mm, 1-g acceleration and a 5 nm positioning noise at $3\ \sigma$. To this day, research focuses on optimizing Halbach arrays and stator coil arrangements [8–10]. In 2012, Lu et al. presented the first cable-free 6-DoF direct drive planar motion system, in which the number of coils increases linearly (vs. quadratically in earlier solutions) with the stroke. The system is, therefore, easily scalable to meters-long X-Y strokes, enabling the technology to be used for workpiece transportation and flexible production [11]. In their prototype, Lu et al. used a 6-DoF stereo-vision-based positioning sensor to provide control feedback. Based on his experience, Lu co-founded Planar Motor Incorporated (PMI) in 2017.

A great research challenge lies in the modeling and controlling of maglev planar motors. Due to the physical functionality of magnetic levitation, the system presents a nonlinear multi-input-multi-output (MIMO) control problem. The general approach is to decouple the system equations to get six single-input-single-output (SISO) system equations that can be controlled by PID or by state controllers [2, 12, 13]. While earlier work focuses on developing analytical models, recent research presents solutions that include numerical models and data-driven control approaches [3, 14].

Besides model accuracy, also model computation time is of importance. In order to realize control for high precision positioning of a maglev planar motor, low latency and high sample times in a range of 5 kHz are required [1, 15].

2.2   Commercially Available Systems

For a few years, there have been several commercially available systems on the market. Beckhoff Automation, for example, offers the XPlanar and PMI the Planar Motor System, which B&R Industrial Automation also distributes as ACOPOS 6D. Mafu Robotics offers the M-Drive, a modified version of the planar motor from PMI, which is operatable in a vacuum environment. A maglev stage for high payloads is offered by Phillips Engineering Solutions. The development of such maglev planar motors is restricted by patents held by PMI (e.g., [16–18]). Table 1 shows the technical specifications of the maglev systems. Most movers are available with square as well as rectangular geometries. Depending on the mover size, the maximum payload is up to 14 kg or 100 kg in the case of the system from Phillips. Furthermore, the movers can reach speeds of up to 10 m/s. The maglev systems offer 6-DoF with a repeatability of the mover position from $\pm 2\,\mu$m (Mafu) to 50 μm (Beckhoff) [19–22].

Not much information regarding industrial applications using maglev planar motors is available in the public domain. Examples can be found for the dispersion of adhesives, and for plasma applications. In these, the needle or the plasma nozzle is fixed and the mover performs all necessary movements underneath [19, 21, 23]. Planar motors are also used for simple assembly tasks. The Eutect GmbH sells assembly lines based on the maglev motor of Beckoff that have the advantage of flexible part geometries as well as flexible process task orders. However, the planar motors are only used for simple pick and place applications but not for precision assembly. Further, they do not use the 6-DoF of the maglev motor but rely on serial robots [24].

Table 1 Technical specifications of the maglev systems commercially available

Specs/manufacturer	Mafu robotics GmbH	Beckhoff automation GmbH	PMI/B&R Automation	Phillips engineering solutions
Mover size [mm^2]	210×210	113×113 to 235×235	120×120 to 450×450	500×500
Max. speed [m/s]	10	2	10	1
Repeatability [μm]	± 2	50	± 5	–
Max. payload [kg]	2	0.4–4.2	0.6–14.4	100

3   Concept for a Maglev Motor Micro-Assembly Station

In conventional high-precision assembly machines, part handling is typically carried out by a portal or gantry system, while alignment and assembly are performed by multi-axis high-precision positioning systems (HPPS). The redundancy of axis and the utilization of high-tech micro- and nanopositioners results in high investment costs. Moreover, process times are increased by the time needed for machine loading and unloading and by the limited motion speed of the HPPS. However, to make optical products economically viable in more applications, an overall cost reduction for machinery and manufacturing is required [6]. Thus, in this chapter, we suggest a solution for an assembly station that uses a maglev planar motor for workpiece transportation as well as an HPPS for assembly. This results in reduced investment costs regarding workpiece handling and assembly. Compared to commercially available 6-DoF HPPS, the investment costs of approx. 15,000 € for a mover and a stator are relatively low. A further advantage of the system is the compatibility with clean and especially vacuum environments, which are, in some cases, mandatory for optics manufacturing. In this context, the small footprint of the station is financially favorable. Compared to conventional assembly machines, which are normally highly specialized, the use of maglev motors enables smart and flexible production.

The following chapter presents the concept for a novel inline assembly station based on a maglev planar motor system. In Sect. 3.1, the components of the station are described. Afterwards, Sect. 3.2 presents the suggested assembly process.

3.1   Components of the Inline Assembly Station

Figure 1 shows the general design of the inline assembly station. The main components are the planar motor, the vertical axis with attached tools (gripper, dispensing unit and distance sensor), an UV light source, and the camera setup. The components, as well as their tasks, are described in the following.

The maglev planar motor, especially the mover (1), is the base of the inline assembly station. The movers carry the wafers on which the components (e.g., laser diodes) are mounted, as well as the components in a workpiece tray. Alternatively, it is possible to realize part feeding with additional dedicated magazine movers. Due to the flexibility of the planar motor system, the magazine movers can overtake other movers and stations and remain in a waiting spot near the assembly station. During the assembly processes, the movers position the components and the assembly location underneath the gripper (3). In addition, the movers perform every alignment movements for the required assembly accuracy. The design of our concept is based on the M-Drive system from Mafu Robotics since it is the only system operational in vacuum environment.

An additional vertical axis (2) provides more flexibility for the assembly station because of the larger travel range (in the z-direction) compared to the movers. This, in turn, enables larger parts to be mounted on the mover. Furthermore, the vertical axis is necessary to measure gripped components with the camera (7). For this purpose, the gripper and the components are positioned on the focal plane above the camera. Additionally, the vertical

1 - maglev planar motor

2 - vertical axis

3 - gripper

4 - dispense unit

5 - UV light source

6 - confocal distance sensor

7 - camera

8 - telecentric lens

9 - beam splitter

Fig. 1 Components of the inline assembly station based on maglev planar motor

axis carries the tools described in the following. The presented concept uses a linear axis with a stepper motor and a lead screw.

There are three tools mounted on the vertical axis. The first one is a micro vacuum gripper. Its purpose is to pick up the components from the magazine and place them onto the wafer. The second tool is a volumetric controlled adhesive dispense unit (4). It is mounted on a slider to move it up and down (waiting position higher than gripper; working position lower than gripper). To cure the adhesive, an UV light source (5) is implemented. The third tool is a distance sensor (6) used to level the mover for the different assembly steps. In the presented concept, a confocal sensor with a resolution of 126 nm is used.

The camera, in combination with a telecentric lens (8), is used to measure the x-/y-positions of the components and the assembly locations. A rotatable beam splitter (9) is mounted in front of the lens to look down onto the mover (component, magazine, assembly location) and up to the tools (gripper, dispense unit). The camera setup is mounted on rail guides to move it in and out of the workspace. The chosen camera setup provides a field of view of 3.32 x 2.22 mm^2 and a theoretical resolution of 0.6 μm/pixel.

3.2 Suggested Process of the Inline Assembly Station

The assembly process is demonstrated by placing a micro laser diode in front of a waveguide on a wafer. The diode has a cuboid geometry with a volume of 300 x 300 x 100 mm^3 (LxWxH). First, the mover carries the wafer and the magazine with the diodes into the assembly station and positions the diode that is to be assembled underneath the distance sensor. By measuring a few surface points, the diode plane is calculated and leveled by a tip-tilt motion of the mover. Next, the mover positions the diode underneath the gripper. Here, the x-/y-position of the diode is measured with the camera and adjusted by a movement to the gripper. The diode is gripped with the vacuum tool, and the assembly location is leveled using the distance sensor. Then, the assembly location is moved under the adhesive needle. The lateral deviation of the needle is corrected using the camera. Following, the adhesive is applied by the needle and a movement of the vertical axis. Next, the gripper is positioned above the assembly location and the x-/y-position is adjusted to the gripped part using the camera. Finally, the diode can be placed by a movement of the vertical axis.

4 Experimental Examination of a Maglev Planar Motor

Apart from a specification for the positioning repeatability of the M-Drive system, currently, there is no technical information regarding the positioning noise and the system's dynamics available. In order to qualify the M-Drive system as a suitable actor for high-precision positioning and our designed assembly station, we conducted several preliminary experiments on a pre-series system. Our experiments aimed to clarify the system's positioning repeatability, positioning noise, and dynamic behavior. To do so, we measured the proximity of the mover to a confocal sensor from Micro-Epsilon (IFS2405-3) in repetitive movements. The principal measurement setup is illustrated in Fig. 2. To account for measurements in the different directions of movement, the sensor was oriented in configurations A and B (highlighted in blue and yellow resp.). A parallel gauge block, which was adhered to the mover, provides a plain measurement surface. The sensor has an operating range of 3 mm starting at a distance of 20 mm from the sensor's head and is specified with a measuring resolution of 126 nm. The sensor noise was measured to be 90 nm at 1000 Hz sample frequency. Due to the status of a pre-series system, an active air or water cooling system for the stator coils was unavailable, so we had to operate the system in a thermally unstable state.

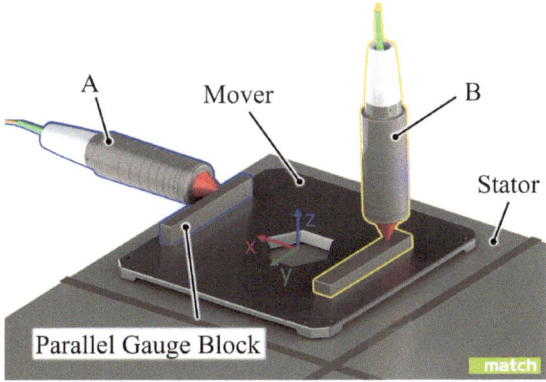

Fig. 2 Experimental setup

In order to identify the system's positioning repeatability, the mover was set up to approach the measuring position multiple times with single-axis movements (x-direction and z-direction resp.), whereby different accelerations and speeds have been tested. We recorded multiple measurement series with similar results. As an example Fig. 3a shows the average positioning deviation from the commanded position for the x-direction for multiple repetitions. The average positioning deviation and positioning noise have been derived from a stabilized state of the mover. Given the measurements from Fig. 3a (19 repetitions), we calculated the positioning repeatability for the x-direction of 1.8 μm at 3 σ. It is noticeable that the deviation of the distance to the nominal position decreases with the number of repetitions. Looking at the course of the measurements in Fig. 3a and the significant trends in the other series of measurements, we assume a superimposition with a thermal drift which is caused by thermal expansion of the system. Consequently, the actual positioning repeatability of the system might be more precise than our experiments imply. Nevertheless, our experiments provide a first inside into the system's capabilities.

Analogous to the x-direction, we investigated the mover's positioning repeatability for z-axis movements. Our results show that the positioning repeatability increases with a decreasing air gap, which can be attributed to the sharp decrease of magnetic force with an increasing air gap. At z = 2.5 mm, we derived the positioning repeatability to be 1.6 μm, and 1 μm at z = 0.5 mm (values given at 3 σ and calculated from 16 repetitions). The acceleration of the z-movements could not be specified because of the pre-series status of the control. Analytically, we approximated the acceleration to be 200 mm/s^2 from the derived velocity.

Looking at the long-term positioning noise in the x-direction, the mover shows an oscillating behavior depicted in Fig. 3b. At rest, the system oscillates periodically with an amplitude of approximately 0.23 μm with a standard deviation of 0.22 μm and a frequency of 0.11 Hz. Whether this systematic disturbance in the measuring is due to non-optimal control parameters or due to other environmental influences is still to be investigated in the future.

Our experiments regarding the positioning repeatability also provide information about the motor's dynamics. Figure 3c and d show the course of a characteristic transient response of x- and z-axis movements. Regarding the x-axis, the mover stabilizes to an acceptable deviation in approx. 0.5 s with a max. positional undershoot of approx. 23 μm. It is notable that the planar motor does not show the typical behavior of a damped oscillation like traditional machine axis do. This could be referred to neglected coupling effects, inaccurate model parameters, or internal filtering of the step excitation. For the z-axis, we have determined the max amplitude and stabilization time to be approx. 6 μm and 0.1 s at z = 2.5 mm. Comparing Fig. 3c and d, the z-axis oscillation appears much more harmonic. For lower z-positions, the motor's undershoot is reduced while the stabilization time and the initial overshoot remain at the same level.

Regarding accuracy requirements for optics assembly, the 1 dB planar alignment tolerance for grating couplers can be specified with ±2.5 μm and for edge coupling with ±0.5 μm [5]. Considering the positioning noise and the thermal influences on our measurements, we are optimistic that with adequate cooling and tuning of the control parameter, the precision

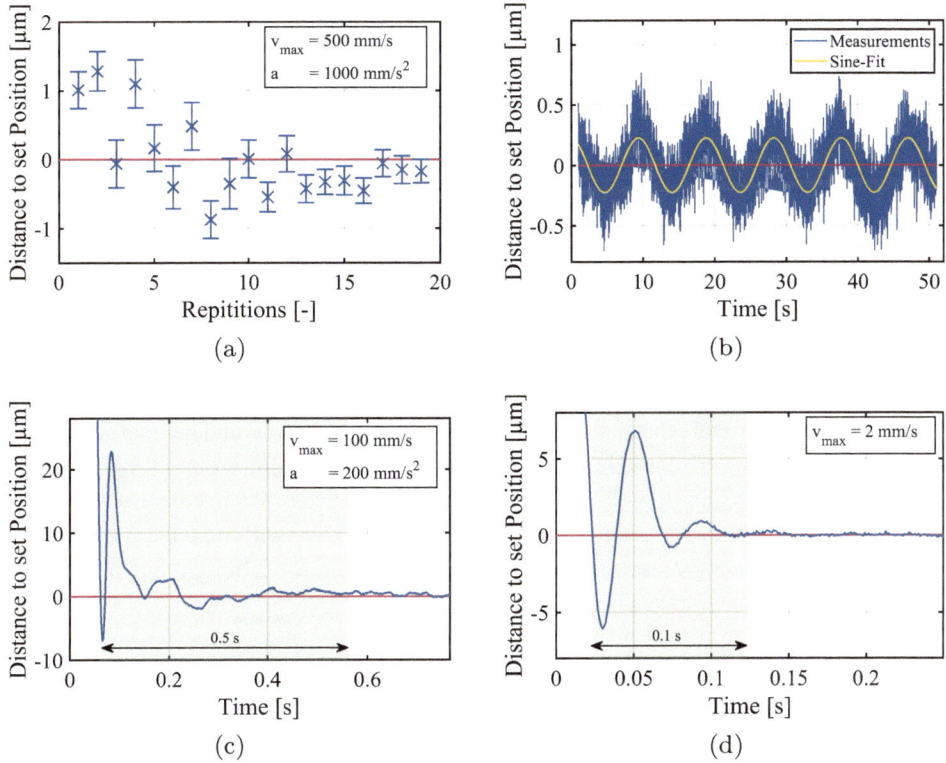

Fig. 3 Positioning repeatability measurements **a** and positioning noise **b** of the mover in x-direction as well as the mover dynamics in x-direction **c** and z-direction **d**

of the planar motor is sufficient for the assembly of optical components. Further, our findings provide information regarding the stabilization time and the z-position-dependent positioning accuracy, which are essential for the operation of the system in assembly processes.

5   Conclusion and Outlook

Although the concept of maglev planar motors is old, commercially available systems have just been introduced to the market a couple of years ago. Given the high positioning precision, the system is not only suitable for flexible workpiece transportation but also as a functional positioning unit within different processes. However, as discussed in Sect. 2.1, controlling maglev planar motors is still a current research topic. While research proposes to use maglev planar motors in lithography processes, we envision using them in micro-assembly of optical systems. Thus, this paper presents a novel precision assembly station that can be integrated into an inline production facility for optical devices or PIC production. The shown concept uses the maglev planar motor to handle, position and align components. An additional vertical axis performs the placing motion to extend the flexibility for larger parts. By using the planar motor for all positioning and alignment tasks, the number of stacked axis is reduced and, thereby also the complexity of the facility.

Due to the novelty of maglev planar motors on the commercial market, there are not much public information about the systems precision and their dynamics. Thus, we conducted some preliminary experiments on a pre-series planar motor regarding the system's positioning repeatability, positioning noise, and the system's dynamics. Although we could not guarantee thermal stability of the stator coils in these first experiments, we determined some key characteristics of the system and identified the motor as viable for our design. In future work, we will present a more sophisticated analysis in a thermally controlled environment. In addition to the properties already examined, we will investigate the repeatability of movements in all 6-DoF, the absolute accuracy, the path accuracy, and the motor's stiffness. Due to the coupling of the axis and the high dynamics, a mature measuring setup or an advanced sensing technology is required to record the system's movement in 6-DoF with high position resolution at a high sample rate. In parallel, we will implement our design of a maglev micro-assembly station and qualify the machine for different assembly tasks on optical devices.

Acknowledgements The research in this paper is funded by the Deutsche Forschungsgemeinschaft (DFG, German Research Foundation) under Germany's Excellence Strategy within the Cluster of Excellence PhoenixD (EXC 2122, Project ID 390833453). The authors thank the German Research Foundation (DFG) for their support. Additionally, we thank our partners Dr. Axel Günther and Prof. Dr.-Ing. Wolfgang Kowalsky from TU Braunschweig and Mafu Robotics for the excellent cooperation.

References

1. Kim, W-j: High-Precision Planar Magnetic Levitation. Doctoral thesis, Massachusetts Institute of Technology, Massachusetts (1997)
2. Jansen, J.W., et al.: Modeling of magnetically levitated planar actuators with moving magnets. IEEE Trans. Magn. **43**(1), 15–25 (2007)
3. Ou, T., et al.: Generation mechanism and decoupling strategy of coupling effect in Maglev planar motor. IEEE/ASME TMECH (Early Access), (2022)
4. Cheng, L., et al.: Grating couplers on silicon photonics: design principles. Emerg. Trends Pract. Issues. Micromach. **11**(7), 666–691 (2020)
5. Carroll, L., et al.: Photonic packaging: transforming silicon photonic integrated circuits into photonic devices. Appl. Sci. **6**, 426–447 (2016)
6. Integrated Photonic Systems Roadmap-International (IPSR-I): IPSR-I - Assembly. https://photonicsmanufacturing.org/sites/default/files/documents/assembly.pdf, last accessed 2023/02/03
7. Trumper, D., Williams, M., Nguyen, T.: Magnet arrays for synchronous machines. IEEE Indust. Appl. Soc. Ann. Meeting **1**, 9–18 (1993)

8. Rovers, J.M.M., Jansen, J.W., Lomonova, E.A.: Design and measurements of the Double Layer Planar Motor. In: IEMDC, pp. 204–211, IEEE, Chicago (2013)
9. Zhang, X., et al.: Analytical model of magnetic field of a permanent magnet synchronous motor with a trapezoidal halbach permanent magnet array. IEEE Trans. Magnet. 7(55), 1–5 (2019)
10. Kleijer, M., Jansen, J. W., Lomonova, E. A.: Optimization of quasi-halbach topologies to maximize the acceleration of moving-magnet planar motors. In: International Conference on Electrical Machines, pp. 942–948. IEEE, Valencia (2022)
11. Lu, X., Usman, I.-U.-R.: 6D direct-drive technology for planar motion stages. CIRP Ann. - Manufact. Technol. 61(1), 359–362 (2012)
12. Zhou, G., et al.: Research on control methods of planar motors based on the modal forces. In: World Automation Congress (24), pp. 1–5, Waikoloa (2008)
13. Reymond, V., et al.: Modeling and control of a magnetically levitated planar motor. In: European Control Conference, pp. 3927–3932. Budapest (2009)
14. Li, X., et al.: Data-driven multiobjective controller optimization for a magnetically levitated nanopositioning system. IEEE/ASME TMECH 25(4), 1961–1970 (2020)
15. Hu, T.: Design and Control of a 6-degree-of-freedom levitated positioner with high precision. Doctoral thesis, Texas A&M University, College Station (2005)
16. Lu, X., et al.: Displacement devices and methods for fabrication, use and control of same. US 2020030995 A1, United States Patent and Trademark Office (2020)
17. Lu, X., Xiao, Y. K.: Systems and Methods for identifying a magnetic Mover. US 20210376777 A1, United States Patent and Trademark Office (2021)
18. Lu, X., Williamson, G.: Stator modules and robotic systems. EP4078791A1, EPO (2022)
19. Beckhoff Automation GmbH Homepage, https://www.beckhoff.com/en-gb/products/motion/xplanar-planar-motor-system/apsxxxx-xplanar-tiles/aps4322-0000-0000.htmlbeckhoff.com, last accessed 2023/01/17
20. Mafu Robotics GmbH Homepage, https://www.mafu-robotics.de/index.php/en/vakuumkammer-27663008.htmlmafu-robotics.de, last accessed 2023/01/17
21. Planar Motors Inc. Homepage, https://planarmotor.com/en/productsplanarmotor.com, last accessed 2023/01/17
22. Philips Engineering Solutions Homepage, https://www.engineeringsolutions.philips.com/news/magnetic-levitation-stage-planar-motor/philips.com, last accessed 2023/01/18
23. Beckhoff Automation GmbH & Co. KG, Floating planar mover simplifies transportation of sensitive workpieces. In Beckhoff Automation GmbH & Co. KG PC-Control, vol 01-2020, pp.18–21 (2020)
24. Eutect GmbH Homepage, https://eutect.de/en/xplanar/eutect.de, last accessed 2023/01/18

Development of a Workflow for the Aggregation and Usage of Data in Digital Twins of Adaptable Assembly Systems

Sarah Zimmer⑩, Jan Molter, Lennard Margies⑩, Martin Karkowski, and Rainer Müller

Abstract

Challenges of the fourth industrial revolution such as shorter product life cycles and increasing product variance require the continuous use of digital technologies in production. Only through these solutions the use of adaptable systems is possible. In this context, the Digital Twin is of crucial importance. The Digital Twin is represented across multiple sectors to meet the technological change. In manufacturing companies its areas of application are versatile. The usage can range from production planning and production monitoring to production optimization. In order to be able to use the advantages of the Digital Twin in the mentioned areas, a targeted data aggregation is necessary. In this paper a cross-application workflow for the efficient planning and usage of Digital Twins is presented. The workflow covers the whole process from the goal and requirement definition up to the implementation and usage of the technology. Particular attention is paid to the aggregation and usage of data in Digital Twins. The developed workflow is evaluated based on an adaptable assembly system.

Keywords

Digital Twin • Data aggregation • Data modeling • Adaptability

S. Zimmer (✉) · J. Molter · L. Margies · M. Karkowski
Center for Mechatronics and Automation gGmbH (ZeMA), 66121 Saarbrücken, Germany
e-mail: sarah.zimmer@zema.de

R. Müller
Chair of Assembly Systems, Saarland University, 66123 Saarbrücken, Germany

S. Ihlenfeldt et al. (eds.), *Annals of Scientific Society for Assembly, Handling and Industrial Robotics 2023*, https://doi.org/10.1007/978-3-031-74010-7_6

1   Introduction

1.1   Motivation

Nowadays, manufacturing companies are forced to be flexible and adaptable due to enormous price pressure, shortened product life cycles, an increasing number of product variants and many other reasons. As part of the fourth industrial revolution, different solutions have evolved to meet these challenges. With the development of new technologies, the aggregation, processing and use of data is gaining particular importance in manufacturing companies. The Digital Twin plays a central role in addressing the challenges mentioned due to its versatile range of applications. Current estimates assume that the market for the Digital Twin will grow by 30 to 40 percent annually until 2025, reaching a value of 7 billion euros [1]. To maximize the potential of the Digital Twin in the future, there are still some challenges that must be addressed today. The first obstacle lies in the term "Digital Twin" itself. There are different definitions existing in literature which can differ significantly, depending on the area of application and the objectives pursued. Additionally, Digital Twins and similar applications, such as product life cycle management systems, can often be indistinguishable from one another.

A further challenge lies in the aggregation and usage of the relevant data in the Digital Twin. Currently there is a lack of goal-driven holistic approaches that support the planning and utilization of Digital Twins across various different fields of application. Existing methods mostly focus on creating and using Digital Twins in specific software programs [2].

1.2   Scope of the Paper

This paper aims to deal with the described challenges. After giving an approach for a definition of the term "Digital Twin" and outlining the main connections of assembly systems, a workflow for the efficient planning and implementation of Digital Twins is developed. Thereby, the main focus is on the aggregation and usage of data in Digital Twins. The workflow is designed to be used in manufacturing, especially in assembly processes. It will be validated in an adaptable assembly system for fuel cell stacks.

2 State of the Art

2.1 Digital Twin

The first description of the concept was provided by Michael Grieves in 2002 as part of his research in product lifecycle management at the University of Michigan. The concept, initially known as the "Conceptual Ideal for PLM", was further developed in the following years, which also changed the name of the model. The approach, that was called "Mirrowed Spaces Model" or "Information Mirroring Model" in the meantime, was not called Digital Twin until a 2011 publication by Grieves and Vickers [3]. Regardless of the various names, the idea of the Digital Twin was ranked among the "Top Trends 2017" in the Gartner Hype Cycle for Emerging Technologies [4]. In connection with the Digital Twin, terms such as Digital Shadow, as well as Digital Model, are often mentioned. The Digital Model is commonly characterized as a representation of an asset from the physical world in the virtual world, that requires manual updating of changes [5, 6]. The so-called Digital Shadow on the other hand, consists of aggregated data traces and models [7]. Through this, Digital Shadows are generally described as a virtual representation of an asset that enables automatic reflection of changes from the physical world to the virtual world [5, 6]. However, a proper taxonomy for the terms Digital Model, Digital Shadow, and Digital Twin is still under discussion today [8].

Due to this circumstances, various sources have been considered to determine the following definition of a Digital Twin, used in this paper:

A Digital Twin is a virtual information construct of a physical product [3], a service system [9], a system behavior [10], a process [11] or other existing or planned elements, which contain a complete description of the respective object. The Digital Twin is characterized by the possibility of a bi-directional data exchange between the virtual and the physical world [3]. To realize an automated data transfer from the physical world to the virtual world, the Digital Twin uses a Digital Shadow, which is a main component of the Digital Twin alongside the Digital Model [5, 9]. The Digital Model in turn contains the master data, geometric models or other core data and represents the basis of a Digital Twin [6]. Only through the combination of Digital Model, Digital Shadow and an extension of the Digital Shadow with an information transfer to the physical world, a complete Digital Twin can be created as shown in Fig. 1.

Independent of the internal structure of a Digital Twin, Digital Twins can be categorized in three main types. Depending on its use it can either be categorized as Digital Product Twin, Digital Production Twin or Digital Performance Twin [12, 13]. The Product Twin is a product focused Digital Twin. It is used in product development and aims to improve different product characteristics before the product is finally being manufactured [14]. Furthermore, it helps to optimize a product, its characteristics or its properties [13]. The Production Twin is a virtual representation of a production system. It is mainly used to enable a virtual commissioning but also supports users control and optimize production

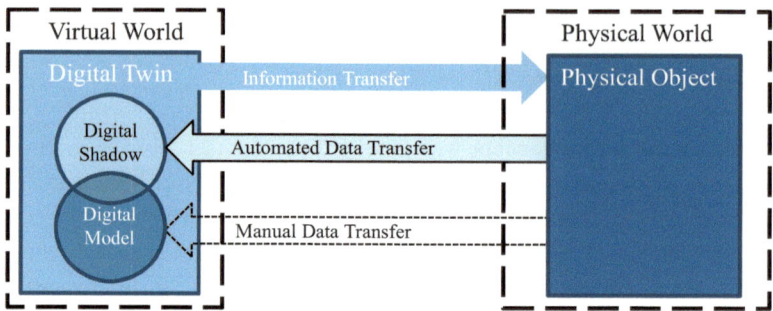

Fig. 1 Interaction between the Digital Twin and the physical world

processes [13]. The Performance Twin works as a protocol of all relevant data and provides the user with the possibility to improve a product, a process or a production system during its life cycle [13, 14]. In the context of this paper the main focus is on the Digital Performance Twin.

2.2   Assembly Systems

Assembly data. When considering assembly systems, it is important to mind the stringent connection between products, processes and the used equipment as shown in Fig. 2.

To assure the customers quality demanded, it is necessary to supervise the underlying assembly processes, the equipment used in these processes and their influence on the final product. Therefore, data and information must be aggregated, analyzed and harnessed during the entire assembly process. At this point it is necessary to differentiate between the terms data, information and knowledge. Data represent abstractions of the real world. They are machine-readable characters that follow a specific syntax, such as numbers or digits. Linking and analyzing data creates information with a subject-dependent meaning. Through the interpretation and processing of information, knowledge is generated [14]. During assembly various types of operational data are accumulated that include task data,

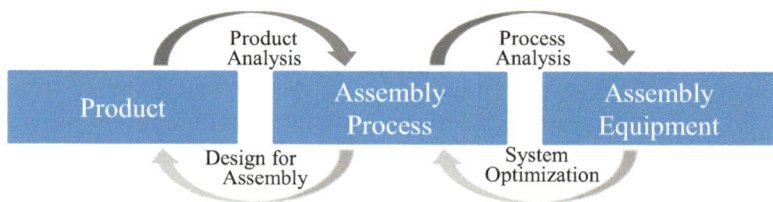

Fig. 2 Dependencies between product, process and equipment in assembly

personnel data, warehouse data, quality data etc. In this context, the focus is most of all on quality data. Quality data provides defect indices, causes of rejects, results of quality analyses and inspection values [15].

Adaptability. An increasingly important factor in the development and construction of assembly systems is adaptability. Adaptability refers to the potential of a system to respond to changes even beyond defined corridors [16]. Adaptable systems thus have no explicit boundaries in their implementation [17], which distinguish them from flexible systems [18]. To ensure responsiveness to change, an assembly system must have characteristics that enable it to adapt. Change enablers in production are universality, mobility, scalability, modularity and compatibility [19].

In the context of this paper, scalability and modularity are of particular importance. Scalability describes the ability to expand and reduce in terms of space and personnel. Modularity characterizes the use of standardized, functional elements that are modularly interchangeable [19].

Planning of an assembly system. There are various approaches for planning assembly systems. The workflow presented in this paper is based on the Lotter and Wiendahl approach. Their planning framework consists of eleven steps, that are shown in Fig. 3. A detailed description of the planning framework is given in Lotter and Wiendahl's "Montage in der industriellen Produktion" [20].

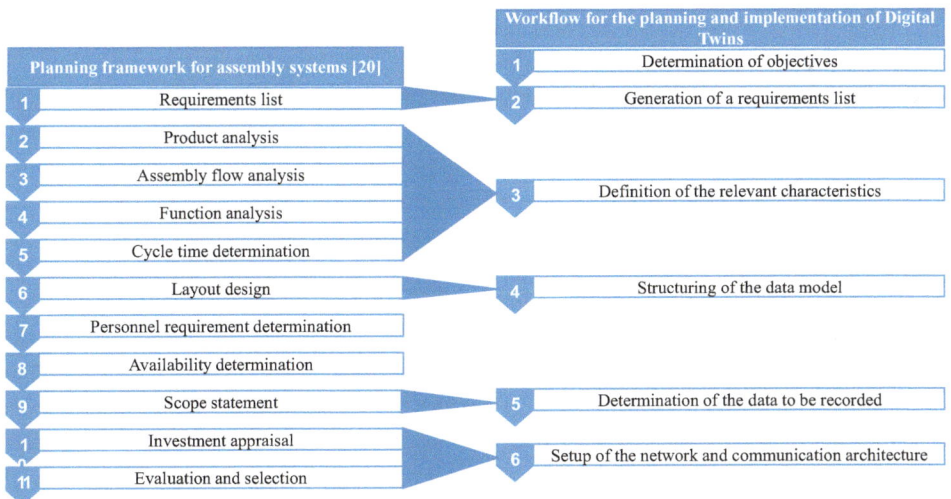

Fig. 3 Derivation of the workflow based on the general planning framework according to Lotter and Wiendahl

3   Aggregation and Usage of Data in Digital Twins

3.1   Development of a Workflow

The aim of the workflow presented here is to facilitate the planning and implementation of Digital Twins for the use in assembly systems. The approach is subject to a holistic consideration in order to efficiently utilize Digital Twins in various fields. It is based on the general planning framework for assembly systems according to Lotter and Wiendahl as described in Sect. 2.2. The derivation of the workflow from the general planning framework is shown in Fig. 3.

In contrast to the general planning framework according to Lotter and Wiendahl, the development of a Digital Twin begins with the **determination of objectives**. Recent studies show, that the second most common barrier to using Digital Twins is companies not knowing where to begin with the implementation [21]. Therefore, clear objectives are indispensable right from the start. It is necessary to differentiate if the subsequent Digital Twin is about an already existing assembly system or one that is in the planning phase. In the case of already physical existing products, processes and equipment, future goals can possibly be derived from experiences. If the assembly system is still in the planning phase, use cases need to be identified in order to tailor the Digital Twin to the underlying system's needs. Objectives for the implementation of a Digital Twin may result from general corporate objectives. Exemplary goals can be cost reduction or quality improvement with the help of the Digital Twin. If possible, goals should be formulated SMART, which means specific, measurable, achievable and relevant.

The second step of the workflow results from the determination of objectives and includes the **generation of a requirements list**. At this point, it is specified which characteristics the Digital Twin must implement in order to achieve the defined goals. A structured guideline to identify, document, analyze and manage requirements is given in Christoph Ebert's book "Systematisches Requirement Engineering" [22]. Requirements for the Digital Twin can be derived from general requirements for the assembly system (see step 1 Planning framework for assembly systems according to Lotter and Wiendahl). If the assembly system under consideration is adaptable, the requirements must also be designed to be adaptable and scalable. They should be regularly reviewed and updated if necessary. The requirements list affects all further steps, which is why it is important to include experts from all affected areas.

Once the objectives and requirements have been determined, the next step is to **define the relevant characteristics** for the considered products, processes and equipment. This provides an important basis for the later aggregation and usage of data. First of all, it is necessary to define critical areas in the assembly. Critical can refer to both, cost and quality. Thereof relevant characteristics and conditions, that should be monitored and optimized with the help of the Digital Twin can be derived. Depending on whether assembly data has already been recorded or not, the procedure in this step differs. If data is available, it should be analyzed first. In terms of quality, regular deviations affecting specific products, processes and equipment can be detected. If the focus is on cost reduction, the data aggregation can at best be used to find the location of the greatest cost causation. Thus, it is possible to identify the relevant characteristics and conditions of products, processes and equipment. Another advantage is that devices recording the respective data in the past can be connected with the Digital Twin in the future. With the help of algorithms, dependencies can be identified and the systems can be optimized. If no data has been recorded or saved so far, carrying out an FMEA in a team of experts is one way of revealing current weaknesses in the assembly. The characteristics to be monitored and optimized in the future can be derived from the products, processes and equipment that have been uncovered as critical after conducting the FMEA.

The next step of the workflow includes the **structuring of the data model**. The data model is an important basis for the Digital Twin. It represents the data to be described and processed as well as their interrelationships. The requirements for the data model should already have been defined in the requirements list, created in the second step. When it comes to an adaptable assembly system, the most important requirement for the data model is also adaptability. The change enablers modularity and scalability in particular therefore play an important role. Just like the modules of the assembly systems, the individual objects of the data model should be easily exchangeable and extendable. It should be able to respond to product or process changes and be extensible to new services and models. The data model must be anchored in a suitable database. In general, a distinction can be made between SQL and NoSQL databases. While SQL databases provide strict specification and thus reduce the error rate, NoSQL databases are flexible but in turn can have problems with consistency.

During the structuring of the data model the next step can already start, the **determination of the data to be recorded**. It can mostly be derived from the definition of the characteristics. The defined characteristics are information about products, processes and assembly equipment. It needs to be determined which data has to be aggregated and how this data needs to be linked and analyzed to obtain the relevant information. With this knowledge it is possible to specify the data to be recorded and thus also the measuring equipment can be defined and allocated to the specific product, process or equipment in the data model. In the context of the Digital Model, the target values for the relevant data and information must be defined so that the recorded data can be aligned. It is also necessary to define which data and information is stored in the Digital Twin and whether any pre-processing is necessary. Next, a linkage between the objects in the data model and the physical objects is required. The data model should be able to receive data from the physical world to process and return to the physical system. In this way, the Digital Twin is created.

The next and final step before implementing and using the Digital Twin is the **setup of the final network architecture**. It must be defined how the Digital Twin communicates with the physical assembly system and vice versa. Interfaces such as MQTT or OPC UA are required. An example is given at the end of Sect. 3.2.

3.2   Implementation and Validation of the Workflow

The workflow was used and validated as part of the research project H2SkaProMo. The goal of the project is the development of a scalable assembly system for the production of fuel cell stacks.

The first step of the workflow was to define the goals to be achieved with the help of the Digital Twin. Ensuring the quality of the product, processes and equipment in the assembly from the early planning phase onwards was the most important objective in this use case. To achieve this goal, a continuous monitoring and optimization of the underlying assembly processes is necessary.

Based on this, a requirements list for the Digital Twin was created. The list consists of numerous demands, some of which are highlighted in the following. First of all, the Digital Twin and thus the data model must have a modular, scalable setup in order to be able to react quickly to changes and adjustments. Furthermore, the Digital Twin should contain consistent information about the entire assembly system. The collected data should enable the user of the Digital Twin to make statements about the past (traceability of mistakes), about the present (descriptive analytics, diagnostic analytics) and about the future (predictive analytics, prescriptive analytics). Additionally, the Digital Twin should be used as a single source of truth in the context of this project. Every user should be able to interact with the Digital Twin in order to get the relevant information for his use case. Therefore, a user-friendly interface is also part of the requirements list.

The next step of the workflow includes a definition of the relevant characteristics. Since there was no previous data from the past that could be evaluated, potentially critical areas of the assembly had to be identified in another way. Here, a FMEA was used to recognize potential faults in the assembly processes that may occur in the future. From these potential faults, specific characteristics were transferred that should be monitored and optimized in the future in order to reach the defined goals.

On the basis of this knowledge the structuring of the data model was conducted. The dependencies between components, processes and assembly equipment were modeled in a modular, scalable way. Figure 4 shows the underlying structure of the data model.

The structure was implemented in the NoSQL database Arango DB, using its flexibility which is needed to create a scalable, modular Digital Twin.

In the next step, the data to be recorded was deduced from the previously defined characteristics. It was evaluated which data need to be aggregated to get the relevant information. The respective information and data were assigned to the individual components, processes and operating resources in the data model. For the final implementation

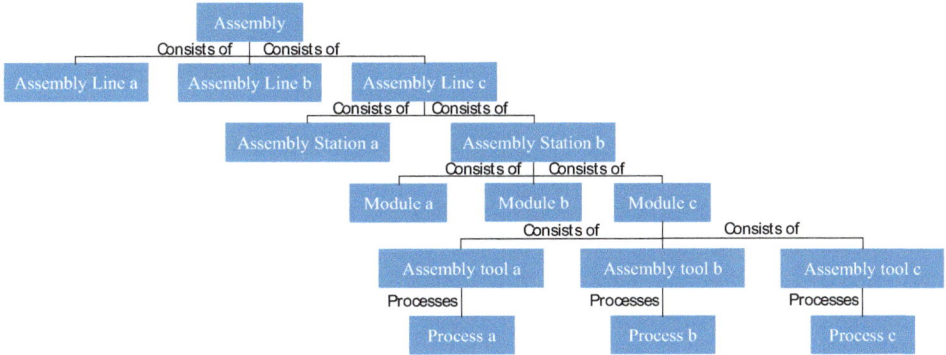

Fig. 4 Modular data model

of the Digital Twin, a communication infrastructure between the Digital Twin and the real world had to be established. The implemented infrastructure is shown in Fig. 5.

By continuously following the workflow, a complication-free implementation of the Digital Twin was possible during the course of the project. The steps were not carried out strictly sequentially but partly iteratively.

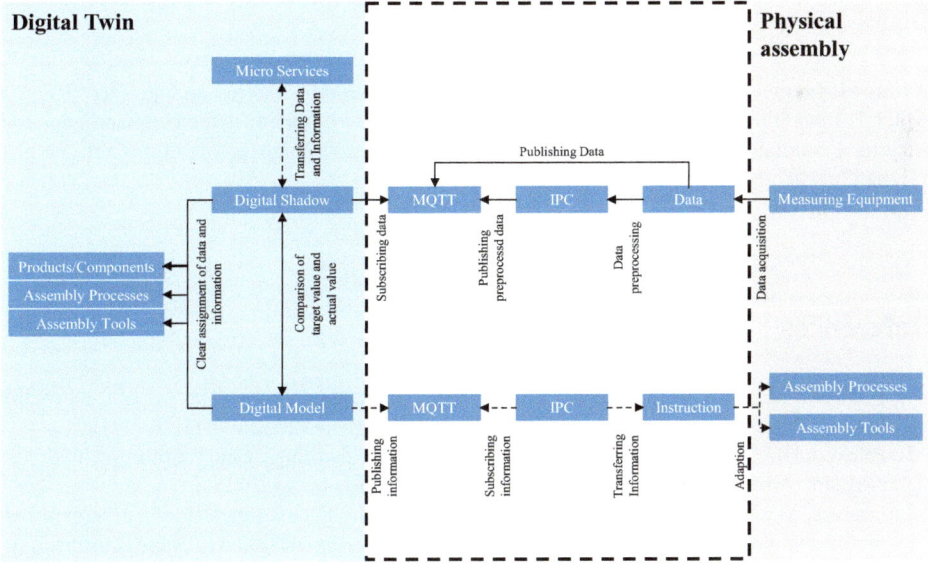

Fig. 5 Communication between the Digital Twin and the assembly system

4   Discussion

The developed workflow consists of six steps: the determination of objectives, the generation of a requirements list, the definition of relevant characteristics, the structuring of the data model, the determination of the data to be recorded and the setup of the final network architecture. It can be used as a guideline for the planning and implementation of Digital Twins. The approach intents to contribute to establishing standards regarding the Digital Twin. The workflow provides an overview, but the individual steps shown can in turn be very complex and need to be researched and standardized. In addition, it must be emphasized that after the creation of this workflow, further research must be carried out on how to optimize an assembly system with the help of a Digital Twin.

The developed workflow focuses on the Digital Performance Twin. While the first steps of an approach for planning Digital Product and Digital Production Twins are not too different from those presented, the final steps before the implementation differ. When considering Digital Product or Production Twins, simulation software must be implemented to virtualize assembly processes. In addition, topics such as virtual commissioning have to be considered. As a consequence, the influence of geometrical parameters is increased in Digital Product and Production twins.

In summary, it can be said that the Digital Twin can already support the planning and optimization of assembly systems today. However, there are still different areas that need to be researched more detailed in the future in order to exploit the highest potential of the Digital Twin.

Acknowledgements This paper was written in the framework of the research project "VProSaar" which is funded by the European Union. This work is also supported by the German Federal Ministry for Economic Affairs and Climate Action (BMWK) with funds provided through the project "H2SkaProMo" (03EN5014I).

References

1. Brossard, M., Chaigne, S., Corbo, J., Mühlreiter, B.: Digital Twins: The Art of the Possible in Product Development and Beyond. McKinsey & Company (2022)
2. Rückert, F.U., Sauer, M.: Die Erstellung eines digitalen Zwillings. Eine Einführung in Simcenter Amesim. essentials. Springer Vieweg, Wiesbaden, Heidelberg (2021)
3. Grieves, M., Vickers, J.: Digital Twin: mitigating unpredictable, undesirable emergent behavior in complex systems. In: Kahlen, F.-J., Flumerfelt, S., Alves, A. (eds.) Transdisciplinary Perspectives on Complex Systems, pp. 85–113. Springer International Publishing, Cham (2017)
4. Gartner Hype Cycle: Top Trends in the Gartner Hype Cycle for Emerging Technologies 2017 (2017). https://www.gartner.com/smarterwithgartner/top-trends-in-the-gartner-hype-cycle-for-emerging-technologies-2017

5. Kritzinger, W., Karner, M., Traar, G., Henjes, J., Sihn, W.: Digital Twin in manufacturing: a categorical literature review and classification. IFAC-PapersOnLine (2018)
6. Zeballos Raczy, J.-P.: Digital models, digital shadows, and digital twins help transform and optimize industrial and business operations. InTech, 20–21 (2022)
7. Brecher, C., Buchsbaum, M., Storms, S.: Control from the Cloud: Edge Computing, Services and Digital Shadow for Automation Technologies (2019)
8. Grieves, M.: Digital Model, Digital Shadow, Digital Twin (2023)
9. Stark, R., Anderl, R., Thoben, K.-D., Wartzack, S.: WiGeP Positionspapier: "Digitaler Zwilling." ZWF—Zeitschrift für wirtschaftlichen Fabrikbetrieb **115**, 47–50 (2020)
10. Schäfer, S.F., Gorke, N.T., Cevirgen, C., Yeong-Bae, P., Nyhuis, P.: Elemente der "Fabrik der Zukunft". Teil 2: Smart Plant—der Digitale Zwilling des Fabrikgesamtsystems. ZWF—Zeitschrift für wirtschaftlichen Fabrikbetrieb 117 (2022)
11. Lindow, K.: Digitaler Zwilling (2020)
12. Krack, M.: Der Digitale Zwilling—Kern der Fabrik der Zukunft. topsoft Fachmagazin, 4–5 (2021)
13. Siemens: Digitaler Zwilling. https://www.plm.automation.siemens.com/global/de/our-story/glossary/digital-twin/24465. Accessed 31 Jan 2023
14. Mertens, P., Bodendorf, F., König, W., Schumann, M., Hess, T., Buxmann, P.: Grundzüge der Wirtschaftsinformatik, 12th edn. Springer, Berlin Heidelberg, Berlin, Heidelberg (2017)
15. Kurbel, K. (ed.): Produktionsplanung und—steuerung im Enterprise Resource Planning und Supply Chain Management, 6th edn. Oldenbourg, München, Wien (2005)
16. Stähr, T.J.: Methodik zur Planung und Konfigurationsauswahl skalierbarer Montagesysteme. Ein Beitrag zur skalierbaren Automatisierung. Dissertation, Karlsruhe Institut für Technologie (KIT) (2020)
17. Cisek, R., Habicht, C., Neise, P.: Gestaltung wandlungsfähiger Produktionssysteme. ZWF—Zeitschrift für wirtschaftlichen Fabrikbetrieb **97**, 441–445 (2002)
18. Abele, E., Liebeck, T., Wörn, A.: Measuring Flexibility in Investment Decisions for Manufacturing Systems. CIRP Annals (2006)
19. Nyhuis, P., Heinen, T., Brieke, M.: Adequate and economic factory transformability and the effects on logistical performance. Int. J. Flex. Manuf. Syst. (2007)
20. Lotter, B., Wiendahl, H.-P., Montage in der industriellen Produktion. Ein Handbuch für die Praxis; mit 18 Tabellen, 2nd edn. VDI-/Buch]. Springer Vieweg, Berlin, Heidelberg (2012).
21. Tara, R.: Are We Ready for Digital Twins? Engineering.com audience survey of perceptions and readiness
22. Ebert, C.: Systematisches Requirements Engineering. Anforderungen ermitteln, dokumentieren, analysieren und verwalten, 6th edn. dpunkt.verlag, Heidelberg (2019)

Conception of a Flexible Modular Method for Automation Recommendation on the Basis of a Potential-Based Assembly Process Evaluation

Michael Hübner, Benedikt Kelm, Fabian Adler, Anne Blum, and Rainer Müller

Abstract

Existing automation analyses are mostly solution-oriented methods that require deep expert and process knowledge. Established methods, such as tolerance chain analyses or variant considerations, are often not considered. However, an objective comprehensive evaluation is only guaranteed by the integration of those methods paired with solution-neutral evaluation criteria. Furthermore, existing automation analyses do not include an evaluation of the temporal optimisation potential of individual processes. This often hinders a target-oriented automation as the focus is too strongly placed on the automatability rather than on the optimal utilization of existing potentials. The flexible modular method presented in this article is intended to enable the user to analyse the conditions of product, process and equipment in a targeted manner by means of a neutral and objective evaluation as well as to make automation recommendations

Certain partial results have been produced within the project "Mittelstand-Digital Zentrum Saarbrücken". The "Mittelstand-Digital Netzwerk" offers comprehensive support for digitalisation with the "Mittelstand-Digital Zentren", the "Initiative IT Sicherheit in der Wirtschaft" and "Digital Jetzt". The Federal Ministry for Economic Affairs and Climate Protection enables free use and provides financial subsidies.

M. Hübner (✉)
ZF Friedrichshafen AG, 97424 Schweinfurt, Germany
e-mail: michael.huebner@zf.com

B. Kelm · R. Müller
Universität des Saarlandes, Lehrstuhl für Montagesysteme, 66121 Saarbrücken, Germany

F. Adler · A. Blum
ZeMA gGmbH, Forschungsbereich Montagesysteme, 66121 Saarbrücken, Germany

© The Author(s) 2025
S. Ihlenfeldt et al. (eds.), *Annals of Scientific Society for Assembly, Handling and Industrial Robotics 2023*, https://doi.org/10.1007/978-3-031-74010-7_7

depending on the optimisation potential. The modular method is designed for assembly processes in existing manual and hybrid workplaces in series assembly. Existing workplaces are analysed regarding their existing optimisation potential. For this purpose, the actual state of the workplace is compared with an ideal state and evaluated in terms of time. The standardised procedure MTM-2 is used for the temporal evaluation of the processes. The evaluation of the automation capability of the processes, using a utility value analysis, is carried out based on product and process-related criteria in four stages. By comparing the available optimisation potential with the technical automatability, automation recommendations are made for individual process steps.

Keywords

Modular method • Process planning • Assembly automation • Process optimisation

1   Introduction

In Germany, the manufacturing industry generates almost a quarter of the gross domestic product and directly or indirectly employs one in three workers [1]. At the same time, labour costs per hour worked are in fourth place worldwide. For comparison, a worker in Poland costs a fourth and in China a fifth per hour [2]. In the course of globalisation, locations where goods that can be traded across regions are increasingly entering into direct competition. As a result, Germany's competitiveness is declining due to high labour costs and a growing shortage of skilled workers. One way to counteract this development and ensure international competitiveness is to increase efficiency through rationalisation measures in production. Assembly, in particular, where the product passes through a large part of its value creation, offers considerable potential for improving economic efficiency [3]. One way to tap this potential is to automate assembly processes.

The identification of automatable assembly processes in industrial practice is often based on expert experience values and primarily examines technical feasibility. However, there is often no detailed analysis with a subsequent comparison of existing potential and technical feasibility. To target the automation of assembly processes, the temporal optimisation potential must be used as a decision-making factor. The state-of-the-art has shown that process steps are often not analysed deeply enough to identify potentials within a process step. The aim of the article is the conception of a modular method which enables a comparison between the temporal optimisation potential and the technical automatability of an assembly process. Depending on the required information density, it is flexible and can be extended by further methods such as a variance analysis. It serves as a guideline for systematic automation analysis in order to identify obstacles to automation in assembly and to develop targeted recommendations for automation solutions. For this purpose, existing methods for automation analysis are first examined and differentiated. Subsequently, the developed conception of the flexible modular method is introduced.

2   Related Work

2.1   Methods for Assembly Automation Analysis

In literature, various methods for evaluating the automation of assembly processes can be found. The automation evaluations presented here are a selection of approaches published in the last 40 years. The focus is on the criteria used for evaluation and serve as the basis for the methodology developed in this work.

Deutschländer [4] defines 15 criteria in a checklist for evaluating the automation of assembly systems, each divided into three forms of expression. A distinction is made between object- and process-related criteria. The sum of the expression of the individual criteria can be divided into the categories "easily realisable", "realisable" and "realisable with great effort". Ross [5] introduces 20 criteria with four forms of expression for evaluating the effort for automating assembly processes. The influence of the criterion on the sub-process, which consists of assembly process, handling and joining, is considered. The sum of the sub-effort values results in an effort value, which allows a decision on the economic automation. In Beumelburg's [6] capability-oriented assembly process planning, four areas are considered: assembly process, ergonomics, component, and parts provisioning. Within these four areas, a criteria catalog containing a total of 22 criteria and a varying number of expressions is introduced for evaluating the automatability. The expression is incorporated into an evaluation vector, which includes various dimensions of an assembly operation. This forms the basis for the capability-oriented division of assembly tasks between humans and robots. Thomas [7] defines different evaluation classes within the framework of an evaluation system for human–robot collaboration. In relation to the automatability evaluation, the classes assembly task, components and technical conditions are relevant. For each of these classes, 9–11 criteria with a weighting factor and an evaluation range from 0 to 10 are introduced. Ermer et al. [8] present 13 criteria, each with five characteristics, as part of a "quick check" for the introduction of hybrid human–robot workplaces. They divide the criteria into the areas of component, supply and removal, potential for improvement, safety and other. The sum of the values of each criterion per process is divided by the maximum sum. The result is a percentage recommendation for automation. Concrete threshold values are not specified.

2.2   MTM-2

With the help of the MTM (Methods-Time-Measurement) process language, a system with process building blocks, the efforts of work processes can be described and evaluated transparently in detail. Process modules are used for this purpose, which represent a standard in terms of content and time. For this reason, the MTM standard performance is used for objective performance evaluation. This represents the permanently achievable performance of an averagely trained person without a drop-in performance due to fatigue. Standardisation makes it possible to record actual conditions and to generate target conditions to compare design alternatives. This makes the design consequences of workplaces and workflows directly measurable and consequently comparable [9, 10].

3   Classification and Differentiation from Existing Methods

Existing automation analyses, as presented in Sect. 2.1, are already focused on specific solutions through their structure and evaluation, like human robot collaboration applications. Such analyses are therefore mostly solution-oriented procedures that require in-depth expert and process knowledge and often ignore other solution approaches. Furthermore, established methods such as tolerance chain analyses [11] or variant analyses are often not included, which makes a comprehensive evaluation more difficult. However, a meaningful evaluation is only guaranteed through the integration of these methods paired with solution-neutral evaluation criteria. In addition, existing automation analyses evaluate complete process steps, such as the placement of a rivet, but not in the finer division of "grasp and release" and "put in place" according to MTM-2. Only this exact division makes it possible to analyse process steps in detail to divide work content between man and machine within a process step. Furthermore, existing automation analyses do not include an evaluation of the temporal optimisation potential of individual process steps. This often prevents solution-oriented automation because the focus is too much on automatability and not on the optimal use of the existing potentials.

4   Conception of the Designed Modular Method

For a simple and targeted identification of automation recommendations, a flexible adaptive conception in the form of a modular method offers the best possibilities for the user. Figure 1 describes the mentioned modular method and is briefly outlined below. The primary methods (potential and technical automatability evaluation) is the focus of this paper and will be discussed in more detail in the sub-chapters.

An analysis of the actual state of the product, the process and the equipment are performed and the assembly priority graph is generated. The actual assembly processes are divided into the movement sequences "grasp and release" and "put in place", based on MTM-2. Subsequently, the product and processes are evaluated in terms of technical automatability using a utility analysis and, accordingly, solution-neutral component as well as process-related criteria. This identifies complexity drivers for a potential automation and points out simplification actions. Furthermore, the sub-processes are evaluated regarding their time potential to identify rationalisation potential. A subsequent comparison of the existing time optimisation potential with the technical automatability makes it

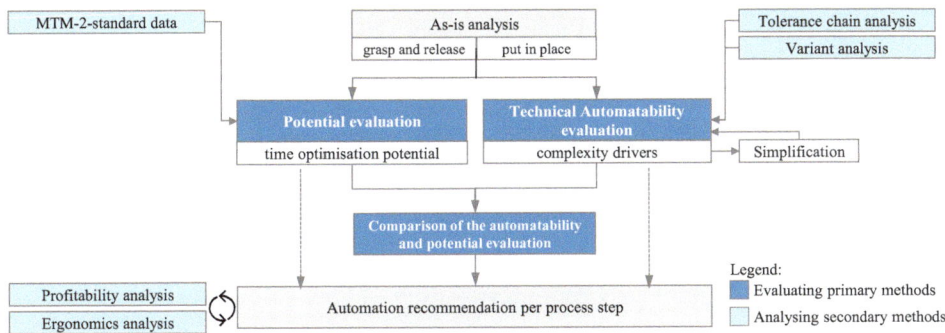

Fig. 1 Conception of the designed modular method for automation recommendation

possible to systematically identify processes with high potential and high or low automatability. From this, targeted automation recommendations can be derived for each process step. A separate identification of the time optimisation potential and the execution of an automatability evaluation as the basis for an automation recommendation are also possible through the selected conception. The modular method is subdivided into necessary evaluating primary methods (dark blue representation) and optional analysing secondary methods (light blue representation). This makes it possible to specify the recommendation more precisely. Secondary methods concretise or support an automation recommendation of the assembly process under consideration.

4.1 Actual State Analysis

At the beginning of the analysis, the present to be evaluated workplace state is to be recorded by documenting the single process steps by using an assembly priority graph. By visualising the assembly process based on an assembly priority graph, the logical and temporal sequence of the individual assembly steps can be displayed transparently [12, 13]. The individual process steps are subdivided into the movement sequences "grasp and release" and "put in place" according to MTM-2 by evaluateing their contents in terms of time. This separation is the basis for the following potential evaluation as well as the technical automation evaluation.

4.2 Subdivision into "Grasp and Release" and "Put in Place"

The execution of a manual assembly activity is essentially limited to the motion sequences of "grasp and release" and "put in place". Here, "grasp and release" according to standard data basis values (MTM-2) includes the motion sequence of reaching, grasping and

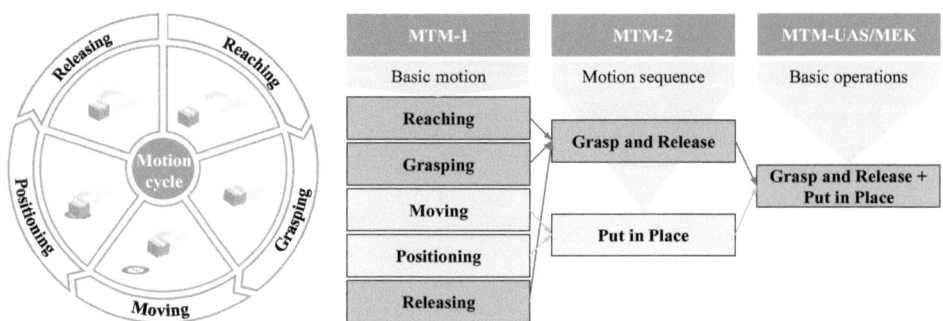

Fig. 2 Illustration and classification of the motion sequences following [9, 10]

releasing and the motion sequence of "put in place" includes the moving and positioning of a component as shown in Fig. 2. The division into "grasp and release" and "put in place" and their subsequent evaluation enables a detailed view of the automatability of the individual process steps. This is a decisive advantage, especially when designing hybrid workplaces through the part-automation of process steps. Specifically, parts of a process step, such as feeding, aligning, and separating, can be automated and the task of positioning can be performed by the employee. For a more ideal approach for the employee, a spiral conveyor can position and orientate the components, while the employee takes over the complex, sensitive placement of the parts to be joined.

The criteria for evaluating the automation capability (see Sect. 4.3) refer either only to the sub-process of "grasp and release" or "put in place" or are equally valid for both sub-processes. The criterion of additional activities (e.g. visual inspection, material supply) is a special case and can occur in both motion sequences. Since additional activities are not required in every process step, their evaluation is optional.

4.3 Definition of the Evaluation Criteria for the Automatability

To be able to evaluate processes regarding to their automatability, criteria are needed for decision-making, based on which the individual process steps can be evaluated objectively and solution-neutrally. To determine such evaluation criteria, automation methods from various authors were examined, compared and supplemented with literary findings from assembly-oriented product design (high influence on the mountability) [14–17]. In the process, solution neutral as well as frequently mentioned criteria were taken over. The evaluation criteria determined in this way are divided into eight component-related and five process-related criteria. In Fig. 3, the criteria with their literary origin can be seen.

To simplify the application and to minimise the required expert and process knowledge, the evaluation criteria used are each divided into four levels. On the one hand, this division

Grasp and release	Put in place	Criterion	Automatability evaluation					Assembly-oriented product design			
			A. Deutschländer [4]	P. Ross [5]	K. Beumelburg [6]	C. Thomas [7]	A.-K. Ermer [8]	P. Konold [14]	S. Hesse [15]	R. Bäßler [16]	W. Eversheim [17]
●	●	Dimensions of the joining component	●	●	○	●	○	●	●	●	○
●	◑	Weight of the joining component	●	○	○	●	●	●	●	●	○
●	●	Dimensional stability of the joining component	●	●	●	●	●	●	●	●	○
●	●	Sensitivity of the joining component	○	●	●	○	●	○	●	●	○
●	●	Geometric tangibility of the joining component	○	●	●	●	○	●	●	●	○
●	●	Number of variant families	●	●	○	●	●	●	○	●	○
●	○	Provision method of the joining component	●	○	●	●	●	○	●	●	○
○	●	Joining accuracy of joining component and base component	○	●	○	○	○	●	●	●	●
○	●	Accuracy requirement of the joining process	●	○	●	○	○	●	●	●	○
○	●	Type of joining movement	●	●	●	○	◑	●	○	●	○
○	●	Required joining forces/torques/sensitivity	○	●	○	○	○	○	○	○	○
○	●	Accessibility joining location	●	●	●	○	○	●	●	●	●
◑	◑	Additional activity (e.g. visual inspection, material supply)	●	●	●	●	●	○	●	○	○

Component-related (rows 1–8); Process-related (rows 9–13).

Rating scale: applicable ● 　 partially applicable ◑ 　 not applicable ○

Fig. 3 Literature comparison criteria nomination

describes the possible states of the criteria sufficiently precisely and, on the other hand, achieves a high degree of selectivity between the levels. This makes it possible for the user to clearly decide on an objective, fact-based evaluation basis between the characteristics of the respective criteria and to evaluate them.

The evaluation of the component-related criterion "sensitivity of the joining component" is described below as an example. As shown in Table 1, four possible component characteristics are proposed to the user. Depending on the characteristic, the user decides which is most similar to the property of the joining component, in this case regarding the sensitivity of the joining component. The joining component is classified in terms of its stability as form stable or rigid (e.g. steel components), deformable with high force (e.g. plastic components), deformable with low force (e.g. cardboard), and unstable or flexible (e.g. seals, cables). The more unstable a component is, the more difficult it is to feed, grip, handle and join. Consequently, the automation capability decreases with increasing lability.

Table 1 Criteria sensitivity of the joining component

Characteristic 1	Characteristic 2	Characteristic 3	Characteristic 4
Form stable/rigid	Deformable with high force	Deformable with low force	Unstable/flexible

4.4   Technical Automatability Evaluation

A utility analysis is used as the basic method for evaluating the technical automatability. For this purpose, criteria were already established in Sect. 4.3 and divided into four levels. The user evaluates each individual process step based on the defined criteria regarding to its technical automatability for "grasp and release" and "put in place" the joining component. The evaluated individual process steps are then combined into a compact number (utility value) that represents the technical automatability regarding "grasp and release" and "put in place" of the respective individual process step.

On one hand, the utility value shows the user the extent to which the evaluated process step is suitable for automation regarding "grasp and release" and "put in place", as well as on the other hand, it points out possible obstacles to automation. Following the evaluation, the process of simplification takes place. This process shows the user possible optimisation approaches to simplify the automation (e.g. forming variant families, modification of the delivery state). This prevents process steps from being categorically excluded due to a supposedly poor automatability, but rather to work on enabling a simpler and more economical automatability. The simplification is the subject of future research and will be published at a later date.

4.5   Potential Evaluation

The evaluation of the time optimisation potential is based on the primary/secondary analysis according to Lotter [18]. The first step is to break down the assembly processes recorded in the actual state analysis into "grasp and release" and "put in place" according to MTM-2. The application of the standardised MTM procedure provides an overview of the actual process times. In the second step, a fictitious ideal state of the workplace under consideration is defined. For this purpose, a limit value for the optimal distance from the supply to the place of assembly is first determined as a function of the size of the assembly object. As shown in Fig. 4, this distance represents the radius of a usually two-thirds circle, the primary area. The secondary area is the space that extends beyond the primary area but is still within the operator's reach.

In ideal conditions, the distance between the supply and the place of assembly is minimized to maximize value creation. Consequently, the entire supply is located within the primary area. Furthermore, it is assumed that the provision of the components takes place individually in an oriented and positioned form. In addition, assembly aids such as hoists or similar can be used in the ideal state. The optimised ideal state is also analysed with MTM-2, resulting in process times for the individual assembly operations. The third step is to determine the time difference between the actual and the ideal state. This difference represents the time optimisation potential for each process considered separately.

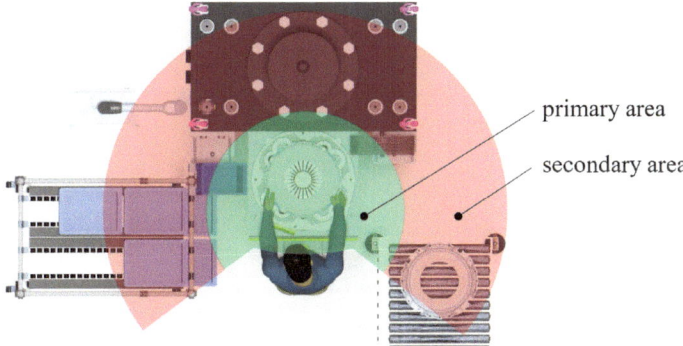

primary area

secondary area

Fig. 4 Comparison of automatability and potential evaluation

The detailed subdivision into "grasp and release" and "put in place" clarifies the exact potential range of the process step.

4.6 Creation of an Automation Recommendation Through the Comparison of the Automatability and Potential Evaluation

The integration of the automatability and potential recommendation allows a comparison between the temporal optimisation potential and the effort of the technical implementation of process automation. The effort for implementation refers to technical complexity. Through this procedure, the usefulness and the technical automatability are evaluated and then compared with each other. Figure 5 shows various assembly processes as a comparison of the two recommendations.

Two extreme constellations can occur. The assembly process "Insert 2 screws" embodies the extreme case of a low potential and a high automation capability. Based on this, a targeted questioning of the added value of automating this process is possible. In contrast to this, the assembly process "Hook in 8 springs" illustrates a high time potential with low automation potential. The indication of this case consists in a more precise examination of the technical effort, since a development of the potential promises a large gain in time. The differentiation between "grasp and release" and "put in place" also makes it clear to the user that it is necessary to take a more detailed look at the process. On this basis, for example, in the case of a high potential for "grasp and release", specific parts of a process step, such as feeding, straightening, and separating, can be tested by automation.

The described approach allows processes with high potential and high or low automation potential to be systematically identified and targeted automation recommendations to be derived.

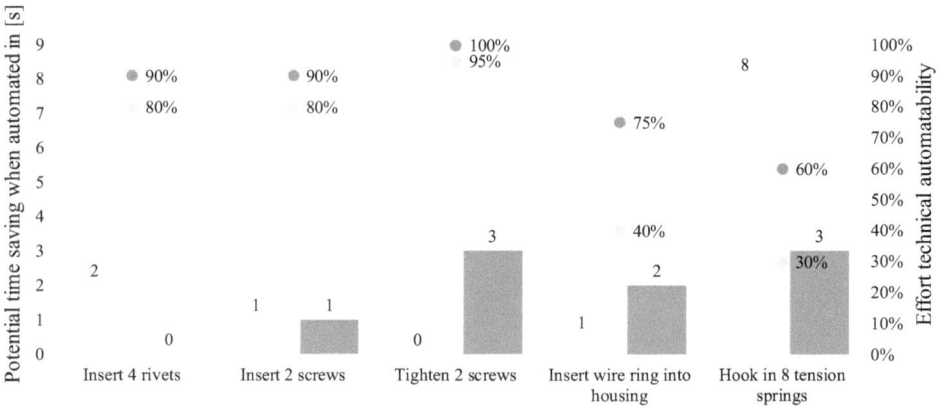

Fig. 5 Comparison of automatability and potential evaluation

5   Summary and Outlook

The article introduces a modular set of methods for solution-neutral automation recommendations for assembly processes. During the actual state analysis, the assembly process is broken down into the parts of "grasp and release" and "put in place". This enables a more precise evaluation of the technical automatability and the time potential of the individual assembly processes. The evaluation of the technical automation potential is concretised by process- and component-related evaluation criteria and supports the identification of complexity drivers. The MTM-2-based potential recommendation shows the temporal optimisation potential for each process considered. A final comparison of the two evaluations makes it possible to systematically identify processes with high potential and high or low automation potential.

The modular method can be used in industrial practice as a guide for a systematic automatability analysis and to uncover automation obstacles in assembly. Through targeted solution recommendations, rationalisation actions can be implemented quickly and thus savings achieved.

Further research is needed in the weighting of the criteria for the evaluation of automation capability, as well as the concretisation of the measures contributing to simplification. Furthermore, the presented methodology should be carried out on practical validation scenarios.

Although the modular method can already be carried out in its current form, the conception of the modular method allows a flexible extension by additional secondary methods. For the simple integration of such secondary methods, clear interfaces must be integrated into the flexible modular method or primary methods during further research work. In addition, it is particularly important to further develop secondary methods such as tolerance chain analysis and variant analysis as well as to increase their practical suitability. In this way, the information density can be increased if necessary and more precise solution recommendations can be made.

References

1. Abele, E., Reinhart, G.: Zukunft der Produktion. Herausforderungen, Forschungsfelder, Chancen. Hanser, München (2011)
2. Schröder, C.: IW-Trends 2/2019. Industrielle Arbeitskosten im internationalen Vergleich. Vierteljahresschrift zur empirischen Wirtschaftsforschung **46**, 63–81 (2019)
3. Mueller, R.: Montagegerecht. In: Bender, B., Gericke, K., Pahl, G., Beitz, W. (eds.) Pahl/Beitz Konstruktionslehre. Methoden und Anwendung erfolgreicher Produktentwicklung. Springer eBook Collection, 9th edn., pp. 725–754. Springer Vieweg, Berlin, Heidelberg (2021)
4. Deutschländer, A.: Integrierte rechnerunterstützte Montageplanung. Zugl.: Berlin, Techn. Univ., Diss., 1989. Produktionstechnik—Berlin, vol. 72. Hanser, München (1989)
5. Ross, P.: Bestimmung des wirtschaftlichen Automatisierungsgrades von Montageprozessen in der frühen Phase der Montageplanung. Zugl.: München, Techn. Univ., Diss., 2002. Forschungsberichte IWB, Bd. 170. H. Utz Verlag, München (2002)
6. Beumelburg, K.: Fähigkeitsorientierte Montageablaufplanung in der direkten Mensch-Roboter-Kooperation. Zugl.: Stuttgart, Univ., Diss., 2005. IPA-IAO Forschung und Praxis, vol. 413. Jost-Jetter, Heimsheim (2005)
7. Thomas, C.: Entwicklung einer Bewertungssystematik für die Mensch-Roboter-Kollaboration. Zugl.: Aachen, Techn. Univ., Diss., 2017. Schriftenreihe des Lehrstuhls für Produktionssysteme, 2017, Band 3. Shaker Verlag, Düren (2017)
8. Ermer, A.-K., Seckelmann, T., Barthelmey, A., Lemmerz, K., Glogowski, P., Kuhlenkötter, B., Deuse, J.: A Quick-check to evaluate assembly systems' HRI Potential. In: Schüppstuhl, T., Tracht, K., Roßmann, J. (eds.) Tagungsband des 4. Kongresses Montage Handhabung Industrieroboter, pp. 128–137. Springer Berlin Heidelberg, Berlin, Heidelberg (2019)
9. Bokranz, R., Landau, K.: Handbuch Industrial Engineering. Produktivitätsmanagement mit MTM, 2nd edn. Schäffer-Poeschel, Stuttgart (2012)
10. Britzke, B. (ed.): MTM in einer globalisierten Wirtschaft. Arbeitsprozesse systematisch gestalten und optimieren, 2nd edn. mi-Wirtschaftsbuch FinanzBuch-Verlag, München (2013)
11. Mueller, R., Esser, M., Janßen, C.: Umfassendes Toleranzmanagement. Eine Notwendigkeit für wirtschaftliche Montageprozesse. wt Werkstattstechnik online (2009). https://doi.org/10.37544/1436-4980-2009-9
12. Bullinger, H.-J., Ammer, D. (eds.): Systematische Montageplanung. Handbuch für die Praxis. Hanser, München, Wien (1986)
13. Ammer, E.-D.: Rechnerunterstützte Planung von Montageablaufstrukturen für Erzeugnisse der Serienfertigung. IPA-IAO Forschung und Praxis, vol. 81. Springer, Berlin, Heidelberg (1985)
14. Konold, P., Reger, H.: Praxis der Montagetechnik. Produktdesign, Planung, Systemgestaltung, 2nd edn. Vieweg Praxiswissen Ser. Springer Vieweg. in Springer Fachmedien Wiesbaden GmbH, Wiesbaden (2003)
15. Hesse, S.: Montagegerechte Produktgestaltung. In: Lotter, B., Wiendahl, H.-P. (eds.) Montage in der industriellen Produktion, pp. 9–78. Springer Berlin Heidelberg, Berlin, Heidelberg (2012)
16. Bäßler, R.: Integration der montagegerechten Produktgestaltung in den Konstruktionsprozeß. IPA-IAO—Forschung und Praxis, IPA, IAO, Stuttgart, Institut für Industrielle Fertigung und Fabrikbetrieb der Universität Stuttgart, vol. 116. Springer, Berlin, Heidelberg (1988)

17. Eversheim, W.: Fertigung und Montage, 2nd edn. Studium und Praxis, Bd. 4. VDI-Verl., Düsseldorf (1989)
18. Lotter, B.: Die Primär-Sekundär-Analyse. In: Lotter, B., Wiendahl, H.-P. (eds.) Montage in der industriellen Produktion, pp. 49–78. Springer Berlin Heidelberg, Berlin, Heidelberg (2012)

Framework for Parallelized Hybrid Strategies in CAD-Based Disassembly Sequence Planning

Sören Münker, Amon Göppert and Robert H. Schmitt

Abstract

Circular economies require efficient disassembly processes for end-of-life products from various product series, variants and with various product states. CAD-based Disassembly Sequence Planning aims to efficiently create feasible disassembly sequences to minimize the needed effort in disassembly execution. Manual, collision-based and feature-based Disassembly Sequence Planning approaches are predominant in research. However, there is no tool covering multiple of these techniques simultaneously. In this paper, we aim to enable the combination of these techniques in a parallelized and hybrid manner. A Service-Oriented Model-View-Controller Framework is presented as an overarching software architecture. The architecture was implemented and tested on an exemplary use case. The validation proved the potential quality and efficiency increase of CAD-based disassembly sequence planning. The framework contributes to choosing the right disassembly sequence planning strategy and thus increasing the efficiency of the circular economy.

Keywords

Disassembly sequence planning · Computer-aided design · Software architecture · Circularity

S. Münker (✉) · A. Göppert · R. H. Schmitt
Laboratory for Machine Tools and Production Engineering (WZL) of RWTH Aachen University, Campus-Boulevard 30, 52074 Aachen, Germany
e-mail: s.muenker@wzl-mq.rwth-aachen.de

R. H. Schmitt
Fraunhofer Institute for Production Technology IPT, Steinbachstraße 17, 52074 Aachen, Germany

S. Ihlenfeldt et al. (eds.), *Annals of Scientific Society for Assembly, Handling and Industrial Robotics 2023*, https://doi.org/10.1007/978-3-031-74010-7_8

1   Introduction

With the rising demand for circular economies, disassembly optimization gains more importance. Crucial for disassembly efficiency is the disassembly structure representation and the planning of disassembly sequences [14]. The application of Computer-aided Deacsign (CAD) not only during product design but also for automated (dis-)assembly planning is an active field of research [4]. However, the applicability of this approach is still low due to the low quality of automatically generated sequence results and high computational demands.

This work aims to develop a framework combining multiple (Dis-)assembly Sequence Planning (DSP) algorithms from recent research and previous work on Assembly-by-Disassembly (AbD) approaches [11, 12] and, thus, make them more applicable for industrial use cases. We propose using parallelized hybrid strategies based on gathered CAD data. *Hybrid* refers to using a combination of different methods (e.g., manual and automatic DSP approaches). Hybridization aims to increase the quality of DSP results. *Parallelized* refers to the ability to perform multiple tasks simultaneously by dividing a large optimization problem into smaller independent subproblems. The parallelization aims to accelerate the DSP process and improve the overall efficiency of CAD-based DSP.

Relevant requirements for a CAD-based DSP software framework with parallelizable components and hybrid strategies are:

- Independence of proprietary CAD software
- Integrateability with (dis-)assembly planning tools or modules (e.g., Job Scheduling software tools)
- Portability of software modules utilized in framework to other frameworks
- Outsourceability of parallelizable computational efforts
- User interface for interacting with CAD and graph environment

This paper first reviews the related work for CAD-based (dis-)assembly planning frameworks and common software architecture patterns. A new framework combining a Service-Oriented Architecture (SOA) pattern and a Model-View-Controller (MVC) pattern is proposed and implemented. The implementation is validated on an exemplary industrial use case.

2   Review of CAD-Based (Dis-)Assembly Planning Frameworks

The "Auto-Assem" system by Xu et al. (2012) is an early framework for combined Assembly Sequence Planning (ASP) and Job Scheduling (JS) with sequential optimisation. The authors' idea was to build a CAD/CAM-like tool with primary usage for assembly planning to assist engineers in making assembly plans. The proposed framework covers five steps: 1. Assembly modelling; 2. ASP; 3. Path planning; 4. Process planning and visualization; 5. Assembly simulation [8]. However, most steps are executed manually rather than automatically.

Li and Lockett (2017) proposed a framework for designing final assembly lines based on product CAD models, including ASP and JS. Their motivation was to overcome the knowledge gap between product designers and final assembly process designers in the aerospace industry. The implementation is done in CATIA V6. Semantic data (e.g., from Failure Mode and Effects Analysis (FMEA) analysis, Bill of Materials (BOM) etc.) are combined in the "Unified Master Data", which is then utilized for sequencing and assembly line design [10]. However, the paper does not provide further details on possible automation of ASP and JS steps.

Beck et al. (2021) developed a framework to integrate (Dis-)assembly Sequence Planning (DSP) modules with robotic path-planning modules. The framework considers information extracted from the product's CAD model. The framework was implemented using the Octopuz Python API, which can automatically read and remodel STEP files for disassembly. A JSON file is used to construct robot programs for any type of robot. The robot-specific execution code can be exported into native robot languages using Octopuz's internal program exporter [6]. However, this approach is not transferable to sequence planning for large-scale assemblies.

Barbu et al. (2022) focus on precise assembly sequence generation based on various optimization criteria without relying on a specific CAD system. The approach includes a reverse engineering algorithm and an additional user interface utilizing a game engine to add the missing information. The user can select and move the product to disassemble parts. The generated information about disassembly paths is saved in a neutral exchange format. This exchange format can then be used for automatically generating assembly animations and videos [5].

Yao et al. (2022) propose a comprehensive ASP system to be applied in practice. The system utilizes CAD files to create Liaison and Interference Matrices representing the mathematical relationships between parts. An adapted Ant Colony Optimization algorithm is then used to generate an optimized assembly sequence based on these relationships. The system can be accessed through a web-based application where users can upload files and interact with the system. The concept and functionality of the system have been validated using a CAD file of an electric motor product as an example [15]. However, the concept lacks modularity, application of hybrid strategies and parallelization of computationally expensive tasks.

The authors' previous work primarily focused on fully automated pipelines to generate Precedence Graph (PG) or AND/OR Graph (AOG) from CAD files. Both approaches utilized a collision-based Assembly-by-Disassembly(AbD) planning approach and were implemented as a CATIA v5 Add-On [11, 12].

Currently, CAD-based DSP approaches are mostly fully manual or fully automated. The automated approaches have feature-based or collision-based approaches with use-case-dependent advantages and disadvantages. There is no approach explicitly considering the parallelization of DSP approaches yet. A framework is needed to ensure the integrability of existing DSP and JS modules, solve the hybrid application of different DSP approaches and enable performance improvement by parallelization. To address these research gaps, in chapter 'Exploratory Pilot Study Investigating Effects of Exoskeletons on Movement Patterns', the parallelizable hybrid DSP approach is introduced, and the accompanied framework is covered in chapter 'Robot-Based Assembly of Hydrogen Tube Fittings for Large-Scale Electrolyzers'.

3   Parallelizable Hybrid Disassembly Sequence Planning

The concept of parallelizable hybrid DSP aims to optimize the disassembly process by identifying which steps can be done in parallel and choosing the right DSP strategy for parallelizable subassemblies. Parallelizable subassemblies can be identified, e.g. through community detection [16], similarity detection or selected manually. Community detection identifies groups of parts or subassemblies that can be disassembled together. In contrast, similarity detection is the process of identifying parts or subassemblies that have similar disassembly requirements.

There are three main (Dis-)assembly Sequence Planning (DSP) strategies to be utilized hybridly: manual DSP, automatic collision-based DSP and automatic feature-based DSP. Manual DSP allows human operators to plan the disassembly process by manually selecting the order of disassembly for the identified communities and similar parts. Automatic collision-based DSP uses algorithms to plan the disassembly process by simulating the movement of parts and identifying potential collisions. It is based on the principle that the disassembly process can be optimized by minimizing the number of collisions between parts. Automatic feature-based DSP uses algorithms to plan the disassembly process by taking into account the characteristics of the parts, such as size, shape, and connections. It is based on the principle that the disassembly process can be optimized by considering the unique characteristics of the parts. The overall concept of parallelizable hybrid DSP is visualized in Fig. 1.

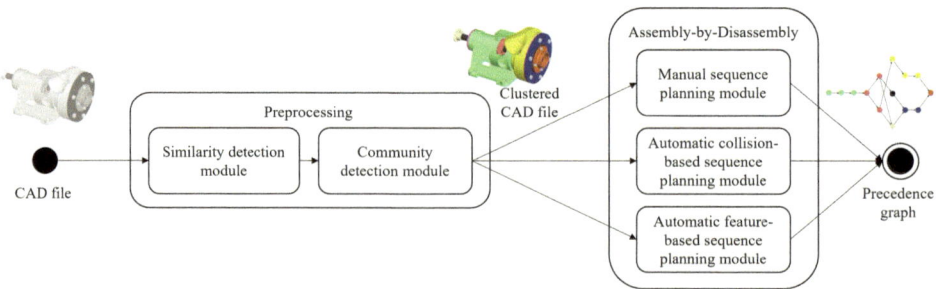

Fig. 1 Exemplary process pipeline of a parallelized hybrid strategy disassembly sequence planning system

4   Service-Oriented Model-View-Controller Framework

Software architecture patterns are design principles that provide a common language and a proven solution for common software development challenges [7]. Different software architecture patterns can be used to design and implement a framework for the parallelized hybrid (Dis-)assembly Sequence Planning (DSP) system. In this chapter, the most common patterns are compared hence their suitability for the problem and a framework with combined patterns is proposed.

4.1   Preselection of Software Architecture Patterns

Many software architecture patterns are intuitively not suitable and can be excluded from consideration (e.g., Peer-to-peer, Mater-Slave, Event-Bus and Broker—all of them focus on communication in networks which is less relevant for user interactive 3D tools). The remaining suitable patterns based on Alebrahim and Heisel (2017) or Bi et al. (2018) [3, 7] are summarized by key characteristics, advantages, and disadvantages in Table 1.

In conclusion, the Model-View-Controller (MVC) pattern is well-suited, as shown in the Table 1. However, it has a disadvantage in limited scalability. It may be an issue for more computationally intensive CAD-based DSP tasks (e.g., for complex engine parts). On the other hand, SOA is a modern approach that offers high scalability and the ability to outsource tasks. However, it requires standardized services with clear interfaces, which may be less suitable for dynamic user intervention during CAD-based DSP analyses because response times could be too high.

The idea of this work is to combine the advantages of both MVC and SOA by integrating both patterns in one framework. This approach is inspired by the "Model-View-Controller-Service-Paradigma" that uses an inner and outer MVC structure to include service-based modules [9]. This combination could provide a more comprehensive solution for software architecture design in CAD-based DSP and is introduced in the following.

4.2   Concept of Service-Oriented Model-View-Controller

The resulting Service-Oriented Model-View-Controller (SOMVC) framework incorporates the advantages of SOA and MVC in a web-based application for CAD-based DSP. It is designed to provide a system-independent way for users to interact with software tools through a browser interface. The main logic of the DSP system is handled by one server, while computationally heavy tasks are outsourced to cloud-based High-Performance Computing (HPC) servers. The framework is illustrated in Fig. 2.

The inner circle of the framework, which is located on the server, is responsible for low-performance tasks. These tasks include manually manipulating the CAD model (e.g.,

Table 1 Comparison of software architecture design patterns

Name	Typical use cases	Advantages	Disadvantages
Layers	• General desktop applications • e-commerce web applications	• Reusability of layers • Potability and exchangeability of layers	• Strict separation of functions in layers hard to achieve • Possibly degradation of performance through multiple layers
Pipes and filters	Linear workflows (e.g., semantic analysis, parsing)	Suitable for multiple transformations on same data	• Not suitable for interactive applications • Filters not directly reusable for other purposes
Model-View-Controller	Interactive applications (e.g., for web applications)	• Well suited for any visual application • Multiple views of same model possible • Easy to develop collaboratively • Easy to maintain • Easy to exchange components	• Structure may lead to overhead in updating each component for every user action • Limited scalability
Client-server	Online applications (e.g. e-mail, document sharing)	• Easy to integrate new servers and functionalities • DevOps possibility with parallel implementation	High dependability on server (single point of failure)
Interpreter pattern	Interpreting programs in dedicated language (e.g., SQL database queries)	High portability (easy replacement of interpreted programs)	• Performance issues for interpreted software • Highly dynamic patterns might lack understandability
Service oriented architecture	Distributed architecture consisting of multiple services with defined interface description	• Maintainability • Independent on platform • Scalability for multiple users or outsourcing computing performance	• High bandwidth server needed • Only integratable with standardized services with clear interfaces

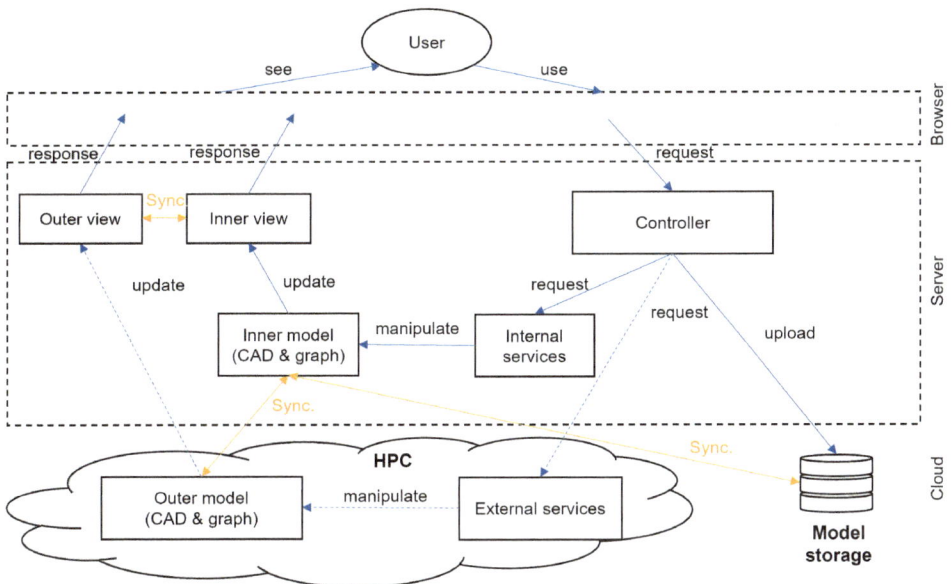

Fig. 2 Block diagram of the proposed service-oriented model-view-controller framework, with HPC = high performance computing

highlighting or excluding parts). The inner model contains the current status of the CAD model and the assembly graph model (e.g., Precedence Graph (PG) or AND/OR Graph (AOG)). The controller has access to all internal services which directly influence the inner model. All changes made by the controller are stored in the inner model, such as which parts are visible and which are invisible. The inner view component is responsible for visualizing the current status of the model, such as displaying different colours for selected parts or only displaying parts with the "visible" attribute.

The outer circle of the framework, which is mainly located on the HPC, is responsible for high-performance tasks such as collision analysis in DSP or graph clustering. The controller can be utilized to request these external services. The external services manipulate the outer model (CAD and graph models). Each model update results in an outer view update.

The user can choose between the inner and the outer view to work on both parallelly. When inner and outer circle processes are terminated, the outer model, inner model and model storage can be synchronized to merge the results of internal and external services. Additionally, the inner and outer views will be synchronized with the model synchronization.

5   Verification and Validation on Exemplary Use Case

The verification and validation of the Service-Oriented Model-View-Controller (SOMVC) framework is an essential step in ensuring that the system works as expected and meets the requirements of chapter 'Latest Challenges in the Development of Scalable Assembly

Systems for Fuel Cell Stacks'. Utilized techniques are unit testing, integration testing and acceptance testing [13]. All tests are executed with an exemplary use case as visualized in Fig. 3. The described use case scenario contains two user interactions: The upload of a CAD model and the Assembly-by-Disassembly (AbD) analysis.

Unit tests are written to test individual system components, such as the internal/external model, the internal/external view, the controller and the internal/external services. These tests ensure that each component works as expected and that the interactions between the components are functioning correctly.

Integration testing examines the interactions between the components of the system. The expected interactions are illustrated in Fig. 3. During the test, it was constantly monitored whether the software adhered to the given diagram. Through iterative programming, adherence was assured.

Acceptance testing is the last validation technique used to test the system from the user's perspective. To ensure objectivity, the concept of face validity was used. Face validity is a technique that involves presenting the system to experts from different fields, such as industry, research projects, working groups, and research institutes. This technique was used to gather feedback from experts in the field of (Dis-)assembly Sequence Planning (DSP) and to ensure that the system meets the needs and requirements of the target audience. Experts from research projects, such as Internet of Construction [2], AdaptAR [1], and REVAMP, were involved in the face validity process. This process helped ensure that the system is practical and usable and fulfils the requirements defined in the introduction. Screenshots of the presented software during face validity are given in Fig. 4. Thus, the resulting framework is a valid solution to increase the efficiency in DSP.

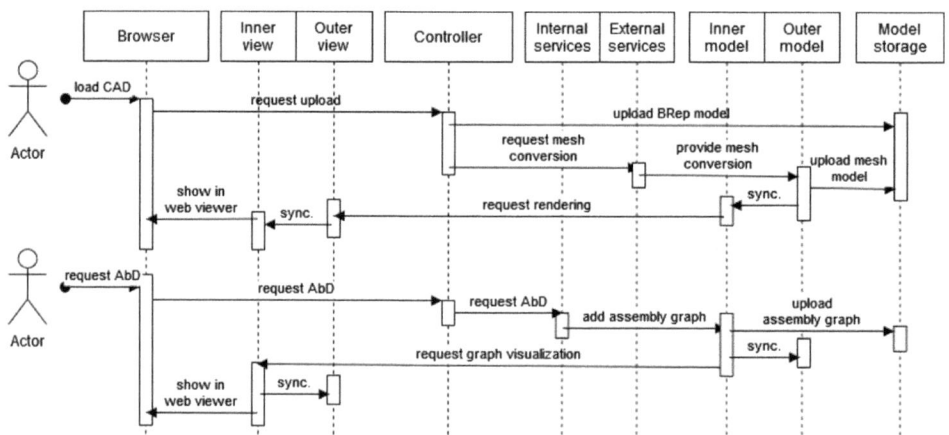

Fig. 3 Use case sequence diagram

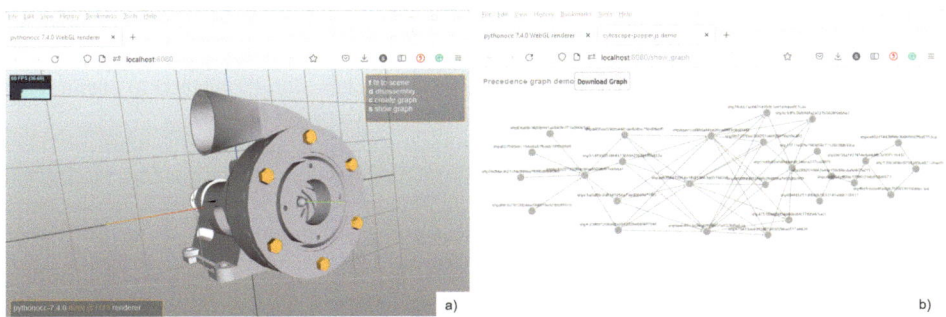

Fig. 4 Browser internal view of **a** the CAD environment and **b** the resulting graph model environment. The user can interact with the controller by keyboard commands

6   Conclusion and Outlook

In conclusion, the Service-Oriented Model-View-Controller (SOMVC) framework enables the parallelized hybrid (Dis-)assembly Sequence Planning (DSP). The use of internal and external processing circles allows for parallelizing DSP strategies (e.g., manual disassembly and automatic collision-based disassembly). Additionally, the integration of different services to identify parallelizable subassemblies, such as community detection and similarity detection modules, can be included. The SOMVC framework was verified and validated through unit testing, integration testing, and acceptance testing, which involved feedback from industry and research experts. These evaluations showed that the SOMVC framework contributes to improving the efficiency and quality of the CAD-based DSP. The contribution will also improve the efficiency and accuracy of remanufacturing planning by making "lot size one" DSP more feasible. Future research can focus on three areas:

1. The integration of the SOMVC framework with other tools and technologies, such as robot motion planning, product design feedback, and assembly instruction tools.
2. Use of artificial intelligence to improve the performance and efficiency of the services. This could include automatically deciding for the right DSP strategy based on labelled assembly topology data.
3. Increase scalability to handle large and complex assemblies with the right High-Performance Computing (HPC) infrastructure.

References

1. Adaptar - adaptive and context-specific technical instructions for the entire product life cycle in an ar environment. https://www.ipt.fraunhofer.de/en/projects/adaptar.html. Accessed 25 Jan 2023
2. Internet of construction - information networks for cross-company collaboration in the production chains of construction industry. http://www.internet-of-construction.com/. Accessed 25 Jan 2023

3. Alebrahim, A., Heisel, M.: Bridging the Gap Between Requirements Engineering and Software Architecture. Springer (2017)

4. Bahubalendruni, M.V.A.R., Biswal, B.B.: Liaison concatenation - a method to obtain feasible assembly sequences from 3d-cad product. Sādhanā **41**(1), 67–74 (2016)

5. Barbu, K., Beck, J., Schäfer, P., Neb, A.: Development of a visual assembly planning system based on neutral files. Proc. CIRP **107**, 446–451 (2022). Leading manufacturing systems transformation - Proceedings of the 55th CIRP Conference on Manufacturing Systems (2022)

6. Beck, J., Neb, A., Barbu, K.: Towards a cad-based automated robot offline-programming approach for disassembly. Proc. CIRP **104**, 1280–1285 (2021). 54th CIRP CMS 2021 - Towards Digitalized Manufacturing 4.0

7. Bi, T., Liang, P., Tang, A.: Architecture patterns, quality attributes, and design contexts: how developers design with them. In: 2018 25th Asia-Pacific Software Engineering Conference (APSEC), pp. 49–58. IEEE (2018)

8. Da Li, X., Wang, C., Bi, Z., Jiapeng, Yu.: Autoassem: an automated assembly planning system for complex products. IEEE Trans. Industr. Inf. **8**(3), 669–678 (2012)

9. Kowarschick, W.: Model-view-controller-service-paradigma. https://glossar.hs-augsburg.de/Model-View-Controller-Service-Paradigma. Accessed 25 Jan 2023

10. Li, T., Lockett, H.: An investigation into the interrelationship between aircraft systems and final assembly process design. Proc. Cirp **60**, 62–67 (2017)

11. Münker, S., Polikarpov, M., Schmitt, R.H.: Virtuelle demontage für xl-produkte/assembly sequence generation for xl-products – virtual disassembly for high numbers of components. wt Werkstattstechnik online **110**(11–12), 833–837 (2020)

12. Münker, S., Schmitt, R.H.: Cad-based and/or graph generation algorithms in (dis)assembly sequence planning of complex products. Proc. CIRP **106**, 144–149 (2022)

13. Sargent, R.G.: Verification and validation of simulation models. In: Proceedings of the 2010 Winter Simulation Conference, pp. 166–183. IEEE (2010)

14. Xiao, J., Anwer, N., Li, W., Eynard, B., Zheng, C.: Dynamic bayesian network-based disassembly sequencing optimization for electric vehicle battery. CIRP J. Manuf. Sci. Technol. **38**, 824–835 (2022)

15. Yao, M., Drescher, B., Stewart, E., Manoj, L., Wittstamm, M., Henke, L., Ghodasara, S., Ge, M.: Computer-aided assembly sequence planning for high-mix low-volume products in the electronic appliances industry. In: Proceedings of the Conference on Production Systems and Logistics: CPSL 2022, pp. 91–100. Publish-Ing., Hannover (2022)

16. Zheng, Yu., Chen, L., Jiang, P., Cheng, H.: A sub-assembly division method based on community detection algorithm. Int. J. Comput. Integr. Manuf. **35**(10–11), 1133–1150 (2022)

Comparison of Global Path Planning Algorithms Regarding Multi Mobile Robot Object Transport Requirements

Henrik Lurz, Tobias Recker and Annika Raatz

Abstract

This paper presents a comparison of multiple global path planning algorithms regarding their usage in multi-robot systems. Most performance-based comparisons only focus on computation time and path length, which are not the most important indicators when transporting an object using a multi-robot formation. Therefore, we decided to add the distance to obstacles and curvature of the path as performance metrics to support the selection of a fitting global path planner for a multi-robot application. In the comparison, we included the PRM, RRT*, Relaxed A* and Voronoi path planning algorithms. Each algorithm performs a specified number of planning tasks and is evaluated by the planning success, path length, obstacle distance and curvature, which are visualized below.

Keywords
Global path planning • Mobile robot • Navigation

1   Introduction

The two-dimensional navigation for a wheeled mobile robot can be separated into global and local path planning. The global path planner calculates a path from a start to a target position while considering static obstacles in the environment. This collision-free path is then used by the local path planner to determine velocities that move a mobile robot along the path while avoiding dynamic obstacles [1]. Therefore, the global path planning process

H. Lurz (✉) · T. Recker · A. Raatz
Institut of Assembly Technology, Leibniz University Hannover, 30823 Garbsen, Germany
e-mail: lurz@match.uni-hannover.de
URL: https://www.match.uni-hannover.de/en/

© The Author(s) 2025
S. Ihlenfeldt et al. (eds.), *Annals of Scientific Society for Assembly, Handling and Industrial Robotics 2023*, https://doi.org/10.1007/978-3-031-74010-7_9

has a major impact on the motion of a mobile robot through the planned path. Global path planning is crucial for the path length and hence the time it takes the mobile robot to reach its target. However, finding the shortest path is not always the most important requirement as the mobile robot must be able to follow the path despite its kinematic constraints.

An example for the necessity to consider kinematic and dynamic constraints is the transportation of an object using multiple differential drive mobile robots. This cooperative object transport is essential when it is not possible to use a single mobile robot due to the object being too heavy, big or delicate. Here, the distances between the mobile robots must remain constant when moving along the path. Any violation of this requirement would result in damaging or breaking the transported object. In this application, the global path planner calculates a path from a start to a target position for the whole formation instead of a single robot to allow for a collision-free transportation. Using the formation path and the position of the robots in the formation allows to derive a path for each individual robot. When deriving the robot path, the position in the formation has a big impact on the movement the mobile robot has to perform to follow the formation. During an orientation change of the formation, a mobile robot's leverage arm to the formation's pivot point can cause the movement to exceed the kinematic or dynamic limits of the mobile robot. This can be caused by rapid formation turning motions or too sharp corners. It is therefore advised to consider the hardware limits of each mobile robot in the global planning phase since the local path planner is responsible for controlling the individual mobile robots during the movement on the pre-planned path and for keeping the distances constant.

In a previous work, we developed the Splined Relaxed A* algorithm for the cooperative object transport using multiple mobile robots [2]. This global path planner uses the Relaxed A* algorithm to find an initial path from the start to a target position. Due to the Relaxed A* algorithm being an optimal path planner, the resulting path is as short as possible. Afterwards, we use Bézier-splines to create and optimize a continuous and smooth path with a curvature-limit, which the formation, consisting of differential drive mobile robots, is able to follow. The Bézier-splines are interpolated using control points from the initially generated path by the Relaxed A* algorithm. The minimal distance to obstacles by the Relaxed A* path and the interpolation of discrete positions cause a significant risk of creating a spline too close to the obstacles. This leads us to the investigation on global path planners that are better suited for generating a path for a multi-robot formation.

2   Related Work

Global path planning algorithms can be divided into three main areas: grid-based, sampling-based and potential-based [3]. Grid-based methods segment the environment into cells with a specified size. Each cell is identified by a two-dimensional coordinate with eight adjacent neighbour cells. Using the cell size, a cell coordinate can be transferred into real world coordinates [4]. Sampling-based algorithms use path planner-dependent query-methods to create a roadmap covering the environment with collision-free trajectories. Using the roadmap, a path planner can calculate the shortest path from the start to the target position [5]. Potential-based methods use attracting and repelling fields to reach the target position. The target position is surrounded by an attractive potential field while all obstacles contain a repulsive potential field [6]. The potential field method features a high complexity due to the bi-operational path model [3] and suffers from the local minima problem [7]. We therefore refrained from using potential-based methods for this comparison. Instead, we included the Relaxed A* and Voronoi path planning algorithm from the grid-based methods. Furthermore, we included Probabilistic Road Map(PRM) and Rapidly-Exploring Random Trees*(RRT*) from the sampling-based methods.

The Relaxed A* algorithm is an evolution of the A* planning algorithm, which features a shorter computation time compared to its predecessor. The Relaxed A* path planner is time-linear and made to efficiently explore large grid environments. The algorithm uses a specified criterion like the Euclidean distance to explore the grid environment in the direction of the target position. After reaching the target position, the explored cells can be backtracked to determine the shortest path [8]. A path of the Voronoi diagram-based path planner is the most accessible and departable. It uses a distance function to skeletonize the free space and creates a pixel-based roadmap that provides the maximal distance to all obstacles in the environment [9]. After creating the roadmap, a path planning algorithm finds the shortest path. The PRM algorithm is a very fast and easy to implement path planning algorithm, which consists of two phases. Firstly, the learning phase randomly generates a specific amount of valid robot positions into the free environment. Nearby positions that can be linked without a collision are connected to form a roadmap. Secondly, the query phase determines the edges of the roadmap that connect the start with the target position [10]. The RRT* path planning algorithm is an evolution of the RRT algorithm and proven to be probabilistically complete, asymptotically optimal and computationally efficient. This algorithm creates a directional graph by sampling new nodes in free space from the start position. Each time a new node is created, the growing tree is rewired to maintain the shortest path to each node from the start position. The node creation is continued until either the target is reached or a total amount of nodes are created without finding the target position [5].

The previously described path planning algorithms create neither continuous nor smooth paths. For that reason, splines can be used to interpolate a determined path to be continuous and smooth, while additionally optimizing the curvature. Like in our previous work [2], we choose to use quintic Bézier-splines for the creation of a continuous path due to their high controllability. Depending on the degree of a Bézier-spline, it uses a specific amount of control points to be created. Two control points are passed through while the others allow for a change in curvature [11]. In this comparison, each path planning algorithm (Relaxed A*, Voronoi, PRM and RRT*) will also be used in conjunction with Beziér-splines.

Multiple global path planning comparisons have been carried out in the past as shown in [3, 8, 12]. On the one hand, these comparisons only consider performance metrics like the path length or computation time which are relatively unimportant for a real-world application in an autonomous mobile robot system. On the other hand, some comparisons only use path planning algorithms of a single working principle. For that reason, we chose to use additional performance metrics like curvature and obstacle distance. Furthermore, we selected multiple path planners using different planning principles.

3   Approach

To compare the selected global path planning algorithms in a real world scenario, we selected the map of the "Deutsches Museum" in Munich [13], which is shown by the black markings in Fig. 1. This environment is relatively big compared to other established testing environments with a width of 168.0 m and a length of 272.3 m. This allows for a path generation of short and long paths ranging from 10 m to over 300 m. Additionally, the map contains multiple tight corridors, complex obstacles and empty spaces that provide a variety of different features the global path planners need to manage.

When calculating a path from a start to a target position, the global path planner relies on the global costmap. The global costmap uses the so-called inflation radius, which is determined by calculating the smallest circle around the mobile robot. In this test series, we used a costmap inflation radius of 0.5 m, based on our mobile robots. The inflation radius is used to enlarge obstacles in the environment and to shrink the navigating mobile robot by the same distance. Doing this allows the global path planner to view the mobile robot as a single point on the map. This procedure is less computationally intensive as the path planner only needs to check a single point for collision. This means that a collision is unlikely if the robot keeps out of the occupied inflated costmap cells.

Using the costmap, 50 position sets are randomly generated in the free space of the map to be used as start and target position for the global path planning. Firstly, this ensures that determining a collision-free path between the start and target position is possible. Secondly, the random selection of points prevents human-bias towards a particular global path planner which leads to a neutral comparison. Thirdly, spreading the random selection over the whole map provides path planning problems that need to avoid different obstacles. All 50 point sets are displayed in Fig. 1 in red.

The comparison of the different paths will be performed using the following performance criteria:

Path planning success: Sampling-based path planning approaches only guarantee to find a path between the start and target position if the path planner can search for an infinite time. But due the time and hardware processing constraints, a timeout is introduced for the calculation. Therefore, not all path planners are able to determine a path from the start to the target position in a given amount of time. However, not finding a path means that the mobile robot is not able to autonomously drive to the target position. Consequently, it is important that a path planner is able to reliably find a path from the start to target position if it exists. We evaluate the reliability by how many paths are successfully planned.

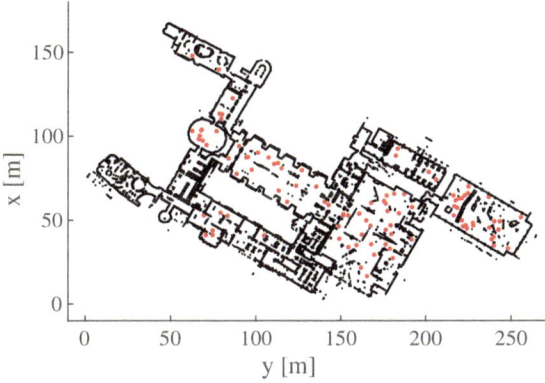

Fig. 1 Map of the "Deutsches Museum" in Munich with all 50 start and target point sets included

Path length: The path length is one of the parameters that influence the time a mobile robot needs to follow a path. Therefore, a typical requirement is to find the shortest and time-optimal path between the start and target position. The total path length is calculated by integrating over the determined path.

Distance to Obstacles: When obstacles intersect the direct link between the start and target position, multiple path planning algorithms calculate the shortest path that is as close to the obstacles as possible, which results in a higher collision probability. It also leads to a smaller error margin for the mobile robot during the actual movement. The most critical position on the path is, where the distance between the path and an obstacle is the lowest. For that reason, we calculated the minimal euclidean distance from all path points to any obstacles.

Curvature: The algorithms of the path planners are vastly different, which leads to different curves along the path. As mentioned at the beginning, the curvature and smoothness of the path is important for a differential drive mobile robot and even more important when moving in a rigid formation. A lower curvature leads to less turning motion and is thus easier to follow. To quantify the necessary turning motion, we derive the target orientation of the robot, which results in the orientation change over the planned path. To evaluate the orientation change we use the root mean square as it amplifies and therefore punishes higher orientation changes.

4   Evaluation

The goal of this chapter is to compare the four global path planners and present the advantages and disadvantages according to the collected data. Additionally, we use Bézier-splines to smoothen each of the four planners. This leads to a comparison of eight path planning algorithms with different properties.

4.1   Path Planning Success

Figure 2 displays the number of paths generated successfully using the 50 randomly generated start and target position sets. The Relaxed A* and Voronoi path planning algorithm successfully calculate all paths and therefore perform the best. This can be explained by the analytical calculation of the path by both algorithms and the guarantee to find a path if one exists. The sampling-based approaches, PRM and RRT*, successfully calculate a path in less than half of the planning attempts. This is likely due to the big environment and the number of points used to create the roadmap in PRM or the tree structure in RRT*. An increase in those values would improve the chance of successfully planning but also extend the calculation time.

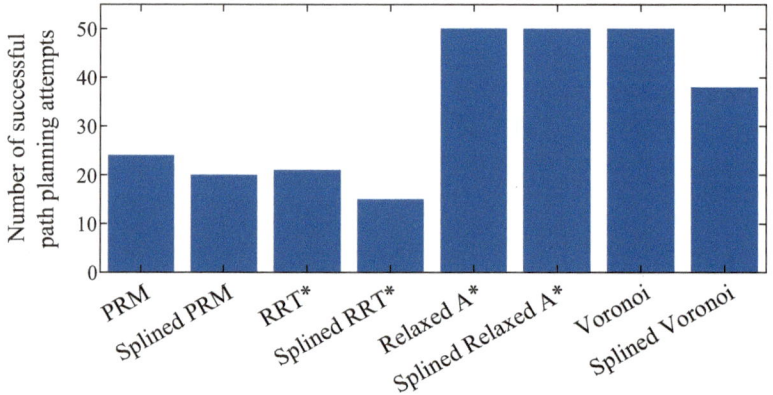

Fig. 2 Overview of all successful path planning attempts of each path planner

All splined versions of the path planners perform worse than the non-splined version, except for the Splined Relaxed A* algorithm. The decrease in performance is due to the sharper turns in the initial paths that the Bézier-spline creation is not able to compensate for. Here, the optimization of the curvature of the spline fails and therefore aborts the path generation as the curvature is a mandatory requirement for the spline.

4.2   Path Length

In order to compare the path lengths of the different path planners, we need to determine the paths where all path planners were successful. This results in 15 out of 50 paths that are included in the path length comparison. We only use the non-splined versions of the path planners for this comparison because the initial path length has been found similar to the splined path length. Furthermore, the low success rate of the splined path planners provides no expressive number of splined paths to compare.

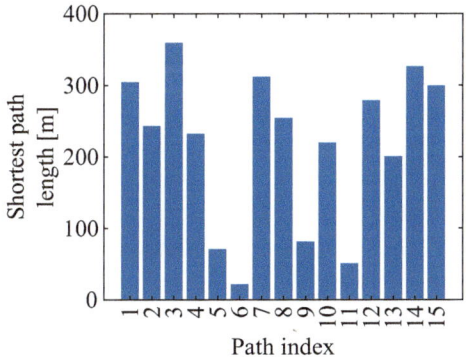

Fig. 3 Shortest path length of the 15 planning attempts of the four path planners

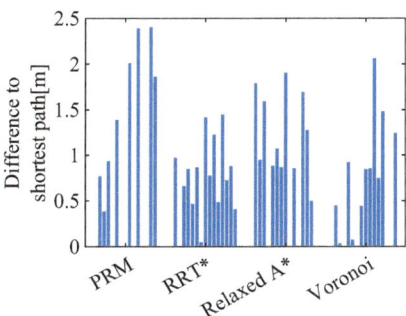

Fig. 4 Difference to shortest path for all 15 planning attempts

In Fig. 3, we show the shortest path length of all four path planners for each of the 15 paths. The different start and target position sets create paths with different lengths each, ranging from 21.5 m to 358.8 m. The different path lengths lead to the conclusion that a longer path length is not the reason for the low success rates shown in Chap. 4.1. Figure 4 shows the difference in path length compared to the shortest path for all path planners and all 15 paths. This means that if the difference to the shortest path is 0 m, then this path planner determined the overall shortest path for the specific start and target position. Figure 4 shows that even on a path length of over 300 m, the difference between each planning algorithm is a maximum of 2.4 m. From the low fluctuation in path length, we can conclude that all path planners create roughly the same path lengths and therefore do not highlight any of the four path planners which find the shortest path.

4.3   Distance to Obstacles

For this comparison, the same paths as in Sect. 4.2 are used. The average of the minimal obstacle distances as well as the smallest and highest minimal distance to an obstacle on any of the 15 paths are shown in Fig. 5. The dashed black line marks the 0.5 m inflation radius. Every path planner where the lower value of the error bar is located below the inflation radius has a high risk of collision. For the PRM, RRT* and Relaxed A* algorithms, the smallest obstacle distance is slightly lower than the inflation radius, which can be explained by the costmap using a cell size of 50 mm. Due to the Voronoi path planner using a roadmap that is based on the maximal distance to any obstacle it is capable of maintaining a distance to the nearest obstacles, which is higher than the inflation radius.

When using the splined variant of the four path planners, only the Splined Voronoi path planner provides paths that are mostly far enough away from any obstacles without causing any risk of collision. Due to the low mean value, the Splined RRT* and the Splined Relaxed A* are very noticeable. Additionally, the smallest minimal distance is almost 0 m and therefore definitely colliding with an obstacle. Consequently, these path planners are not usable in an application since the interpolation of the Bézier-spline is not optimized and versatile enough for such a complex environment. The Splined PRM path planner is better compared to the Splined RRT* and Splined Relaxed A*. However, the smallest minimal obstacle distance still provides a high risk of collision.

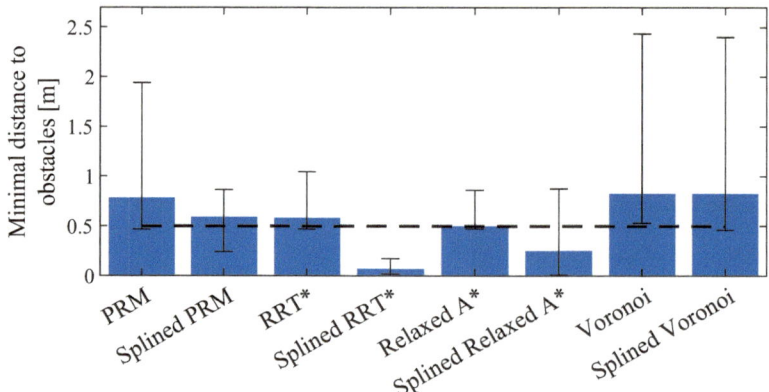

Fig. 5 Minimal distance to an obstacle over the planned path

4.4   Curvature

In Chap. 3, we outlined the importance of the curvature of the path for any formation of mobile robots. A too high curvature or turning motion can lead to an unfeasible motion by the differential drive mobile robots. In this comparison, we want to show the differences in curvature between the path planning algorithms and the improvements a spline can provide to the path. In Figs. 6 and 7, we depict the necessary turning motion a mobile robot needs to perform in order to be able to follow the path. Here, we present the turning angle for each path planner and all 50 paths. In Fig. 6, the turning motion is calculated using the root mean square over the whole path. In Fig. 7, we show the maximal turning angle on the path which will be the most critical location on the path. We already mentioned the low success rate of some path planning algorithms in Sect. 4.1, which is reflected in the results by the grey marked areas where no path was found.

Figure 6 clearly shows the difference between the splined and non-splined version by a much lower average turning motion along the path. The difference between the non-splined path planning algorithms is minor, whereas the Relaxed A* path planner performs slightly better. This algorithm benefits from the high number of path points due to the long straights in the planned paths. The other three path planning algorithms use a roadmap that expels a turning motion when passing an intersection, which results in a higher average value.

Figure 7 shows that the Relaxed A* path planner has the highest turning angles in its path. This is due to the grid-based pathfinding algorithm that only contains 0°, 45° and 90° turns. The other three path planning algorithms perform equally but better than the Relaxed A* algorithm. Figure 7 clearly visualizes the importance of using splines to smooth out the path to reach minimal turning angles along the path. The PRM, Relaxed A* and Voronoi maximal turning angles significantly improve when using the splined variant. Only the Splined RRT* path planner does not feature any improvement to the initial one. This can be explained by the inability of the spline creation and optimization to reach the desired curvature in the critical locations on the path.

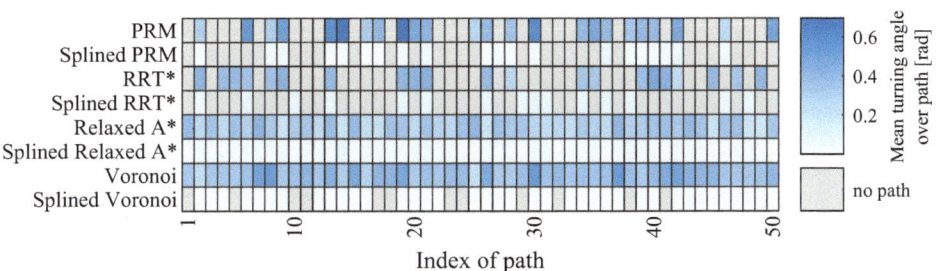

Fig. 6 Mean orientation change on the path

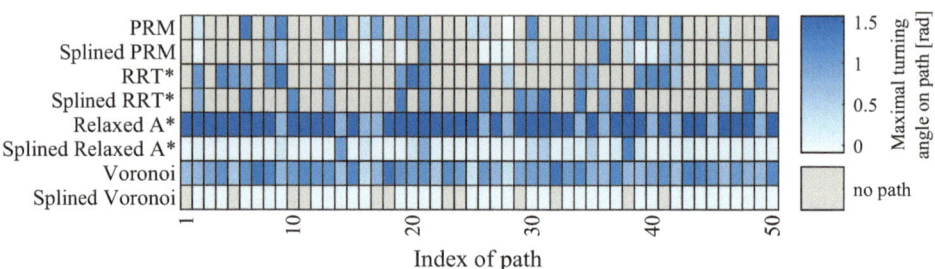

Fig. 7 Maximal orientation change on the path

5   Conclusion and Outlook

We have presented a comparison between the PRM, RRT*, Relaxed A* and Voronoi path planning algorithms. We have also included an existing method for creating continuous and smooth paths, which was applied to all mentioned path planning algorithms. This results in eight series of measurements that were investigated using the criteria planning success, path length, obstacle distance and curvature. We conclude that the sampling-based path planning algorithms are only suited for a big and complex map if time requirement allows the use of a high amount of planning points. The analytically solvable algorithms Relaxed A* and Voronoi succeeded all path planning attempts in the map. The path length comparison on this map provided the result that it is not a meaningful metric as the fluctuation in path length was negligible. The obstacle distance investigation showed that the usage of splines has to be thought through and improved, otherwise the distance to the obstacles is too small. The curvature comparison showed that some path planning algorithms like Relaxed A* and Voronoi are more suitable to be used with splines as these indicate a bigger improvement. The results provide a foundation for a selection on which path planning algorithm to choose for a multi-robot system that moves as a formation. In the future, we will further investigate the spline generation and optimization due to its lower success-rate than the default path planning algorithms.

Acknowledgements The authors gratefully acknowledge the partial funding by the Deutsche Forschungsgemeinschaft (DFG—German Research Foundation)—Project no. 414265976. The authors would like to thank the DFG for the support within the SFB/Transregio 277—Additive manufacturing in construction. (Subproject B04).

References

1. Marin-Plaza, P., Hussein, A., Martin, D., de la Escalera, A.: Global and local path planning study in a ROS-based research platform for autonomous vehicles. J. Adv. Transp. 1–10 (2018)

2. Lurz, H., Recker, T., Raatz, A.: Spline-based path planning and reconfiguration for rigid multi-robot formations. Proc. CIRP **106**, 174–179 (2022)
3. Victerpaul, P., Saravanan, D., Janakiraman, S., Pradeep, J.: Path planning of autonomous mobile robots: a survey and comparison. JARDCS **9**, 1535–1565 (2017)
4. LaValle, S.M.: Planning Algorithms. Cambridge University Press (2006)
5. Karaman, S., Frazzoli, E.: Sampling-based algorithms for optimal motion planning. IJRR **30**, 846–894 (2011)
6. Lee, M.C., Park, M.G.: Artificial potential field based path planning for mobile robots using a virtual obstacle concept. IEEE/ASME **2**, 735–740 (2003)
7. Milos, S.: Roadmap methods versus cell decomposition in robot motion planning. In: ISPRA, vol. 6, pp. 127–132 (2007)
8. Ammar, A., Bennaceur, H., Chaari, I., Koubaa, A., Alajlan, M.: Relaxed Dijkstra and A* with linear complexity for robot path planning problems in large-scale grid environments. Soft. Comput. **20**, 1–23 (2015)
9. Choset, H., Burdick, J.: Sensor based planning, part i: the generalized Voronoi graph. In: IEEE International Conference on Robotics and Automation (1995)
10. Kavraki, L.E., Švestka, P., Latombe, J.-C., Overmars, M.H.: Probabilistic roadmaps for path planning in high-dimensional configuration spaces. IEEE Trans. Robot. Autom. **12**, 566–580 (1996)
11. Piegl, L., Tiller, W.: The NURBS Book, 2nd edn. Springer, Berlin (1997)
12. Chaari, I., Koubaa, A., Bennaceur, H., Ammar, A., Alajlan, M., Youssef, H.: Design and performance analysis of global path planning techniques for autonomous mobile robots in grid environments. In: IJARS, vol. 14 (2017)
13. The Cartographer Authors, https://github.com/cartographer-project/cartographer_ros/blob/master/docs/source/data.rst

Material-Adapted Gripping and Handling of PEO-Based Cell Components for All-Solid-State Battery Cell Stacking

Do Minh Nguyen, Timon Scharmann, and Klaus Dröder

Abstract

Lithium-ion batteries (LIBs) immensely contribute to the electromobility's success for achieving climate change goals. As LIBs are forecasted to succumb to optimization limits in the coming decade, next generation battery technologies, such as all-solid-state batteries (ASSBs), gain noteworthy attention for meeting ever-increasing cell performance requirements. By deploying solid electrolytes (SEs), compared to liquid electrolytes in current LIBs, ASSBs benefit from enhanced safety against flammability and allow for the usage of lithium metal anodes for higher energy densities. Here, polymer solid electrolytes, such Polyethylene oxide (PEO), are widely used for their high flexibility and hence beneficial processability properties compared other SEs. However, their adhesive behavior poses challenges when conducting handling and stacking processes with conventional grippers during cell assembly. In this research, we present a parameter study on ASSB handling and stacking with PEO-based cell components aiming to promote process understanding and point out optimization potentials. An experimental design for testing different grippers is devised by which deposition accuracy is systematically assessed in relation to the holding force, gripper speed, and placement distance. Within this evaluation, the electrostatic gripper with a Polytetrafluoroethylene (PTFE) dielectric provides adequate position and orientation accuracies in almost all experiments while showing improved accuracies with higher holding forces.

D. M. Nguyen (✉) · T. Scharmann · K. Dröder
Institute of Machine Tools and Production Technology, Technische Universität Braunschweig, 38106 Braunschweig, Germany
e-mail: do-minh.nguyen@tu-braunschweig.de

Battery LabFactory Braunschweig, Technische Universität Braunschweig, 38106 Braunschweig, Germany

© The Author(s) 2025
S. Ihlenfeldt et al. (eds.), *Annals of Scientific Society for Assembly, Handling and Industrial Robotics 2023*, https://doi.org/10.1007/978-3-031-74010-7_10

Parameter settings achieving higher overall deposition accuracies for all tested grippers are identified. This research provides insights into the establishment of stacking processes for realizing an industry-scale ASSB production.

Keywords

All-solid-state-battery • Battery production • Cell stacking • Electrostatic gripper • Deposition accuracy • Polymer electrolyte

1   Introduction

Facing rapid decarbonization in the automotive sector, lithium-ion batteries (LIBs) establish themselves as a key technology driving the transition towards electromobility. As current LIBs are subject to performance issues due to insufficient energy and power densities along with unfavorable charge rates, new battery generations, such as all-solid-state batteries (ASSBs), rise to prominence for overcoming these obstacles and meeting requested demands. In ASSBs, solid electrolytes (SEs) are used instead of liquid electrolytes for significantly improving system safety as the risk of leakage and flammability is reduced [1]. Additionally, this enables the usage of lithium metal anodes (LMAs) for tackling the aforementioned performance demerits of LIBs [2]. Compared to ceramic SEs with a brittle nature, polymer SEs, such as Polyethylene oxide (PEO), not only exhibit high stability against lithium but also show great processability as a result of their flexibility [1]. Despite this, conducted material tests with PEO-based cell components at our facilities showed its high mechanical sensitivity along with increased deflection tendencies and a noteworthy adhesive behavior, especially in contact with itself. Both factors become a potential challenge in ASSB cell assembly with its multitude of handling operations and the desired battery cell quality, for which process adaptions in accordance to new materials are required.

Current research on handling of ASSB cell components is scarce and focuses on the handling of LMA which exhibits a similar mechanical sensitivity and adhesion towards the machinery to PEO [3–5]. As handling of polymer-based SEs has not been addressed specifically yet, the need for gripping concepts aiming for an efficient handling of solid electrolytes becomes apparent. This publication presents a parameter study investigating the handling of PEO-based cell components in ASSB cell stacking with material-adapted grippers. This experimental approach, being derived from production-relevant material properties as well as set requirements on the desired handling system, is finally validated considering specific process parameters.

2   Handling and Gripping in Battery Cell Assembly

In battery cell assembly, handling operations mainly occur during cell stacking in which electrodes and separator/electrolyte are built into a compound cell stack. Three different processes have been established for cell stacking and are shown in Fig. 1.

In cell winding (Fig. 1a), continuously fed electrode and separator webs are wound around a winding core [6]. Occurring small bending radii induce high material bending stress making this process unsuitable for brittle or stiff materials which concerns ceramic SEs [7, 8]. When being coated on both sides, the adhesive behavior of polymer-based SEs poses challenges in the web-based material provision as the material cannot be easily unwound. In z-folding (Fig. 1b), grippers deposition single sheet electrodes on the continuously fed separator web which is then alternatingly folded over the electrodes' edge and thereby forms a meander shape [6]. As with cell winding, the listed factors deem z-folding inappropriate for processing of SEs. In contrast, in single sheet stacking, electrodes and separator are provided as single sheets and alternatingly stacked onto each other by grippers [6]. Advantages such as high material and format flexibility along with low material bending stresses distinguish single sheet stacking as the most preferred ASSB cell stacking process in future production systems [7, 9].

Grippers as the main component in interacting with the cell components can be divided according to their working principle.

Figure 2 gives an overview on different grippers used in battery cell assembly with their respective working principle. The shown concepts belong to force-closed gripping principles for ensuring material-sensitive handling of such limp foils [10].

Pneumatic grippers (Fig. 2a–c), especially vacuum suction grippers, are most commonly used as state-of-the-art handling solutions in LIB cell assembly and were already

(a)   (b)   (c)

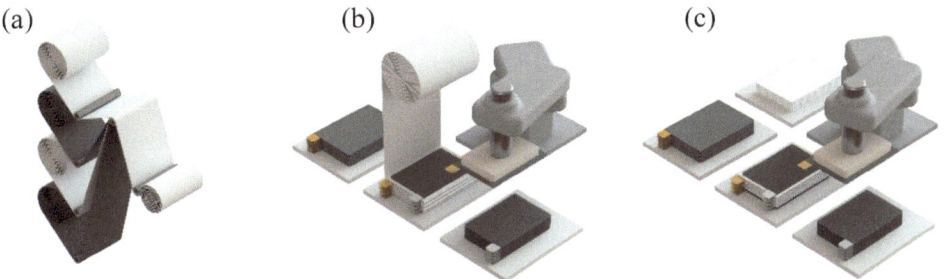

Fig. 1 Cell stacking methods, **a** Winding process with continuous electrode and separator web, **b** Z-folding process with single sheet electrodes and continuously fed separator web, **c** Single sheets stacking process with single sheet electrodes and separators

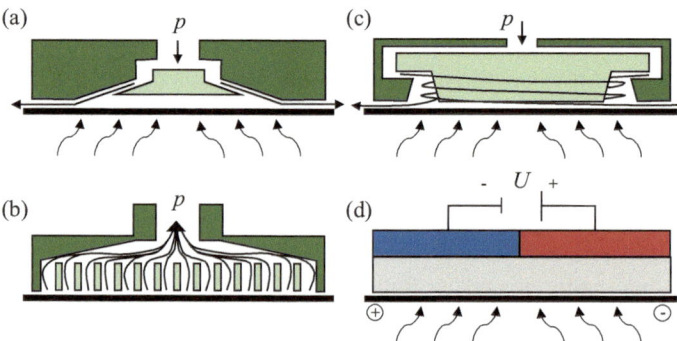

Fig. 2 Grippers principles used in battery cell production, **a** Bernoulli gripper, **b** Vacuum suction gripper, **c** Cyclone gripper, **d** Electrostatic gripper [11]

intensively investigated towards their applicability in ASSB cell stacking by different researchers [3, 4, 10, 11]. Vacuum suction grippers (Fig. 2b) create a holding force via vacuum between gripper and electrode which is ideally evenly distributed over the entire gripper surface via numerous holes for surface load reduction. Central disadvantages lie in the necessity for a process medium and the potential material adhesion due to electrostatic charge and surface damage resulting from mechanical contact. In contrast, Bernoulli grippers (Fig. 2a) allow for contact-free material handling by means of compressed and fast-flowing air streams creating overpressure and thus a lifting force. However, uncontrolled and gap-dependent vibrations potentially hinder safe electrode handling [4, 11]. Another gripping principle utilizing overpressure is the cyclone gripper (Fig. 2c). Here, a circularly rotating air flow generates negative pressure in the gripper's center resulting in an overpressure with fast air flow between gripper and electrode. The resulting lifting force also exhibits a dependency on the gap but with distributed fixation of the electrode and comparably low air consumption [12]. In contrast, the electrostatic gripping principle (Fig. 2d) utilizes the attractive Coulomb force generated by an electric field acting between the gripper and workpiece for creating a lifting force. The gripper features two bipolar conductive electrodes charged with AC or DC power which are spatially separated and electrically insulated from the opposing workpiece by a dielectric. This dielectric prevents a charge equalization and dielectric breakdowns at high voltages. Assuming an ideally homogenous electric field influencing infinitesimally small charges in the workpiece, an electrostatic gripper potentially achieves a more homogonous force distribution over its surface compared to vacuum suction grippers. In recent experimental handling investigations with LMAs in our institution, an electrostatic gripper prototype with a Polytetrafluoroethylene (PTFE) dielectric surpassed established pneumatic grippers considering the achieved deposition accuracy. Here, the usage of AC power is advised for potentially minimizing adhesion due to the periodic polarization change. In this research, the following grippers are investigated:

- Vacuum suction gripper with a METAPOR contact surface (VSG-METAPOR),
- Vacuum suction gripper with a PTFE contact surface (VSG-PTFE),
- Bernoulli gripper (BG),
- Electrostatic gripper with a PTFE dielectric (ESG-PTFE), and
- Electrostatic gripper with a DLC dielectric (ESG-DLC).

METAPOR refers to a microporous polymer with a homogonous pore distribution for enabling lower surface loads over its smaller holes compared to the PTFE material used in the VSG-PTFE. For sufficient material adaption, the entire handling system must fulfill requirements considering material properties, the assembly process, and established handling principles. Konwitschny et al. as well as Fröhlich et al. listed the following criteria for ASSB cell component handling systems [4, 5]:

1. Damage-free component processing with avoidance of particulate contamination,
2. Adjustable and homogenously distributed lifting forces with high reproducibility for repeated handling processes,
3. Fast build-up and breakdown of a lifting force to ensure fast handling cycles,
4. Reliable and sufficient deposition accuracy,
5. Material separation when gripping from loaded stacks,
6. Avoidance of material adhesion after deposition, and
7. Possible lifting force generation in conditioned atmospheres with low media consumption.

This research focuses on the analysis of the deposition accuracy during gripping as a quantifiable measure for evaluating and comparing different gripping concepts. It is characterized by the pose accuracy AP_i as well as the pose repeatability RP_i according to ISO 9283 [13]. While this norm describes pose accuracy as "the deviation between a command pose and the mean of the attained poses when approaching the command pose from the same direction", it depicts pose repeatability as "the closeness of agreement between the attained poses after n repeat visits to the same command pose in the same direction" [13]. Both aspects are divided into components considering position in x, y, z-axes (position accuracy and repeatability) and the orientation around those axes (orientation accuracy and repeatability) while being calculated as follows [13]:

Pose accuracy:

Position accuracy:

$$AP_{x,y} = \sqrt{(\bar{x} - x_c)^2 + (\bar{y} - y_c)^2 + (\bar{z} - z_c)^2} \qquad (1)$$

$$\bar{x} = \frac{1}{n} \sum_{j=1}^{n} x_j \quad \bar{y} = \frac{1}{n} \sum_{j=1}^{n} y_j \quad \bar{z} = \frac{1}{n} \sum_{j=1}^{n} z_j \qquad (2)$$

Orientation accuracy:

$$AP_\varphi = \left(\frac{1}{n} \sum_{j=1}^{n} \varphi_j - \varphi_c \right) \tag{3}$$

Pose repeatability:

Position repeatability:

$$RP_l = \bar{l} + 3S_l \tag{4}$$

$$\bar{l} = \frac{1}{n} \sum_{j=1}^{n} l_j \tag{5}$$

$$l_j = \sqrt{(x_j - \bar{x})^2 + (y_j - \bar{y})^2 + (z_j - \bar{z})^2} \tag{6}$$

$$S_l = \sqrt{\frac{\sum_{j=1}^{n}(l_j - \bar{l})}{n-1}} \tag{7}$$

Orientation repeatability:

$$RP_\varphi = \pm 3S_\varphi = \pm 3\sqrt{\frac{\sum_{j=1}^{n}\left(\varphi_j - \bar{\varphi}\right)^2}{n-1}} \tag{8}$$

where x_c, y_c, z_c and φ_c are coordinates or angles of the command pose and x_j, y_j, z_j and φ_j are those of the j-th attained pose. As mentioned in [14], increasing deposition accuracy values can be correlated to a decrease in discharge capacity which is undesired. According to Mooy, deposition accuracy should not exceed 0.5 mm in position and 0.5° in orientation accuracy as well as repeatability in order to guarantee sufficient cell performance which is assumed for the following sections [15].

---

3   Experimental Design and Setup

Considering the specimen type, experiments are performed on cathodes consisting of lithium iron phosphate (LiFePo$_4$) and carbon black as an active material where PEO with lithium bis(trifluoromethanesulfonyl)imide (LiTFSI) is applied as a SE. These are manufactured and provided by the Institute of Particle Technology of the Technische Universität Braunschweig in using a solvent-free processing via melt granulation, subsequent extrusion, and direct calendering. Details on the conducted dry electrode processing routes

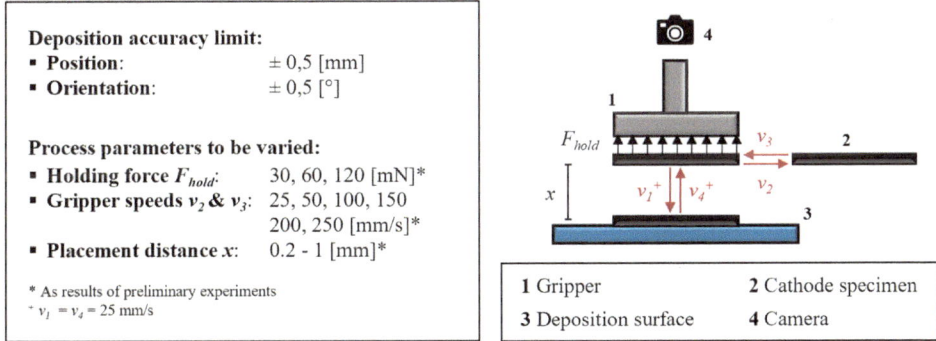

Deposition accuracy limit:
- **Position**: ± 0,5 [mm]
- **Orientation**: ± 0,5 [°]

Process parameters to be varied:
- **Holding force** F_{hold}: 30, 60, 120 [mN]*
- **Gripper speeds** v_2 & v_3: 25, 50, 100, 150 200, 250 [mm/s]*
- **Placement distance** x: 0.2 - 1 [mm]*

* As results of preliminary experiments
⁺ $v_1 = v_4 = 25$ mm/s

| 1 Gripper | 2 Cathode specimen |
| 3 Deposition surface | 4 Camera |

Fig. 3 Schematic test setup, including component details, movement sequences (red), and varied process parameters

are further discussed by Helmers et al. [16]. The cathodes feature a coating thickness of 60 μm on 20 μm thick aluminum substrates and are cut to the overall dimension of 50 × 70 mm².

Figure 3 depicts the devised setup for the experiments aimed to minimize disturbances while ensuring high result reproducibility.

Here, the tested grippers, mounted on a vertical articulated robot arm (FANUC LR Mate 200iD/4S), grab the PEO-based cathodes from the deposition surface. The cathodes are then exposed to vertical and horizontal robot movements and finally released onto their initial position on the lighted deposition surface. A camera with a ring light and an image processing software are then used to visually analyze the difference in x- and y-coordinates and orientation (rotation around the z-axis) in comparison to a reference for calculating the deposition accuracy. With a distance of 70 cm between camera lense and deposition surface as well as a camera resolution of 5,742 × 3,648 px, the camera is capable of detecting displacements errors with tolerances of 0.034 mm. The Python-based image processing software consists of a Gaussian filter for noise reduction along with edge smoothing and a Canny algorithm for edge detection via brightness contrast. This is devised in accordance to the setup thoroughly described in [3]. Following the camera-based analysis, the specimens can then be grasped again for another iteration of the handling process. All movements are carried out as linear movements to ensure ideal congruence between gripper and specimen during pickup and release. The measured values are then used to calculate the total deposition accuracy as the superposition of position accuracy and repeatability as well as orientation accuracy and repeatability with formula (1)–(8). The deposition accuracy is systematically evaluated in relation to the holding force F_{hold}, the gripper speed v_2 and v_3 and the placement distance x. The variation steps shown in Fig. 3 are derived from pretests, including a holding force normalization for correlating the holding force with the applied pressure or voltage. For emphasizing the effects of lateral acceleration on the specimens during handling, both vertical movements are kept at the same speed ($v_1 = v_4 = 25$ mm/s). Overall, the experimental design consist of 75 experiments per gripper with 30 iterations each according to ISO 9283 [13]. A reference is taken before each experiment and used in every iteration for the evaluation. With a chosen horizontal movement length of 700 mm, the average duration of one iteration is around 22 s. Experiments are conducted in the dry room of the Battery LabFactory Braunschweig of the Technische Universität Braunschweig featuring a room temperature of 20 °C and a dew point of -40 °C.

4   Results and Discussion

During testing, the ESG-DLC is not able to achieve sufficient gripping of the PEO-based cathodes with the chosen AC voltages, so that therefore these results are omitted. When testing the BG, sufficient specimen handling is not achieved at higher pressures and placement distances due to occurring specimen vibrations which hinder a safe transportation of the cathodes after they have been gripped. These results are therefore also omitted.

Figure 4 shows the total deposition accuracies as the superposition of position and orientation accuracy (columns) and position and orientation repeatability (red error bars) for the remaining tested grippers. An excerpt of the gripper speed and placement distance variation is shown for the purpose of clarity.

Both tested vacuum suction grippers do not meet the proposed total position accuracy and repeatability limits for the majority of the experiments with very high position repeatability values of $AP_{x,y} = 1.5$–2 mm. The VSG-PTFE only achieves acceptable position accuracies and repeatabilities during the placement distance variation, starting from 0.6 mm with lower holding forces of $F_{hold} = 30$ mN. Deviations to the set orientation accuracy and repeatability are lower compared to the position accuracy and repeatability while only being sufficient at lower gripper speeds of $v_{2,3} = 25$ mm/s with values of $AP_\varphi = RP_\varphi \approx 0.25°$. In contrast, the VSG-METAPOR meets set orientation accuracy and repeatability throughout all tested holding forces and gripper speeds at lower placement distances of $x = 0.2$ mm. Only the position accuracy seems to improve with increasing holding forces at $v_{2,3} = 250$ mm/s. As partial specimen adhesion before deposition is observed at the highest occurring accuracy and repeatability values, both grippers' poor performance can be attributed to occurring mechanical contact and therefore electrostatic charge within the specimens with increasing holding forces or decreasing placement distances. Both factors impair an adequate deposition as detected during placement distance variations with $x = 0.2$ mm and holding force variations with $F_{hold} = 120$ mN for both grippers. Considering optimized parameter settings, both grippers show their best results at the lowest gripper speed of $v_{2,3} = 25$ mm/s and lower holding forces of $F_{hold} = 30$–60 mN. While the lowest deposition accuracy value is achieved at higher placement distances starting at $x = 0.8$ mm for the VSG-PTFE, the optimal placement distance for VSG-METAPOR is $x = 0.4$ mm. In summary, the results delivered by both vacuum

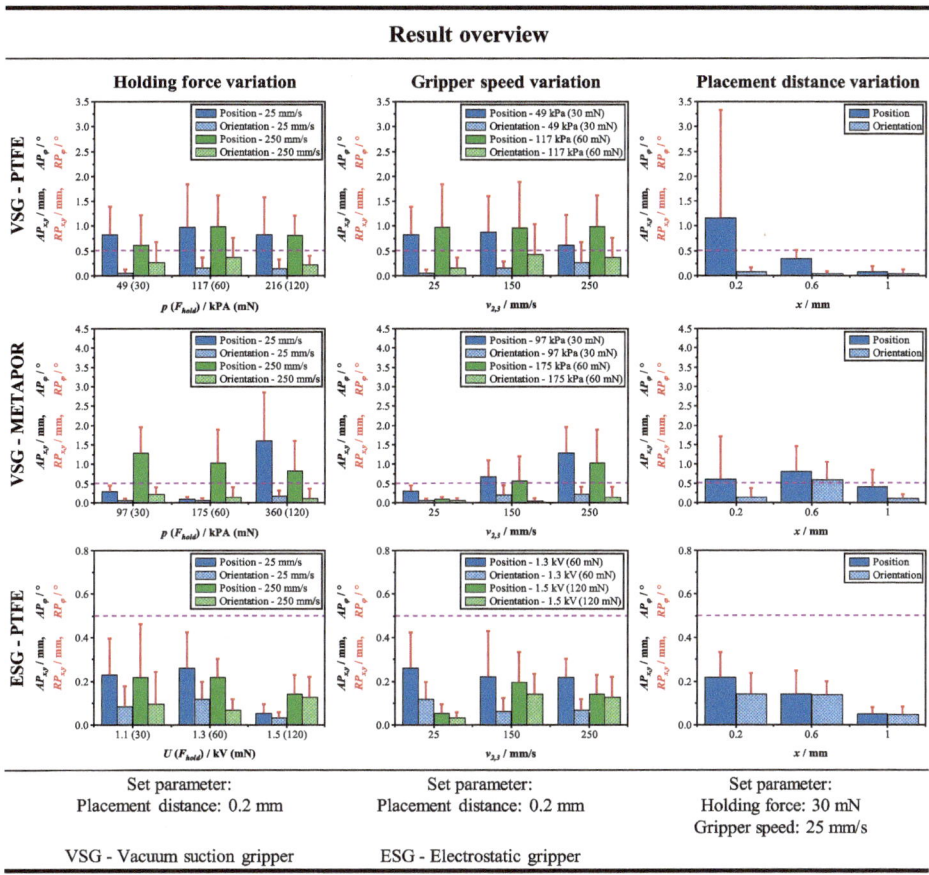

Fig. 4 Result overview of total the deposition accuracy for the tested grippers with set parameters

suction grippers are unsatisfactory for handling PEO-based cathodes during ASSB cell stacking.

In comparison, the ESG-PTFE with AC power reaches sufficient total deposition accuracy in almost all experiments with combined values of around 0.3 mm considering the position and 0.2° considering orientation (Fig. 4). The combined value of position accuracy and repeatability exceeds the desired limit only during one parameter configuration by 0.13 mm. During the holding force and gripper speed variation, an elevated holding force of 120 mN has shown to improve the accuracy and repeatability for both position and orientation. As seen with the vacuum suction grippers, a decrease in placement distance appears to correlate to higher accuracy and orientation values which can be attributed to the higher likelihood of adhesion. Also, experiments with placements distances higher than $x = 1$ mm exhibit worse gripper performance with higher deviations in accuracy and repeatability from the reference. Calculated position and orientation repeatability values are slightly higher than the respective accuracy values, especially at lower holding forces of $F_{hold} = 30$ mN and grippers speeds of $v_{2,3} = 25$–150 mm/s. For optimized deposition accuracy, the combination of a higher holding force of $F_{hold} = 120$ mN, a lower gripper speed of $v_{2,3} = 25$ mm/s, and a high placement distance of $x = 1$ mm is desired. As sufficient deposition accuracy is achieved even with elevated gripper speeds of $v_{2,3} = 250$ mm/s when testing the ESG-PTFE, we propose the usage of such speeds for minimizing stacking cycle time. Therefore, the usage of ESG-PTFE for ASSB cell stacking could close the gap towards an industry-orientated ASSB cell assembly. Nevertheless, specimen adhesion onto the machinery is occasionally apparent even after power shutdown which points to a remaining polarization of the gripper or specimens. Component adhesion could be tackled by applying different dielectrics in the gripper design or through the usage of ion blowers. Further testing with DC power should be considered.

5   Conclusion and Outlook

The main goal of this research is the thorough investigation of ASSB handling and stacking process with adhesive and mechanically sensitive PEO-based cathodes for realizing sufficient and optimized ASSB cell assembly. It serves as a main contribution for process development towards industry-scale ASSB production. This is achieved by a systematic experimental assessment of significant process parameters during cell stacking and their influence on the deposition accuracy. Despite identifying optimized parameter configurations for each tested gripper, the electrostatic gripper with a PTFE dielectric has proven to be the most suitable option for the tested PEO-based cathodes. Nevertheless, the remaining specimen adhesion in several iterations and even after power shut down leaves room for improvement in the experimental setup. Conducted pull-off tests with PEO-based cathode material under different specimen temperatures suggest the beneficial reduction of the PEO's adhesive behavior with specimen cooling, so that a sufficient specimen cooling system could be beneficial in future experimental setups. Additionally, tests with DC power and voltages above 1.5 kV could be of interest. As the negative influence of particulate contamination during electrode handling on the resulting cell performance is apparent [11], investigations on specimens' surface damages after gripping should additionally be included for thorough gripper evaluations. The proposed experimental design with improvements will be used for future gripper evaluations with other SE types, e. g. sulfide-based SEs, and compound components, such as cathode-separator-electrolyte-compounds, for more realistic use cases of industry-orientated ASSB production.

Acknowledgements The authors acknowledge the support of the German Federal Ministry of Education and Research in funding the research project "FB2-Prod" (03XP0432A).

References

1. Schmaltz, T., Wicke, T., Weymann, L., et al.: Solid-State Battery Roadmap 2035+. Fraunhofer ISI, pp. 7–9, 16–19, 40–41 (2022)
2. Lin, D., Liu, Y., Cui, Y.: Reviving the lithium metal anode for high-energy batteries. Nat. Nanotechnol. **12**, 194–206 (2017). https://doi.org/10.1038/nnano.2017.16
3. Fröhlich, A., Masuch, S., Dröder, K.: Design of an automated assembly station for process development of all-solid-state battery cell assembly. In: Schüppstuhl, T., Tracht, K., Raatz, A. (eds.) Annals of Scientific Society for Assembly, Handling and Industrial Robotics 2021, pp. 51–62. Springer International Publishing, Cham (2022)
4. Fröhlich, A., Gresens, D., Vervoort, B., et al.: Design and evaluation of a material-adapted handling system for all-solid-state lithium-ion battery production. Procedia CIRP **93**, 143–148 (2020). https://doi.org/10.1016/j.procir.2020.03.040
5. Konwitschny, F., Schnell, J., Reinhart, G.: Handling cell components in the production of multi-layered large format all-solid-state batteries with lithium anode. Procedia CIRP **81**, 1236–1241 (2019). https://doi.org/10.1016/j.procir.2019.03.300
6. Kampker, A., Hohenthanner, C.-R., Deutskens, C., et al.: Fertigungsverfahren von Lithium-Ionen-Zellen und -Batterien. In: Korthauer, R. (ed.) Handbuch Lithium-Ionen-Batterien, pp. 237–247. Springer Berlin Heidelberg, Berlin, Heidelberg (2013)
7. Schnell, J., Günther, T., Knoche, T., et al.: All-solid-state lithium-ion and lithium metal batteries—paving the way to large-scale production. J. Power. Sources **382**, 160–175 (2018). https://doi.org/10.1016/j.jpowsour.2018.02.062
8. Schilling, A., Schmitt, J., Dietrich, F., et al.: Analyzing bending stresses on lithium-ion battery cathodes induced by the assembly process. Energy Technol. **4**, 1502–1508 (2016). https://doi.org/10.1002/ente.201600131
9. Duffner, F., Kronemeyer, N., Tübke, J., et al.: Post-lithium-ion battery cell production and its compatibility with lithium-ion cell production infrastructure. Nat. Energy **6**, 123–134 (2021). https://doi.org/10.1038/s41560-020-00748-8
10. Schröder, R., Glodde, A., Aydemir, M., et al.: Increasing productivity in grasping electrodes in lithium-ion battery manufacturing. Procedia CIRP **57**, 775–780 (2016). https://doi.org/10.1016/j.procir.2016.11.134
11. Fröhlich, A., Leithoff, R., von Boeselager, C., et al.: Investigation of particulate emissions during handling of electrodes in lithium-ion battery assembly. Procedia CIRP **78**, 341–346 (2018). https://doi.org/10.1016/j.procir.2018.08.322
12. Li, X., Li, N., Tao, G., et al.: Experimental comparison of Bernoulli gripper and vortex gripper. Int. J. Precis. Eng. Manuf. **16**, 2081–2090 (2015). https://doi.org/10.1007/s12541-015-0270-3
13. Internationale Organisation für Normung: Manipulating Industrial Robots—Performance Criteria and Related Test Methods, vol. 9283, 2nd edn. International standard ISO (1998)
14. Leithoff, R., Fröhlich, A., Dröder, K.: Investigation of the influence of deposition accuracy of electrodes on the electrochemical properties of lithium-ion batteries. Energy Technol. **8**, 1900129 (2020). https://doi.org/10.1002/ente.201900129
15. Mooy, R.J.M.: Beitrag zur Produktivitätssteigerung in der Vereinzelung, Positionierung und Orientierung von Elektrodenfolien durch eine kontinuierliche Materialbewegung. Technische Universität Berlin (2019)
16. Helmers, L., Froböse, L., Friedrich, K., et al.: Sustainable solvent-free production and resulting performance of polymer electrolyte-based all-solid-state battery electrodes. Energy Technol. **9**, 2000923 (2021). https://doi.org/10.1002/ente.202000923

Framework for Automatically Generating Robot Programs of Multi-Heavy-Duty-Robot Systems

Shuxiao Hou, Mohamad Bdiwi, and Steffen Ihlenfeldt

Abstract

In today's car body shop, the number of cases where multiple robots share workspaces is growing rapidly. This trend means more productivity—but it poses a challenge to program the increasingly complex robotic systems. We present a framework for automatically generating robot programs of multi-heavy-duty robot systems in car body manufacturing. The framework was verified in a simulation environment. The results show that the presented framework can reduce the programming effort and production cycle time.

Keywords

Robot motion planning • Industrial robot • Multi-robot-system

1 Introduction

With the rapid evolution of agile manufacturing, industrial robots are widely used in the car body shop, such as in spot welding, flow-drill screwing, handling and other manufacturing processes. Due to space limitations in the production line and the flexibility of the

S. Hou (✉) · M. Bdiwi · S. Ihlenfeldt
Fraunhofer IWU, Reichenhainer Straße 88, 09126 Chemnitz, Germany
e-mail: shuxiao.hou@iwu.fraunhofer.de

M. Bdiwi
e-mail: bdiwi.mohamad@iwu.fraunhofer.de

S. Ihlenfeldt
e-mail: steffen.ihlenfeldt@iwu.fraunhofer.de

S. Ihlenfeldt et al. (eds.), *Annals of Scientific Society for Assembly, Handling and Industrial Robotics 2023*, https://doi.org/10.1007/978-3-031-74010-7_11

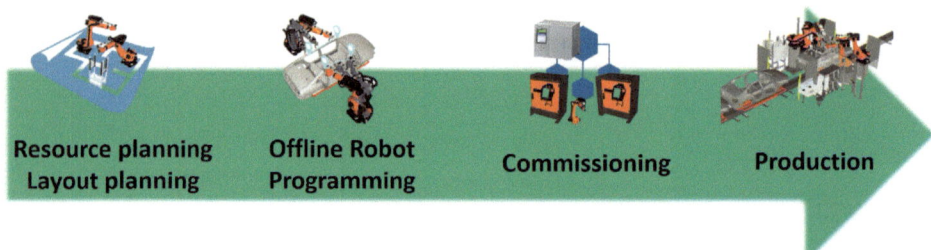

Fig. 1 Typical engineering process from planning to deploying of a robot system

agile production process, robot-based manufacturing stations have become more complex. This complexity poses new challenges for the planning and programming industrial robot systems.

Figure 1 shows a typical engineering process from planning to deploying a new robotic system. After selecting the robots and other equipment and determining the work cell layout, the robot programmer begins to program the collision-free robot motion offline in a simulation environment. The manually generated robot programs are then transferred to the robot controller. Simulation-based offline programming enables early creation and validation of the robot programs before the system is not even physically exists. Furthermore, offline programming allows the robot to continue working while the programmer changes the program, reducing the downtime required if the robot is being programmed live.

However, manual offline programming the motion of multiple robots in the complex production line is very time-consuming, as the engineer must take many various factors into account when programming the robot motion:

- Avoidance of collisions between the robot and static interfering contours in the environment.
- Definition of interlocking areas and logic to avoid collision between multiple robots.
- Avoidance of singularity of robot kinematics.

After the initial robot motion is generated, the specified cycle time should be achieved by iteratively manual optimization of the robot motion.

Figure 2 shows a production cell with four robots for a spot welding application, where 15 welding spots (colored ball in Fig. 2) are set. More than 100 program instructions are created for this application. The manual effort increases rapidly in a car body manufacturing production line with several hundred robots and several thousand welding spots. Agile production requires flexible replanning of the process flows with high productivity. Even seemingly minor changes in the production process, such as altering the order of welding spots, can necessitate replanning the paths and interlocking logic of the involved robots, leading to multiple iterations in order to optimize cycle time. Therefore, manual

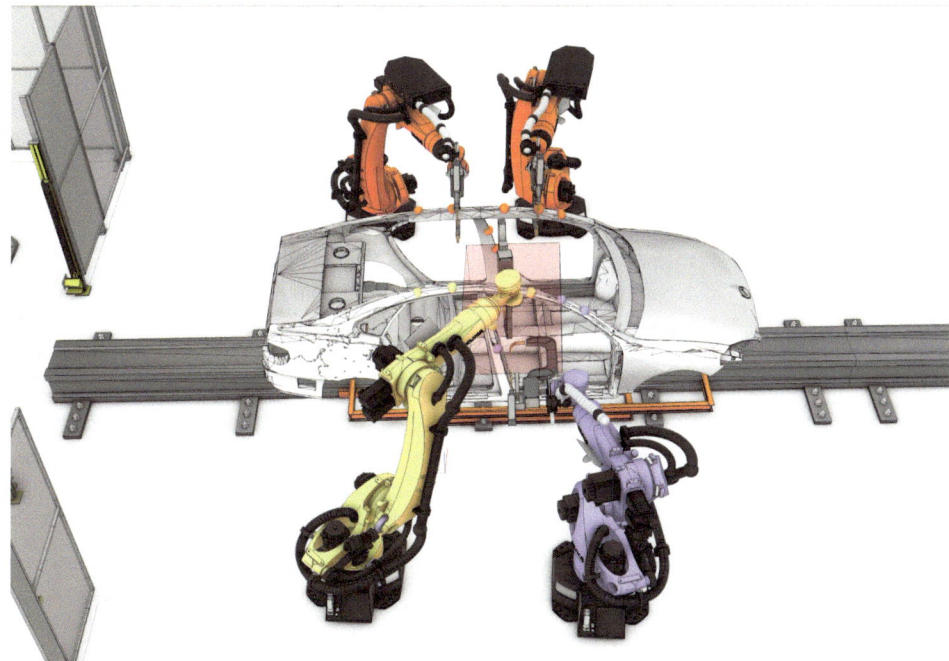

Fig. 2 Production cell with four robots for spot welding application

robot programming can no longer meet these requirements and automatically generating robot programs is currently an urgent need.

2   Related Works

2.1   Automatic Path Planning Methods

Automatic path planning methods can be divided into two types. They are sampling-based methods and optimization-based methods.

In the sampling-based path planning methods, a few samples (the red points in Fig. 3a) are generated in the workspace. Connecting the samples generates a collision-free trajectory from the start to the goal configuration. The core of sampling-based path planning methods is generating collision-free robot configurations, so-called "samples". In [1–8], different sampling-based path planning methods with different sampling strategies are developed. Due to the random nature of the sample generation, the generated robot trajectories may contain redundant points, which increases the executing time of the robot motion.

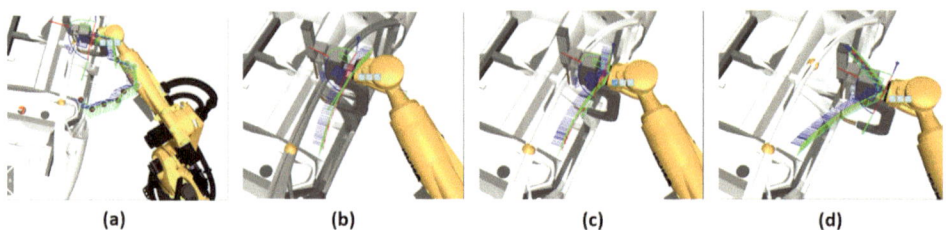

Fig. 3 **a** Sampling-based path planning method, **b–d** Opitimization-based path planning method

Optimization-based path planning methods generate an optimal motion considering various constraints and criteria such as dynamics, smooth motion and collision avoidance. The optimization usually starts with a direct connection between start and goal configurations (Fig. 3a). Since the direct connection usually leads to the collision between robots and obstacles in the environment, several iterative optimization loops are required to generate a collision-free path. For example, during the optimization, the robot avoids the obstacle (B-pillar of the car body) step by step (Fig. 3b, c). However, the optimization technique has limited success in the path planning of the high-dimensional robot. This can be mainly attributed to two reasons: the high computational cost of optimization problems with complex cost functions or constraints and the local minima in the computation of the non-convex optimization problems [9–12].

2.2 Gap Between Automatic Path Planning Methods and Deployment in Practical Applications of Multi-Heavy-Duty Robotic Systems

Most sampling-based and optimization-based methods can only generate the robot motions in the form of low-level motion commands, such as the interpolated position or angle of the robot joints (such as a series of interpolation points along the robot trajectory in Fig. 3). However, the control systems of most heavy-duty robots only provide the possibility to program the robot motions with high-level motion commands, e.g., point-to-point and linear motions. Therefore, the low-level motion commands generated by the methods described above cannot be directly deployed in most control systems of heavy-duty robots.

Furthermore, the interlocking of multiple robots should be implemented in a higher-level control system to coordinate the motion of multiple robots. The methods mentioned above cannot generate interlocking logic and are only used to plan the motion of a single robot.

3 Framework for Automatically Generating Robot Programs of Heavy-Duty Robot Systems

In order to automatically generate the control programs for robot systems—especially the multi-robot systems with multiple heavy-duty robots—a framework was developed in the course of a research project. The workflow of the presented framework (Fig. 4) consists of four functional modules. The modularized workflow enables the combination of the advantages of sampling-based and opting-based path planning methods. Compared to the path planning methods presented above, the framework generates the robot motion as high-level robot motion commands and interlocking signals, which can be directly transferred to the robot control system. The details of the function modules are presented in this section.

3.1 Model Manager

In the "Model Manager" module, the cad data relevant to motion planning is imported and processed in a suitable data structure. The information includes the geometry and kinematics of the robots, safety concepts and geometry of obstacles in the environment.

3.2 Trajectory Generator

The relevant information described above is then passed to the module "Trajectory Generator". In this module, sampling-based trajectory planning algorithms have been integrated to generate multiple trajectories (initial trajectories) that fulfill the derived boundary conditions (Fig. 5a). Furthermore, the presented framework allows the user to flexibly set how many initial robot paths should be generated and which ones should be optimized in the subsequent module. The following boundary conditions are considered:

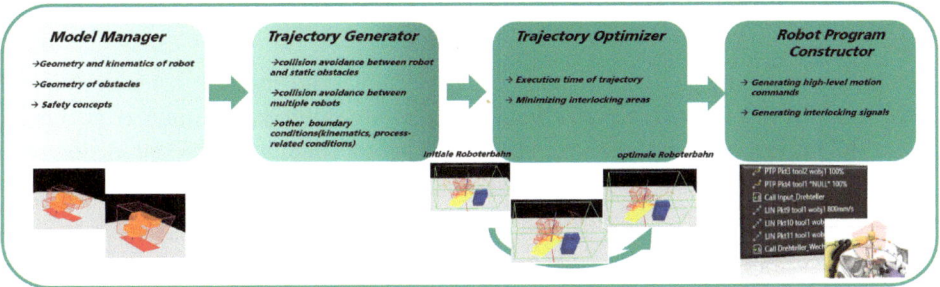

Fig. 4 Workflow of the presented framework

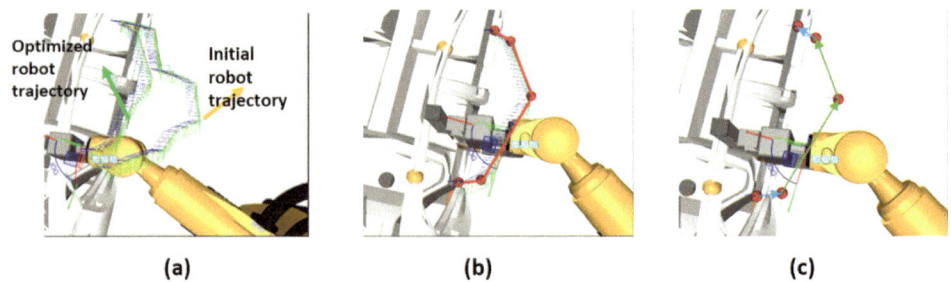

Fig. 5 **a** Optimization of the robot motion based on a valid initial robot trajectory. **b** Division of the optimized robot motion into sub-segments. **c** Approximation of the robot motion with high-level motion commands: Linear motion shown as blue arrows and point-to-point motion shown as green arrows

- Collision avoidance with static obstacles: It should be ensured that the generated trajectories do not collide with static disturbance contours in the environment.
- Collision avoidance with other robots: The motions of all other robots are considered as dynamic obstacles. If the workspaces of several robots overlap, the overlapping areas are calculated and defined as interlocking areas.
- Other boundary conditions: The kinematic, process-related and safety-related boundary conditions are taken into account.

3.3   Trajectory Optimizer

The initial robot trajectories are optimized in the module "Trajectory Optimizer" according to various criteria. The following criteria are considered:

- Execution time of trajectory: It is always necessary to reduce the production cycle time, which can be achieved by optimizing the executed robot motion
- Interlocking logic: The unnecessary waiting time of the robots are optimized by minimizing the interlocking areas.

The optimization in this function module starts from valid initial robot trajectories that already fulfill the derived constraints. Therefore, the optimization time is significantly reduced. Furthermore, local minima can be avoided by simultaneously optimizing several initial robot trajectories.

3.4   Robot Program Constructor

Since the optimized trajectories are described in form of interpolated joint angles, they should be converted into high-level motion commands in module "Robot Program Constructor". The trajectory is first divided into several segments (red line in Fig. 5b). The interpolated joint angles on each segment are approximated by a high-level motion command. Suppose the robot motion of an approximated high-level motion command does not collide with other obstacles and other robots. In that case, the segment is replaced by the high-level motion command (Fig. 5c). Otherwise, the segment is divided further and the approximation should be performed again.

In addition to the high-level movement commands, the locking signals are derived from the optimized interlocking areas. The time points at which each robot enters and leaves overlapping areas are calculated.

4   Result

The presented framework was verified in a typical spot welding application with a multi-robot system in the body-in-white shop. The confined space and the complex couture of the car body make programming the robot motion and interlock logic difficult. The proposed framework eliminates the need for manual planning, resulting in a complete savings of effort in that regard. Furthermore, Table 1 compares the manually programmed robot programs with the automatically generated robot programs. The automatically generated robot programs have reduced the execution time of robot trajectories on the one hand and the waiting time of the robots in the interlocking areas on the other hand.

This waiting time reduction is attributed to optimizing interlocking areas through our framework. As depicted in Fig. 6, our framework automatically generates three smaller interlocking areas (represented by the red cube) during the optimization phase, as opposed to a single large interlocking area (Fig. 2). In Fig. 2, the robot on the right had to wait 2.4 s until the robot on the left completely exited the large interlocking area. However,

Table 1 Comparison of robot programs generated by the presented framework and manual programming

	Presented framework		Manual	
	Robot left (s)	Robot right (s)	Robot left (s)	Robot right (s)
Wait time	0,0	1,1	0,0	2,4
Execution time of robot motion	7,8	8,6	8,9	9,1
Cycle time	8,7	9,7	8,9	11,5
Total cycle time	**9,7**		**11,5**	

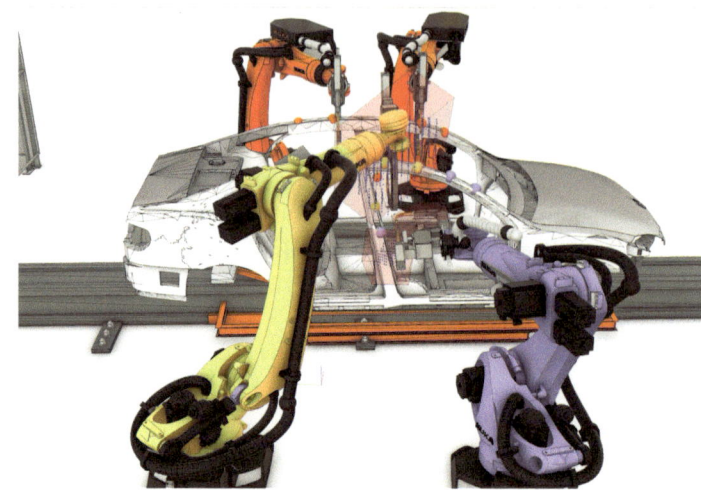

Fig. 6 Automatically generated interlocking areas (red cubes)

with our framework, the robot on the left occupies the first small interlocking area in just 1.1 s. It then proceeds to the second small interlocking area, allowing the robot on the right to enter the first small interlocking area. When the robot on the right enters the second small interlocking area, it does not need to wait as the robot on the left has already left the second small area. As a result, the wait time for the robot on the right is reduced from 2.4 to 1.1 s.

5   Conclusion

This paper presented the urgent need for automatically generating robot programs for heavy-duty robot systems. Then we analyzed various state-of-the-art automatic path planning methods and issues in the deployment of planned robot motion in control systems of heavy-duty robots. In order to solve the problems, a framework for automatically generating robot programs of multi-heavy-duty robot systems was developed. Compared to state-of-the-art path planning methods, the presented framework generates complete robot programs, which include high-level control commands and the interlocking logic to coordinate multiple robots. The generated robot programs can be directly deployed into the robot controllers of different manufacturers. The framework is not limited to a specific path planning method thanks to the modularization of the workflow. Other new path planning algorithms can be integrated into the framework in the future.

In a simulation environment, the framework was verified with a real industrial application. The results showed that the framework could significantly reduce the engineering effort for programming the robot systems on the one hand and improve the quality of the robot motions in terms of cycle time on the other hand.

Further work includes the verification of the framework in real robot systems. In addition, the framework can be extended to more applications by considering various manufacturing technologies.

Acknowledgements The development of the presented framework was supported by the SAB Research Funding Project "Effiziente Online-Anlagensteuerung mit einem modellbasierten Steuerungssystem".

References

1. Kavraki, L.E., Kolountzakis, M.N., Latombe, J.-C.: Analysis of probabilistic roadmaps for path planning. Trans. Robot. Autom. **14**, 166–171 (1998)
2. Bohlin, R., Kavraki, L.E.: Path Planning using lazy PRM. In: IEEE International Conference on Robotics and Automation, pp. 521–528. San Francisco (2000)
3. Karaman, S., Frazzoli, E.: Sampling-based algorithms for optimal motion planning. Int. J. Robot. Res. **30**(7), 846–894 (2011)
4. La Valle, S.M.: Rapidly-exploring random trees a new tool for path planning. Iowa State University, Technical Report No. 98-11 (1998)
5. Kuffner, J., La Valle, S.M.: RRT-connect: an efficient approach to single-query path planning. In: Proceedings of the 2000 IEEE International Conference on Robotics and Automation, pp. 995–1001 (2000)
6. Karaman, S., Frazzoli, E.: Sampling-based algorithms for optimal motion planning. Int. J. Robot. Res. **30**(7) (2011)
7. Jinwook, H., Lee, D.D., Isler, V.: Neural cost-to-go function representation for high dimensional motion planning. In: Workshop: Motion Planning with Implicit Neural Representations of Geometry. ICRA (2022)
8. Yuan, C., Liu, G.: An efficient RRT cache method in dynamic environments for path planning. Robot. Auton. Syst. **131** (2020)
9. Ratliff, N., Zucker, M., Bagnell, J., Srinivasa, S.: CHOMP: gradient optimization techniques for efficient motion planning. In: IEEE International Conference on Robotics and Automation (2009)
10. Zucker, M., Ratliff, N., Dragan, A.D., Pivtoraiko, M., Klingensmith, M., Dellin, C.M., Bagnell, J.A., Srinivasa, S.S.: Chomp: covariant Hamiltonian optimization for motion planning. Int. J. Robot. Res. (2013)
11. Schulman, J., Ho, J., Lee, A.X.: Finding locally optimal, collisionfree trajectories with sequential convex optimization. Proc. Robot. Sci. Syst. **9**, 1–10 (2013)
12. Schulman, J., Duan, Y., Ho, J., Lee, A., Awwal, I., Bradlow, H., Pan, J., Patil, S., Goldberg, K., Abbeel, P.: Motion planning with sequential convex optimization and convex collision checking. Int. J. Robot. Res. (2014)

Analysis of Surface Properties for Process-Reliable Handling with Dry-Adhesive Microstructures

Mirja Louisa Krüger and Kirsten Tracht

Abstract

The aim of this paper is to investigate the influence of different surface parameters on the adhesion behavior of dry-adhesive microstructures. Selective laser melting surfaces, which are partly reworked in a solid rolling process, are used to investigate the influence of different surface parameters. The surface parameters are determined using a laser scanning microscope. Experiments with three different test masses and dry-adhesive microstructures are carried out on the surfaces to determine the pull-off stress. The experiments show that the surface parameter root-means-square roughness has an influence on the adhesion behavior of the dry-adhesive microstructures. It becomes clear that the size relationship between the root-means-square roughness and the tip diameter of the dry-adhesive microstructures has a crucial role in the adhesion behavior. Moreover, increasing the pull-off stress reduces the number of surfaces where adhesion takes place. Furthermore, it can be seen that when the average element length of the ripple is reduced and the total height of the ripple, the arithmetic average roughness and the mean roughness depth are increased, the adhesion behavior is reduced. In addition, increasing the pull-off stress leads to a reduction in the number of surfaces to which adhesion occurs. The reduction in adhesion behavior can be explained by the impediment of contact formation between the mushroom heads of the dry-adhesive microstructures and the surface.

M. Louisa Krüger (✉) · K. Tracht
University of Bremen, Bremen Institute for Mechanical Engineering, Badgasteiner Str. 1, 28359 Bremen, Germany
e-mail: krueger@bime.de

© The Author(s) 2025
S. Ihlenfeldt et al. (eds.), *Annals of Scientific Society for Assembly, Handling and Industrial Robotics 2023*, https://doi.org/10.1007/978-3-031-74010-7_12

133

Keywords

Handling • Biological inspired design • Dry-adhesive microstructure

1 Introduction

Some reptiles and insects can easily cling to various surfaces and move forward without having sticky feet. The adhesion and forward movement is made possible by the microstructures on their feet, whose appearance resembles small hairs, called Spatulae [1]. In this respect, it is interesting to note that adhesion and forward movement are also possible on vertically aligned surfaces or even upside down. Based on these findings from the natural sciences, the geometry of the microstructures is reproduced with the help of technical materials in order to be able to create a dry-adhesive bond in technical processes. The dry-adhesive microstructures can be used as grippers for handling objects [2]. This novel gripper technology is characterized by its reversible adhesion properties. Furthermore, the grippers can be used without electricity, in vacuum and without compressed air. Handling objects with smooth surfaces poses no problems whatsoever for the dry-adhesive microstructures. Rough surfaces, on the other hand, can cause more problems for handling. In addition, handling can be made more difficult by wear and contamination.

The experiments conducted in this paper are intended to determine to what extent the surface properties have an influence on the adhesion behaviour of the dry-adhesive microstructures and which surface parameters have a positive or negative effect on the adhesion behavior.

2 Background

2.1 Dry-Adhesive Microstructures

Dry-adhesive microstructures are artificially structured adhesives made of elastomers that have a mushroom-shaped structure on the surface. The dry-adhesive microstructures are made of individual small mushroom heads that can adhere to various surfaces due to the large contact area and the Van der Waals forces that occur. In addition, the adhesion of the dry-adhesive microstructures can be dissolved without residue and afterwards the microstructure can be attached to another surface. For this reason, the dry-adhesive microstructures are suitable for a variety of handling applications.

The development of dry-adhesive microstructures has been guided by the findings of the last two decades. Arzt et al. [1] divided the research findings into two areas. The first area is the fibril level where the focus is on the behavior, understanding and development of fibrillar elements. On the fibril level size, shape, elastic modulus and moisture play a crucial role in terms of adhesion performance [1].

In the second area, the array level, it was observed that a large number of individual fibrils do not behave like a single fibril. This is because the elastic connection of the fibrils through the support layer, the load distribution on the individual fibrils and the detachment of the individual fibrils from the contact surface play a decisive role [1].

However, the dry-adhesive microstructures developed from these findings were based on those of the leaf beetle in their microstructure and do not correspond to the ibrillary structures of the gecko [3].

2.2   Use of Dry-Adhesive Microstructures with Different Surface Parameters

For the dry-adhesive microstructures, the surface roughness can be an obstacle in terms of adhesion, which is created by the intermolecular interactions at a short distance. This is because, depending on the degree of roughness of the contact surface, the formation of the interactions can be hindered [1].

Gorumlu and Akask [4] have already experimentally investigated the adhesive force of mushroom-shaped microstructures with a diameter of 10 μm and a height of 20 μm on rough surfaces. It was concluded that they adhere worse to a rough surface with a square deviation Rq of 373 nm and higher roughness values with an adhesive force of 1 N/cm^2 than to a surface with a square deviation Rq of 2 nm with an adhesive force of 10 N/cm^2 [4].

Materzok et al. [5] also studied dry-adhesive adhesion on flat and rough surfaces and developed a simulation. They could find out that the pull-off stress of the simulation matches very well with the pull-off stress of the experiments. The pull-off stress was in the range between 8 and 20 nN. However, in the simulation and the experiments, only single dry-adhesive microstructures were investigated, which adhered at an angle. The length of each dry-adhesive microstructure was 440 nm and the head had a rectangular shape with edge lengths of 175 nm and 20 nm [5].

Huber et al. [6] found in their experiments with dry-adhesive microstructures that they exhibited very low pull-off stress between a Rq of 100 nm and 300 nm. Based on these observations, they suggested that at a Rq value of the magnitude of the dry-adhesive microstructures, only partial contact between the dry-adhesive microstructures and the surface is possible, since the roughness peaks are too small for the spatula to adhere to them [6].

Bauer et al. [7] have also conducted experiments with dry-adhesive microstructures and they investigated the influence of different surface geometries on the adhesion force. Squared deviation was used to determine the surface geometries. In the investigations by Bauer et al. [7], the dry adhesive microstructures were pressed onto flat, rough luminium and it became clear that the adhesion was reduced by 35% to 50% compared to measurements on flat, smooth glass substrate. This leads to the assumption that increasing the roughness reduces the adhesion.

In addition, it was found in the experiments that adhesion is limited by the onset of buckling and that the mushroom-shaped microstructures adhere better to wavy, rough surfaces and the pull-off force is greater than with flat tip structures [7].

3   Materials Used for the Pull-Off Experiments

3.1   The Dry-Adhesive Microstructure

For the experiments conducted in this study, dry-adhesive microstructures from the company Klettband Technik called Gecko®-Tape were used. These dry-adhesive microstructures replicate the structure of the leaf beetle's foot, shown in Fig. 1 as a laser scanning microscope image.

The dry-adhesive microstructures are applied to a silicone rubber film. The silicone rubber film has a thickness t of 0.34 mm. The mushroom-shaped microscopic adhesive elements are arranged on one of the surfaces. The mushroom-shaped microstructures have

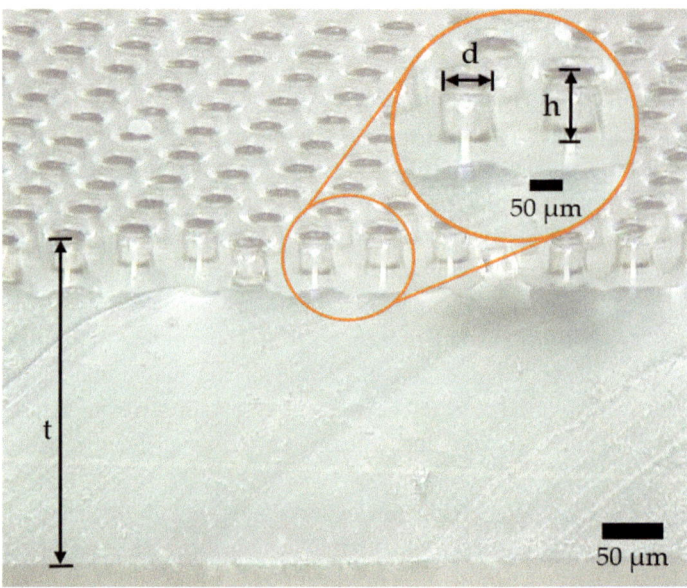

Fig. 1 Laser scanning microscope image of the dry-adhesive microstructure, taken by T. Brunkhorst, H. Dierks, H. Siesenis and C. Tuitje

a height h of approx. 60 μm and a tip circle diameter d of approx. 38 μm, to be seen in Fig. 1. There is an applied adhesive layer on the other side of the tape.

The tape has the ability to adhere to any smooth and even surface, even if this surface is wet, greasy or soapy, the adhesive strength of the dry-adhesive microstructures is not reduced. The Van der Waals forces that form create electrostatic interactions, resulting in an inter-molecular attraction comparable to that of a magnet. The dry-adhesive microstructures can be removed without leaving any residue and can be used in a vacuum. It is also reusable and can be cleaned by washing it with water [8].

The dry-adhesive microstructure tape can be used as grippers, since low-energy adhesion and release is possible. The challenge with the dry-adhesive microstructures is the wear and contamination behavior. Due to dust and other small particles, the dry-adhesive microstructures can quickly become contaminated and the adhesive force is significantly reduced. In addition, it must be found out on which surfaces the dry-adhesive microstructures adhere well and thus handling is possible.

3.2 The Selective Laser Melting Surfaces

In this study, cubes made from AlSi 316L with an edge length of 50 mm were used. The cubes were produced by selective laser melting. The surface structure of the cubes was varied by choosing different hatch distances h_d of 150 μm for cube no. 1 and 120 μm for cube no. 2. The layer thickness of 50 μm, the applied laser power of 235 W as well as the scanning speed of 700 mm/s were kept constant for both cubes. The two SLM cubes were deep rolled on two sides. The deep rolling process was carried out with a ball diameter d_b of 6 μm, a deep rolling pressure p_r of 100 bar, a rolling speed v_r of 100 mm/min and a stepover s_f of 0.1 mm [9]. The contact pressure tests were carried out on the SLM-printed and the deep-rolled surfaces.

The SLM printed and deep-rolled cubes were examined using the VK-X100K/X200K 3D laser scanning microscope from Keyence in accordance with DIN EN ISO 4287. During the examination, the surface parameters of the 10 surfaces were determined using the software "VK Analyse Modul".

The choice of the surface parameter can be traced back to the investigations of Gorumlu and Aksak [4], Materzok et al. [5], Huber et al. [6] and Moreira Lana et al. [10]. The investigations of Moreira Lana et al. [10] used the roughness parameters arithmetic average roughness Ra and the mean roughness depth Rz to describe the surfaces. For the ripple parameters, Bauer et al. [5] used the average element length of the ripple WSm and the total height of the ripple Wt in their experiments. In addition, the investigations of Huber et al. [6] showed that the Rq has an influence on the adhesion at the same order of magnitude as the dry-adhesive microstructures. The surface parameters of the individual surface areas can be taken from Table 1. Surface no. 5, 7, 9 and 10 are deep rolled areas of the SLM surfaces. The other surfaces have not been not post-processed.

Table 1 Surface parameters of the 10 different surfaces

Surface no	Cube no	Post-processing	Rq (μm)	WSm (μm)	Wt (μm)	Rz (μm)	Ra (μm)
1	1		57	1054	817	1182	35
2	2		42	1136	495	713	31
3	1		41	953	178	380	32
4	2		35	980	185	372	27
5	1	Deep rolled	21	1521	4	174	17
6	2		20	689	239	353	15
7	2	Deep rolled	20	1719	53	139	15
8	1		19	658	262	362	14
9	2	Deep rolled	5	827	39	69	5
10	1	Deep rolled	5	629	56	97	3

The different values for the individual surface parameters of the 10 different surfaces are due to the fact that two cubes were printed with different hatch distances, the surfaces are located on different sides of the cube and were partly post-processed by deep rolling. The surfaces 1, 3, 5, 8 and 10 are on cube no. 1, the surfaces 2, 4, 6, 7 and 9 are on cube no. 2. The surfaces 1, 2, 3, 4, 5 and 6 are on the top side, i.e. parallel to the layer orientation, and the surfaces 7, 8, 9 and 10 are on one side of the cube, i.e. perpendicular to the layer orientation. Due to this distribution of the surfaces over the different sides of the two cubes, they have different surface parameters. It can be seen, however, that WSm and Ra of the surfaces arranged perpendicular to the layer orientation are lower than parallel to the layer orientation.

4   Methodical Approach of the Pull-Off Experiments

The 10 different surfaces described in Chap. 3, the dry-adhesive microstructure also described in Chap. 3 and three test masses were used for the experiments. The test masses have a mass of 0.4 g (test mass 1), 1.25 g (test mass 2) and 2.6 g (test mass 3). For the selection of the masses, it was assumed that the object that is to be handled in further investigations with the dry-adhesive microstructures with a size of 5 mm × 5 mm has a thickness of 10 mm, a width of 5 mm, a depth of 5 mm and consists of 50% iron and 50% aluminum. The object to be handled in the future will be manufactured using a sintering process, therefore the packing density of the mixture is assumed to be 95%. The weight of the object to be handled is 1.25 g (test mass 2) under these assumptions. In order to investigate the influence of the change in weight, the weight was doubled (test mass 3)

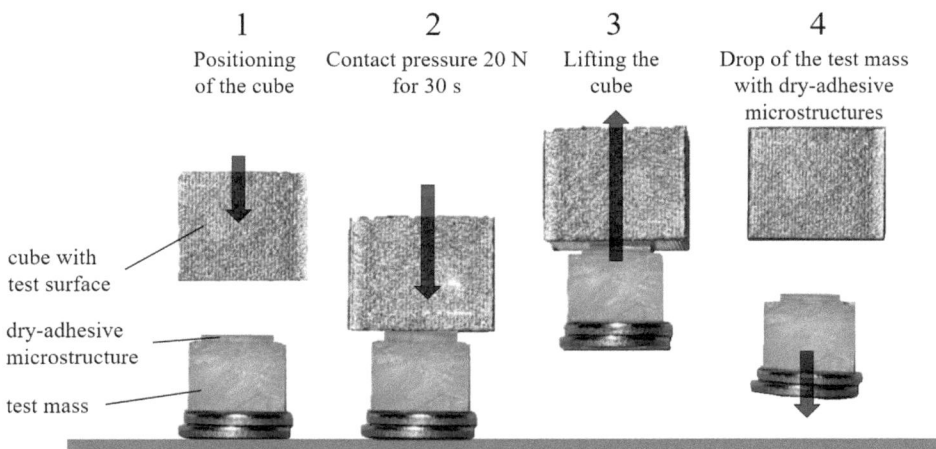

Fig. 2 Schematic representation of the experimentation

on the one hand and reduced to one third (test mass 1) on the other. Test mass 1 is also the weight that the object to be handled would have on Mars.

For each experiment, a new dry-adhesive microstructure, as described in Chap. 3, with dimensions of 5 mm × 5 mm was used.

The test procedure is shown schematically in Fig. 2. In step 1, the surface of the cube is positioned over the dry-adhesive microstructure attached to the test mass. In step 2, the surface of the cube is pressed with a force of 20 N for 30 s onto the dry-adhesive microstructures, which are attached to the test mass.

Afterwards, the cube is lifted with the adhering test mass, see step 3. For the determination of the adhesion time, the time until the detachment of the test mass was recorded, as shown in step 4. If the dry-adhesive microstructures continued to adhere to the respective test mass for more than 30 s, the experiment was terminated. According to the DIN standards for tensile tests, the experiments were repeated five times on the 10 different surfaces, each time with a new, unused dry-adhesive microstructure. For each of the 10 surfaces, five adhesion times were determined from which the mean value for the adhesion time was calculated. The experiments will be used to investigate whether the dry-adhesive microstructures are suitable as a gripping object. In order to handle the gripped objects, they should adhere to the dry-adhesive microstructures for 30 s if possible. With the help of the 10 different surfaces, a first assessment for the adhesion behavior of the dry-adhesive microstructures can be performed.

5   Results

In the first part of the experiment, test mass 1 with a weight of 0.4 g was used. The pull-off stress was therefore 15.7 mN/cm^2. The test mass adhered for 30 s to eight of the ten surfaces. In the second part of the experiment, test mass 2, which has a weight of 1.25 g, was used. The pull-off stress in these experiments was 49 mN/cm^2 resulting in three times the pull-off stress compared to test mass 1. It can be seen that when the pull-off stress is tripled, compared to test mass 1, seven of the ten surfaces adhere for 30 s. In the last part of the experiment, test mass 3, which has a weight of 2.6 g, was used. The pull-off stress in these experiments was 102 mN/cm^2, resulting in a doubling of the pull-off stress compared to test mass 2. Here it can be seen that only three of the ten surfaces adhere for 30 s.

The results of the experiments regarding the root-means-square Rq in relation to the mean value of the adhesion time are shown in Fig. 3. The figure shows on the one axis the Rq in blue, can be found in Table 1, for the 10 surfaces and on the other axis the mean value of the adhesion time for the test mass 1 in green, the test mass 2 in orange and the test mass 3 in red.

Figure 3 shows that test mass 3 adhered only to the deep rolled surfaces for 30 s. On surfaces 1, 2, 3, 4 and 10, which are outlined in red in Fig. 3, test mass 3 adhered with the help of the dry-adhesive microstructures only for less than 30 s. Moreover, test mass 1 and 2 adhered to the surfaces with a higher Rq for 30 s and did not adhere to the non-post-processed surfaces 6 and 8, despite a lower Rq value.

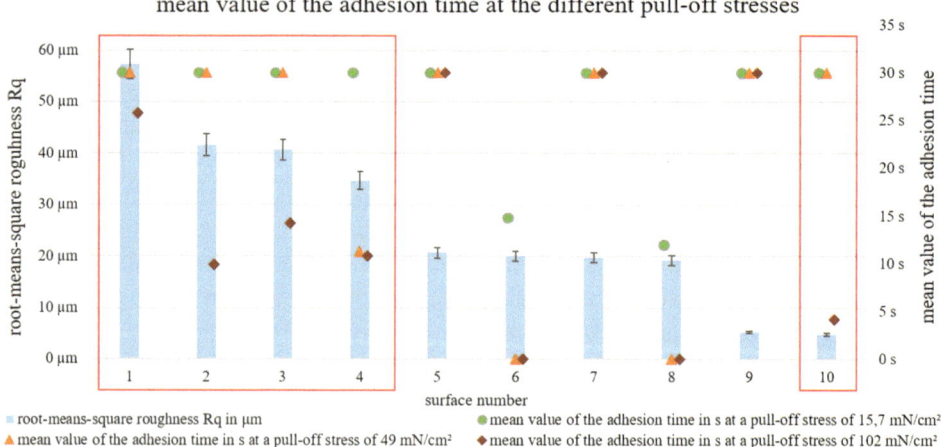

Fig. 3 Root-means-square roughness Rq and the adhesion time at the different pull-off stresses of 15.7 mN/cm^2, 49 mN/cm^2 and 102 mN/cm^2 on surfaces no. 1 to 10

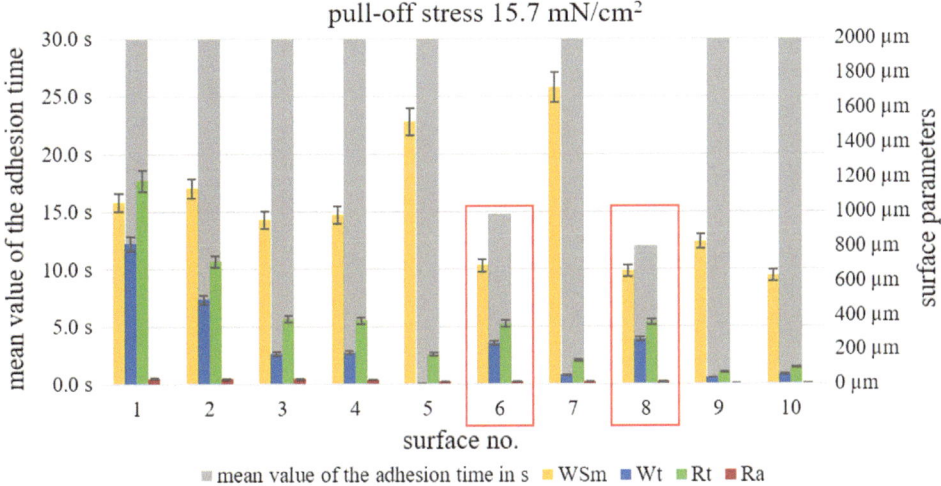

Fig. 4 Mean value of the adhesion time at a pull-off stress of 15.7 mN/cm^2 and the surface parameters on surfaces no. 1 to 10

Figures 4, 5, 6 show on the one axis the mean value of the adhesion time (grey) and on the other axis the different surface parameters of the SLM surfaces WSm (yellow), Wt (blue), Rz (green) and Ra (red) for the experiments with the 3 test masses. In the first part of the experiments, shown in Fig. 4, with the test mass 1 the dry-adhesive microstructures adhere on eight of the ten surfaces. It is noticeable that WSm is less than 700 μm, outlined in red in Fig. 4, for the two surfaces to which the test mass did not adhere for 30 s. No correlation can be found for the other surface parameters in this part of the test.

In the second part of the experiment, shown in Fig. 5, it can be seen that when the pull-off stress is tripled, compared to test mass 1, the dry-adhesive microstructures adhere on seven of the ten surfaces for 30 s. The dry-adhesive microstructures did not adhere only to the surface no. 4, 6 and 8, outlined in red in Fig. 5, where WSm is below 1000 μm. No correlation can be found for the other surface parameters in this part.

In the last part of the experiment, shown in Fig. 6, test mass 3, which has a weight of 2.6 g, was used. The pull-off stress in these experiments was 102 mN/cm^2, resulting in a doubling of the pull-off stress compared to test mass 2. Through this the dry-adhesive microstructures only adhere on three of the ten surfaces adhere for 30 s, outlined in red in Fig. 6. Figure 6 shows that the dry-adhesive microstructures did not adhere or only partially adhered to surfaces with a WSm below 1150 μm, a Wt above 55 μm, a Rz above 175 μm and a Ra above 17 μm.

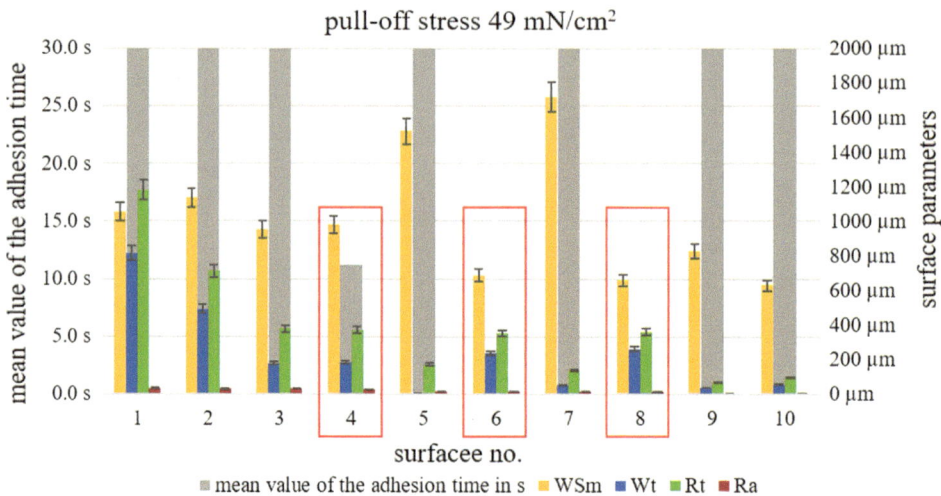

Fig. 5 Mean value of the adhesion time at a pull-off stress of 49 mN/cm^2 and the surface parameters on surfaces no. 1 to 10

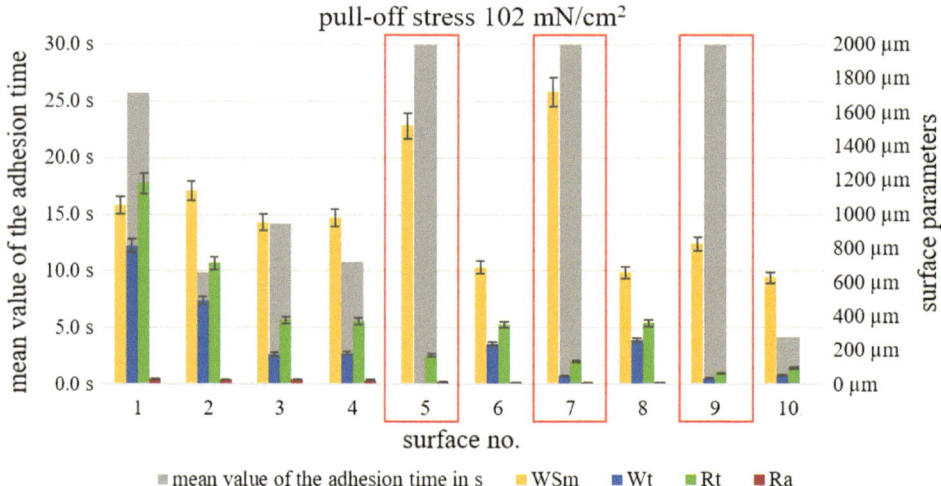

Fig. 6 Mean value of the adhesion time at a pull-off stress of 102 mN/cm^2 and the surface parameters on surfaces no. 1 to 10

6   Discussion and Conclusion

From the experiments carried out, it can be concluded that the surface parameters have an influence on the adhesion time and adhesion force, since the adhesion time on some surfaces decreased with an increase in the test mass. It is noticeable in the experiments with the pull-off stress of 102 mN/cm^2 (test mass 2) that WSm is below 1150 μm for the surfaces on which the dry-adhesive microstructures did not adhere for 30 s. With a reduction of the pull-off stress, WSm below which adhesion no longer occurs also decreases. This indicates that WSm has an influence on the adhesion of the dry-adhesive microstructures. The influence of WSm can be attributed to the adaptation behavior of the dry-adhesive microstructures, because this is limited due to the geometry of the mushroom heads. It also becomes clear that the assumption made by Huber et al. [6] that the dry-adhesive microstructures adhere badly to surfaces with a Rq in the same size scale as the tip diameter of the dry-adhesive microstructures could be confirmed. This is very evident on the surfaces no. 1 to 4. Both test mass 1 and 2 adhere to these surfaces for the maximum 30 s. In addition, test mass 3 adheres to the four surfaces for at least nine seconds. To the non-post-processed surfaces 6 and 8 with the lower Rq, all 3 test masses adhered significantly worse or not at all. The reduced number of formed contacts reduces the adhesion force and consequently less force is needed to release the contacts formed between the mushroom heads and the surface [11].

The results of the experiments show that the adhesion time decreases with increasing pull-off stress for surfaces with a low WSm value or high Wt, Rz and Ra values. At a pull-off stress of 102 mN/cm^2, the dry-adhesive microstructures did not adhere for 30 s for surfaces with a WSm below 1150 μm, Wt above 55 μm, Rz above 175 μm and Ra above 17 μm. In addition, these showed that the adhesion time decreases with increasing pull-off stress for surfaces with Rq higher than the tip diameter of the dry-adhesive microstructures. In addition, it can be seen that the dry-adhesive microstructures adhered very well to the deep-rolled surfaces. When using the dry-adhesive microstructures as grippers for handling objects, care should therefore be taken to ensure that the Rq is not in the same size range as the tip diameter of the dry-adhesive microstructures. In addition, for best adhesion WSm should be as high as possible and Wt, Rz and Ra as low as possible. When the surface parameters are observed, gripping is possible with these dry-adhesive microstructures and the dry-adhesive microstructures can be used as a gripping technology.

Acknowledgements The Federal State of Bremen and the University of Bremen support the 'Humans on Mars Initiative' with funding for seven seed projects until the end of 2024. The initiative employs more than 20 early career researchers and supports undergraduate education related to space exploration. https://www.uni-bremen.de/en/aerospace-at-the-university-of-bremen/humans-on-mars.

References

1. Arzt, E., Quan, H., McMeeking, R.M., Hensel, R.: Functional surface microstructures inspired by nature—From adhesion and wetting principles to sustainable new devices. Prog. Mater. Sci. (120), 100823 (2021)
2. Hensel, R., Moh, K., Arzt, E.: Engineering micropatterned dry adhesives: from contact theory to handling applications. Adv. Funct. Mater. (28), 1800865 (2018)
3. https://www.ingenieur.de/technik/forschung/extrem-haftende-klebefolie-geometrie-kontaktes-entscheidend/, Last accessed 2023/01/13
4. Gorumlu, S., Aksak, B.: Sticking to rough surfaces using functionally graded bio-inspired microfibres. R. Soc. Open Sci. **4**(6), 1–11 (2017)
5. Materzok, T.; Boer, D. de; Gorb, S.; Müller-Plathe, F.: Gecko adhesion on flat and rough surfaces: simulations with a Multi-Scale molecular model. Small (18), 2201674, (2022)
6. Huber, G., Gorb, S.N., Hosoda, N., Spolenak, R., Arzt, E.: Influence of surface roughness on gecko adhesion. Acta Biomater. **3**, 607–610 (2007)
7. Bauer, C.T., Kroner, E., Fleck, N.A., Arzt, E.: Hierarchical macroscopic fibrillar adhe-sives: in situ study of buckling and adhesion mechanisms on wavy substrates. Bioinspir. Biomim. **10**(6), 1–17 (2015)
8. https://www.klettband-technik-schultz.de/shop/Gecko-r-Nanoplast-r-haftet-auf-Glas-durch-Van-der-Waals-Krafte-c142502668, Last accessed 2023/05/16
9. Wielki, N., Meyer, D.: Potential of deep rolling as a finishing process directly after SLM to generate beneficial surface and subsurface properties. In: 19th International Conference of the European Society for Precision Engineering and Nanotechnology, Bilbao. (2019)
10. Moreira Lana, G., Sorg, K., Wenzel, G.I., Hecker, D., Hensel, R., Schick, B., Kruttwig, K., Arzt, E.: Self-Adhesive silicone microstructures for the treatment of tympanic membrane perfora-tions. Adv. NanoBio. Res. **1**, 2100057 (2021)
11. Meiners, F., Tuitje, C., Hogreve, S., Tracht, K.: Model-based prediction of the detachment of microspheres from dry-adhesive gripper surfaces by bending. Procedia CIRP **115**, 101–106 (2022)

Highly Modular Postprocessing by Robot Kinematics in an Additive-Subtractive Manufacturing Process for Polymer Components

Michael Baranowski, Nikolai Krischke, Isabelle Lorei,
Simon Ole von Werder, Marco Friedmann, and Jürgen Fleischer

Abstract

Additive manufacturing processes have the potential to economically address the trend towards greater individualization and shorter product availability driven by globalization and digitization. In addition, future products will need to meet regulatory requirements for reparability to prevent increasing resource scarcity. By repairing products, an extension of the product life cycle can be achieved and, in particular, material resources can be saved. In the context of a shortage of skilled workers, future manufacturing processes must be as flexible as possible to meet a wide range of requirements. In the field of electromobility, the high volatility of the market presents special challenges for production engineering. To meet all these challenges, a highly flexible machine for the additive-subtractive manufacturing of highly functional polymer components was developed. With the help of this process chain with inline process control and an industrial robot for the integration of subcomponents, it is intended to make fully automated post-processing, repair and modification of multi-material components using the fused filament fabrication process possible.

M. Baranowski · N. Krischke (✉) · I. Lorei · S. O. von Werder · M. Friedmann · J. Fleischer
Karlsruhe Institute of Technology, Wbk—Institute of Production Science, Rintheimer Quer-Allee 2, 76131 Karlsruhe, Germany
e-mail: nikolai.krischke@kit.edu

M. Baranowski
e-mail: michael.baranowski@kit.edu

© The Author(s) 2025
S. Ihlenfeldt et al. (eds.), *Annals of Scientific Society for Assembly, Handling and Industrial Robotics 2023*, https://doi.org/10.1007/978-3-031-74010-7_13

Keywords

Additive-subtractive process chain • Fused filament fabrication • Additive repair • Post-processing

1   Introduction

There is an increasing demand for customized products, for which additive manufacturing processes are particularly suitable due to their high flexibility. The fused filament fabrication (FFF) process is characterized by comparatively simple machine technology, low production costs and short production times. However, the main disadvantages are the rough surfaces, insufficient dimensional accuracy and the need for support structures in case of overhangs. This results in a high level of manual and time-consuming post-processing effort and a loss in part quality. Especially for surface-critical applications, this can exceed the manual effort of conventional manufacturing methods [1]. As a result, the application is limited to prototyping, and the potential in the area of series production of individual components remains unused. At the beginning of the COVID-19 pandemic, for example, many protective masks were additively manufactured [2]. The raw material cost of the FFF process is comparatively low since hardly any waste is produced. Under these conditions, series production of individualized parts using 3D printing is in theory economically feasible. However, the time and cost of manual post-processing is high [3].

Due to critical requirements many complex vehicle components are designed as functional, hybrid polymer parts, which are, however, difficult to access for repairs of the functional interior (e.g. electronics). This, though, contradicts the "right to repair", which is stated in a resolution of the EU Parliament as a requirement for manufacturers "…to design their products in such a way that they last longer, can be repaired reliably, and their parts are easily accessible and removable…" [4, 5]. The aim is a long, resource-saving lifetime with the possibility of reusing and modifying individual components. To achieve this in a resource-efficient manner, it must be possible to reclaim existing parts and adapt their functions. In general, the FFF process is suitable as a primary approach, but it must be extended to include additional process elements [6]. These include the fully automated integration of inserts (load introduction elements, electronics, etc.) into the printing process using a robot [7]. Subtractive pre- and post-processing significantly improve dimensional and surface quality. In combination with a non-planar additive manufacturing strategy, the potential of this process chain for the production of complex structures has been demonstrated [8, 9].

Commercially available products are highlighted that offer a combination of additive manufacturing and post-processing steps are highlighted below. Combined printing and milling systems already exist for large components, such as CEAD's AM Flexbot [10] and Thermwood's LSAM [11]. These products have no sensors, which makes automated quality control and reactions to print errors impossible. The effector (extrusion or

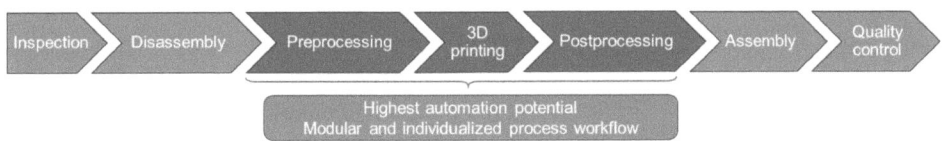

Fig. 1 Exemplary workflow for manufacturing, repair and upgrading of 3D printed parts through an additive-subtractive manufacturing process combined with various pre- and postprocessing

milling) cannot be changed automatically. In addition, both systems only allow an initial printing and subsequent milling of the surface. Integration of other components or connecting parts is not possible. The companies Rösler Oberflächentechnik GmbH [12] and Additive Manufacturing Technologies Limited [13] offer a variety of machines for automated post-processing of additively manufactured parts from powder bed fusion processes. These machines can remove powder residue and support structures from parts as well as smoothen their surfaces. Automated robotic part loading is available in some cases. Support structures can only be removed sequentially and individually. Thus, only partial automation is possible. As a result, due to increasing price pressure, customers continue to prefer the less expensive mass-produced parts.

Automated additive manufacturing processes currently only exist as research projects and for powder bed fusion. A few semi-automated individual modules are available [14]. A comparison with the state of the art clearly shows that a system for the fully automated production of ready-to-use FFF parts does not yet exist. To achieve this functionality, modules from different manufacturers would have to be combined at a high cost and part handling would have to be fully automated.

In the context of extended functional integration and the repairability of additively manufactured polymer components, a process chain as shown in Fig. 1 offers significant advantages. The combination of different manufacturing principles enables the efficient production of highly functionalized, customized, hybrid polymer components. Furthermore, an increase in the number of variants can also be realized after the manufacturing process or during the life cycle, as the process enables modification or targeted repair of used components. Based on the requirements and an existing 3D printer, fully automated post-treatment and additive repair are addressed in two new projects. In this paper, the existing system is presented first, followed by a conceptual implementation of the extended process chain and finally two use cases for post-processing and repair.

2   Basics

In the context of the increasing demand for individualized functional components and the associated need for flexible production machines, this chapter presents a hybrid additive-subtractive fused filament fabrication (FFF) process with a milling spindle and handling robot (2.1). This machine concept aims to maximize the degree of automation and minimize the use of manual activities by workers. In 2.2, the current workflow of digital process planning is presented using the so-called "HybridPlaner".

2.1 Additive-Subtractive 4 K-FFF-Process

As part of the "InnovationsCampus Future Mobility", a novel additive-subtractive manufacturing process for the production of function-integrated polymer components was developed. The basis of this additive-subtractive manufacturing process is an FFF module with four nozzles (multi-material printing—4K) and four axes of motion (x, y, z, C). An AMB 1050 FME-P DI 230V milling spindle from AMB-Elektrik GmbH was mounted on the extruder portal to reduce the manufacturing inaccuracy of about 0.3 mm which is typical for the FFF process. This eliminates play or high process forces during the integration of functional components by increasing the manufacturing accuracy to 0.02 mm. The machine also has a KUKA KR6 HA handling robot from KUKA GmbH for integrating the functional components during FFF printing. Figure 2 shows the basic process flow of the 4K FFF system.

In the first step, the basic structure is produced using the FFF process. The cavity, i.e. the negative shape of the part to be integrated, is already provided in the CAD model and is not filled with a support structure during the printing process. In a layer defined by the user, the printing process is paused at the machine's zero point and thus the measurement of the cavity is started using a 2D camera. An external image processing program is used to compare the target dimension with the actual dimension. The difference or the remaining oversize is transferred to the control of the FFF module and converted into a milling cycle (circular pocket, rectangular pocket, etc.). At the end of the milling cycle, the pocket is checked again. If the pocket is too large, i.e. the dimension of the pocket

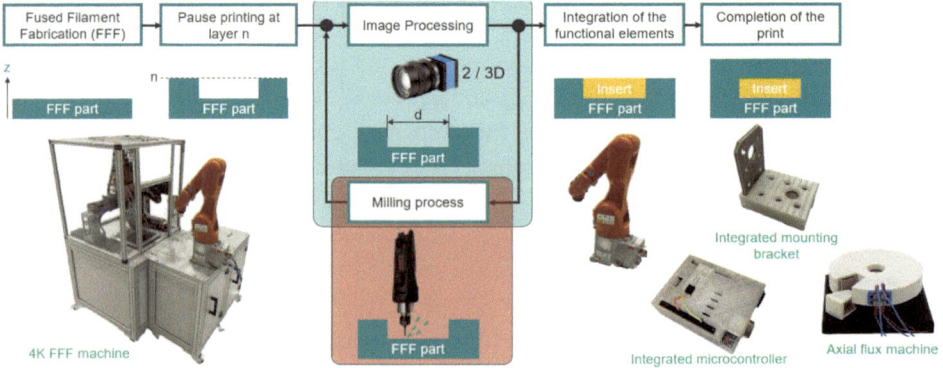

Fig. 2 Process flow of function integration

is out of tolerance, the machine goes into standby mode. The user receives an e-mail notification of the current machine status. If the cavity is within the specified tolerances, the industrial robot is activated. The robot picks up the insert to be placed using a gripper (suction gripper, parallel jaw gripper or similar) on the passive module provided.

The entire CNC code with printing code, milling commands and position data for the robot are generated with the help of the so-called "HybridPlanner"—see Sect. 2.2. After the insert is placed, the printing process is continued and thus the insert is completely embedded in the FFF structure. Fig. 2 (right hand side) shows three components that were produced using the 4K FFF process. These components are FFF parts with embedded steel brackets as force introduction or fixing elements to other components. An Arduino MEGA with an FFF-printed protective case is also shown. Furthermore, by processing conductive filament as well as by integrating temperature sensors and threaded inserts as connection points for the power supply, it was possible to produce complete sensor circuits [9]. Such sensor circuits are used in [8] to measure the winding temperature of an axial flux machine for condition monitoring.

2.2   Process Control and Process Planning

The developed 4K FFF machine is controlled by a Beckhoff programmable logic controller (PLC) with a computer numerical control (CNC) extension. With the help of this CNC extension, it is possible to process CNC codes containing the print and milling data of the FFF component to be produced following DIN 66,025. The target positions of the inserts in the FFF part, which are stored in the CNC code, are transmitted bit by bit to the robot controller (KUKA KRC4) via a serial interface. The inline process control can be realized during the printing process based on trigger points (variables) in the CNC code or faults such as the well-known spaghetti effect can be detected at an early stage after each printed layer [8]. For process planning of the individual process steps (printing, milling, handling, image processing), the MATLAB-based application "HybridPlaner" was developed.. The development of the process planning tool is motivated by the fact that up to now there has been no universal software for hybrid additive manufacturing planning or software that is adapted to the system concept of the 4K FFF system [15]. The Hybrid-Planer provides the operator with a graphical user interface (HMI) that allows him to interactively plan the manufacturing process with all necessary steps. Once planning is complete, the HybridPlaner automatically generates the corresponding CNC code. This code is then exported and imported into the HybridPlaner, where it is immediately translated from the MARLIN-specific code into a DIN-compliant CNC code. This is followed by the planning of the required imaging, milling and handling processes through a process strategy defined by the manufacturing expert. Each added or planned production step is automatically inserted by the HybridPlaner into the prepared CNC code of the FFF process at the appropriate places. The result of the process planning and output of the

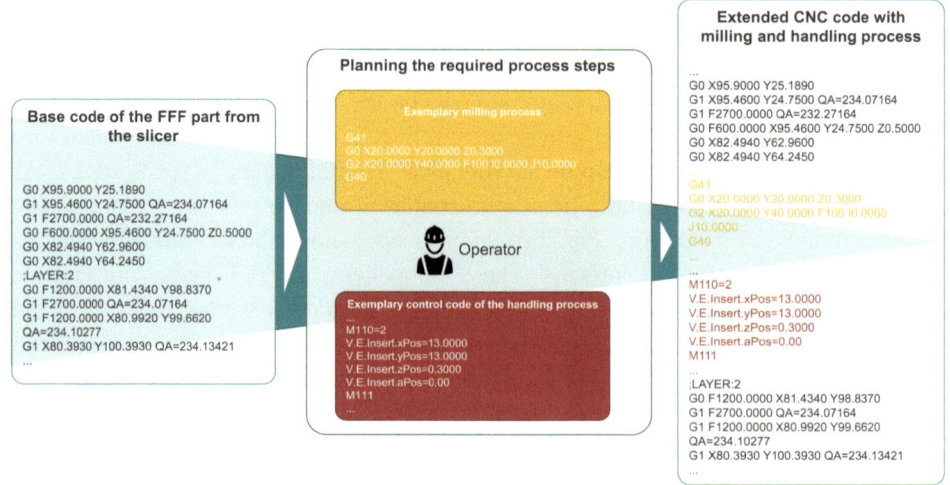

Fig. 3 Basic principle of the extension of the base G-code with the sub-process steps

HybridPlaner is an extended file with the CNC code executable by the 4K FFF system. Figure 3 shows the control code generation in the Hybrid Planner schematically using exemplary code excerpts.

The software helps the user decide which layer of the part to plan a sub-process on by providing a graphical visualization of the individual printing layers and a layer-by-layer display of the CNC code. The actual planning of the various sub-process steps itself takes place in separate, analogously structured program areas. In each of these windows, the user enters all the data required to define the sub-process (geometry, process parameters, machine commands, …) via an input mask, from which the control code is automatically created and displayed for the user on the graphical user interface (HMI). At the same time, the defined sub-process is graphically animated so that the user can understand, evaluate and, if necessary, iteratively adapt the result of his planning. Challenges and limits of the current process chain are:

- Due to the use of three different software tools (CAD, CUR, HybridPlaner), there is still a high manual planning effort.
- Expert knowledge is required for all sub-processes.
- If the manufacturing processes to be planned become more complex, the planning concept of HybridPlaner reaches its limits. If, for example, a complete component contour is to be post-processed with the help of a milling process, the HybridPlaner does not offer the necessary range of functions and the program-side support as CAM solutions available on the market.

- The existing robot module cannot be decoupled from the FFF module due to the prevailing safety logic. In other words, the subsystems cannot be operated separately or autonomously. This has so far made flexible use impossible.

3   Conceptual Process Extension of the 4 K FFF Process

With an increasing shortage of skilled labor, there is an urgent need for flexible and highly automated production systems. In particular, the automated post-processing of FFF parts is nowadays still dominated by manual tasks (loading and unloading of parts in the cleaning bath, painting…). Also in the context of increasing resource efficiency, repairing parts can extend the product life cycle and thus avoid cost- and material-intensive new production. Based on these facts, the existing 4K FFF machine at the wbk Institute for Production Science will be extended in terms of software and mechanics to enable automated repair and post-processing of polymer components. The following points describe the concept of the extended process chain and the intended control architecture.

3.1   Extended Postprocessing Workflow

A modular manufacturing solution will be built for the fully automated mass production of individualized, 3D-printed components. Any combination of the individual steps is possible, as intermediate postprocessing steps can also be realized. A robot kinematic system on a 7th axis serves as the core handling element. Both the multi-material 3D printer described above and a wide variety of post-processing modules can be docked onto this linear axis module with a KUKA KR6 robot (see Fig. 4).

Among other processes, these modules include the following:

- Insert supply and in-process integration by a robot
- Support structure removal using an alkaline or water-based immersion bath
- Optical 3D quality part measurement
- Vibratory grinding for surface smoothing
- Laser ablation

Both the planning of the entire production run and the individual process steps have to be converted into an automatic planning workflow based on the CAD data. As a result, the user should only have to create the CAD model and retrieve the finished, fully postprocessed part. The current approach to realizing these concepts is shown in Fig. 4 as a rendered CAD model.

It shows the existing multi-material 3D printer with an integrated milling spindle. It also shows the central handling module with the attached robot. Standardized modules

Fig. 4 Rendering of the extended process chain. A robot on a linear axis is used to transfer the 3D printed parts from the FFF printer module between the various other processing modules

can be attached around this core element. The modules have an electrical connection, compressed air and an EtherCAT connection to the central switch. With a breadboard-like hole pattern on the top plate, various postprocessing machinery can be mounted. A simple storage solution for a variety of grippers or as well as a buffer for workpieces during production can be realized too.

3.2   Concept of the Digital System Architecture

The following Fig. 5 shows the planned control topology.

The challenges and limitations of the previous system described in Sect. 2.2 are to be made possible through a targeted change in the control architecture. The focus of this

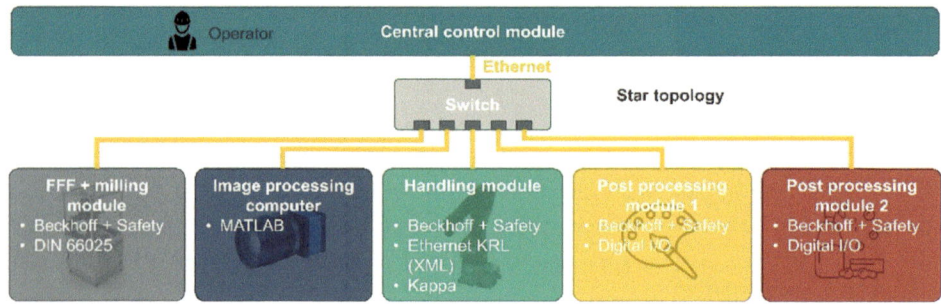

Fig. 5 Control architecture with modules and module-specific structure of the control system

consideration should be a system concept that is as flexible as possible, in which the process chain is configured according to the required process steps.

Topology: In the intended process chain, all systems are connected to a star topology using Ethernet. Starting from the central control module, all subsystems are connected in a star topology using a switch. A major advantage of a star topology is the relative simplicity of expansion and a high transmission rate. In addition, a failure of one component does not lead to a failure of the entire system. Troubleshooting is simple compared to other topologies.

Central control module: The central control module or, in other words, the process control computer primarily forms a human–machine interface to the existing PLSs of the subsystems. Depending on the part and the required process sequence, the relevant production process is to be started and monitored on the connected module. Furthermore, user-friendly planning of the required part-dependent process steps is to be made possible within the central control module and thus the control codes of all modules with required iterations for inline process control are to be generated.

Subsystems: All modules (except image processing) contain their CPU or programmable logic controller (PLC) with the necessary inputs/outputs to process the module-specific tasks. In addition, each module has its safety logic, so that an integrated or autonomous operation can be guaranteed. The control commands of the modules and their sequence of processing are orchestrated by the central control module.

Image processing: The necessary steps for the inline measurement of the components employing a 2D or 3D camera are supported by an external computer and image processing software in MATLAB. The main task of this computer is the processing of the image data including the interpretation of the measurement results and the decision on the further sequence of the process. Such key points are integrated into the control code of the subsystems by the central control module.

Increased system flexibility: Depending on the component, not all modules or subsystems are required. Modules that are not required should be placed in a nearby park or operated autonomously at another location. If it is necessary to connect the modules to form an overall system, the module-specific safety logic must be linked. With the help of the TwinSAFE Loader from Beckhoff, flexible linking of the safety controls via customizing is to be ensured [16]. This means that with the help of a central safety controller (e.g. FFF module), the safety controllers of the other modules are connected or disconnected in a user-defined manner. This way of linking the safety logic ensures flexible and, in particular, safety-compliant operation following the Machinery Directive 2006/42/EG.

4 Use Cases

The following section presents two projects, which are based on the extended process chain. Section 4.1 presents a process chain for automated post-processing of FFF parts. Section 4.2 describes the concept of a process chain for the automated repair or modification of broken components.

4.1 Automated Postprocessing of FFF Parts

Aiming at a fully automated generation of 3D-printed objects using FFF printing technology, the presented concepts are realized in the project *Auto-FFF* funded by the central innovation program for SMEs of the Federal Ministry for Economic Affairs and Climate Action (BMWK). Various use cases demonstrate the bandwidth of necessary processes for the production of additively manufactured parts. These include challenges in the actual 3D printing by adding functionalized structures and additional components. But also, in the subsequent removal of previously required support structures and the finishing of surfaces through processes such as grinding, dyeing or coating.

Due to the high manual effort involved in these very process steps, automation offers significant cost and time savings. Up to this point, the contradictions between individuality and automation have canceled each other out, but 3D printing provides a new and more efficient market for individualized mass-produced products. However, to meet the quality demands of consumers, the finishing processes should also be automated.

Figure 6 shows a total of four different use case parts. The housing components require threaded inserts to fasten other elements. In addition, printed support structures must be removed. A smooth and suitably colored outer surface of the housing is desired. The necessary tolerances for the pin connections must also be complied with. The second use case involves a rounded surface. Here, the staircase effect must be compensated during post-processing to achieve the required tolerances as well as roughness values. In a third use case of an RFID tag, different colored materials have to be printed simultaneously. The RFID electronics installed between the two parts are inserted during the printing process. In the last of the four use cases, a connector adapter is produced in large quantities of more than 2500 units. Due to the geometry, support structures have to be removed subsequently. In addition, a colored coating must be applied. The goal is the continuous automation of manufacturing processes for the ready-to-use production of individualized parts and their handling within the interlinked process steps.

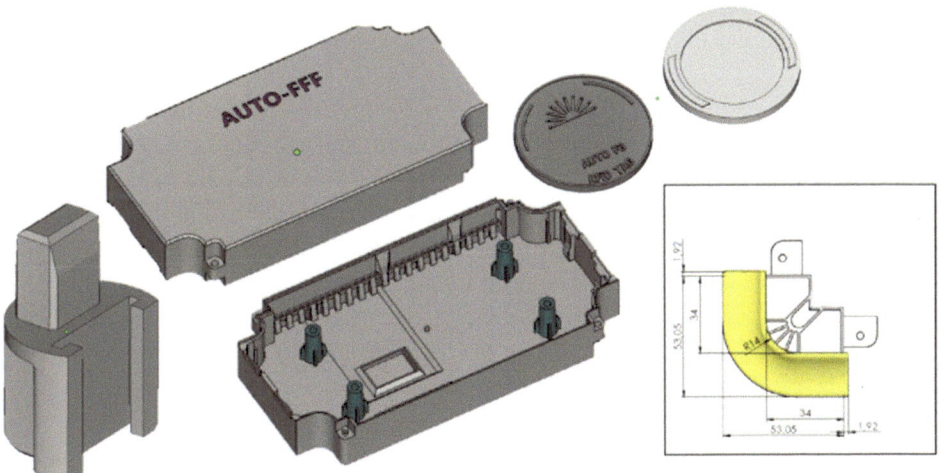

Fig. 6 Collection of parts from various applications with different requirements

4.2   Repair and Modification of Additively Manufactured Parts

The RESTORE project, which is funded by the InnovationCampus Mobility of the Future, has the goal of developing the necessary fundamentals for remanufacturing process chains for functional, hybrid polymer parts. Based on the vision of a sustainable and resource-saving society, which would like to define the long usability of products through a right to repair, the intention is to expand an existing component-flexible machine in terms of hardware and software. In the future, this will enable a high degree of flexibility in the remanufacturing of various components, as the required thermal, mechanical subtractive and additive process steps of the machine will later be defined exclusively via digital process planning. The product to be processed is to be opened, disassembled, repaired or adjusted for the intended purpose by the machine and reassembled for further use (remanufacturing). This remanufacturing strategy offers essential advantages in the context of resource efficiency and reparability. By using the extended process chain, transport routes are eliminated, handling operations are simplified and iterative and combined use of the different manufacturing principles is made possible. Using a highly hybridized polymer component in the form of an adaptable and repairable axial flux machine (AFM), a repair process (e.g. the replacement of a defective motor winding) and a modification process (e.g. insertion of a pole shoe with higher packing density and thus higher torque) are to be demonstrated. Essential research contents of this project are:

- Creation of a development guide for components to be repaired
- Creation of a failure catalog and failure categories
- Enabling the machinery for the repair / modification process

- Derivation of process strategies for damage-minimized disassembly/assembly
- Demonstration of technical feasibility based on the axial flux machine

To analyze and enable the remanufacturing process of the axial flux machine, all process steps are analyzed based on a function plan according to DIN 2860. * MERGEFORMAT Fig. 7 shows the principle sequence of the process chain for the case of the replacement of a defective pole shoe. The axial flux machine is first fed into the process in a fully ordered manner and clamped in place. With the help of a laser process to be developed, a minimally invasive opening of the electric motor is to be achieved. After removing the housing cover, the defective pole shoe is cut out. After the seat of the pole shoe has been prepared (e.g. milling process), the new pole shoe is inserted. Finally, the previously removed housing cover is joined and the complete axial flux machine is assembled. During the entire process, inline process control data will be collected and analyzed so that elements of the AFM are always correctly gripped and handled. The following challenges are to be addressed within the scope of this research project:

- Development of suitable gripping and clamping technology
- Derivation of software-based data processing including all test steps
- Experimental determination of a minimum damage process strategy
- Proof of the technical feasibility of the system (repair and modification process)

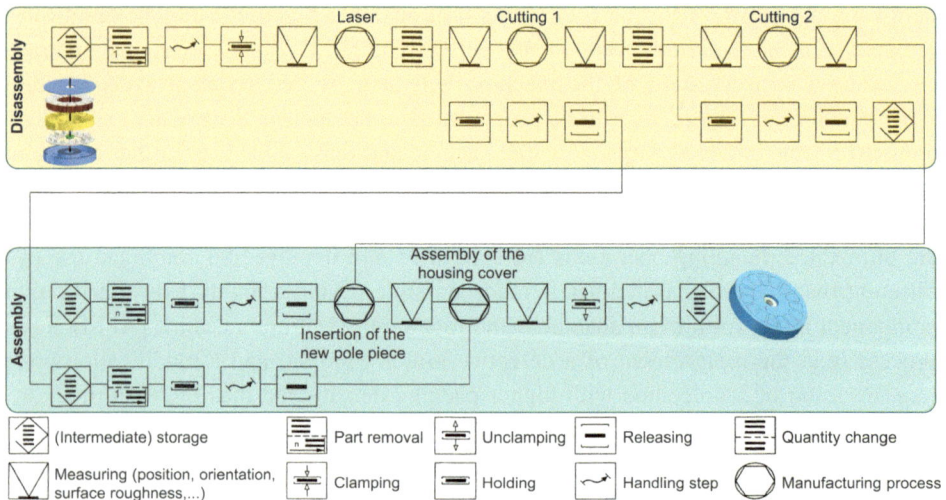

Fig. 7 Exemplary flow chart according to DIN 2860 for the replacement of a defective pole shoe of an axial flux machine

5   Conclusion and Outlook

Additive and especially automated manufacturing of highly integrated components is playing an increasingly important role in today's world. In the context of a growing shortage of skilled workers, increasing individualization and, in particular, the growing right to repair products at the end of the product life cycle, production technology is facing major challenges. In addition to the fully automated production of highly integrated components (Auto-FFF), the automated repair (RESTORE) of components is also becoming the focus of research and development. In two affiliated projects, a fully automated production of FFF components with post-processing and the repair of plastic components are to be made possible in the future based on an existing machine concept with flexible control architecture (star topology). The next steps deal with the implementation of the existing machine concept and the realization of a flexible control architecture.

Acknowledgements The authors would like to thank the Baden-Württemberg Ministry of Science, Research and the Arts for the financial support of the projects within the "InnovationCampus Future Mobility". The Project "Auto-FFF" is supported by the Federal Ministry for Economic Affairs and Climate Action (BMWK) based on a decision by the German Bundestag.

References

1. Beaman, J.J., Bourell, D.L., Seepersad, C.C., Kovar, D.: Additive manufacturing review: early past to current practice 142 (2020)
2. Wesemann, C., Pieralli, S., Fretwurst, T., Nold, J. et al.: 3-D printed protective equipment during COVID-19 Pandemic 13 (2020)
3. Dizon, J.R.C., Gache, C.C.L., Cascolan, H.M.S., Cancino, L.T. et al.: Post-Processing of 3D-Printed Polymers 9, p. 61 (2021)
4. Ballardini, R.M., Flores Ituarte, I., Pei, E.: Printing spare parts through additive manufacturing: legal and digital business challenges **29**, 958 (2018)
5. Europäisches Parlament. Recht auf Reparatur: Recht auf Reparatur Entschließung des Europäischen Parlaments vom 7. April 2022 zu dem Recht auf Reparatur 2022/2515(RSP).
6. Charul Chadha, Albert E. Patterson, James T. Allison, Kai A. James, and Iwona M. Jasiuk, 2019. *Repair of high-value plastic components using fused deposition modeling*, Austin, Texas, USA.
7. Baranowski, M., Netzer, M., Gönnheimer, P., Coutandin, S. et al.: Functional integration of subcomponents for hybridization of fused filament fabrication, p. 319 (2022)
8. Springmann, M., Matkovic, N., Schäfer, A., Waldhof, M., et al.: Flexible and High-Precision integration of inserts by combining subtractive and non-planar additive manufacturing of. Polymers **926**, 268 (2022)
9. Baranowski, M., Schubert, J., Werkle, K.T., Schoner, S. et al.: Additive-subtractive manufacturing of multi-material sensor-integrated electric machines using the example of the transversal flux machine, p. 1
10. CEAD Group. AM Flexbot: Automated large scale additive manufacturing. https://ceadgroup.com/solutions/robot-based-solutions/. Accessed 31 January 2023.

11. Thermwood. LSAM—Large Scale Additive Manufacturing. https://thermwood.de/maschinen/lsam/. Accessed 31 January 2023.
12. Rösler Oberflächentechnik GmbH. AM Solutions - 3D postprocessing solutions. https://www.solutions-for-am.com/de-de/post-process. Accessed 31 January 2023.
13. Additive Manufacturing Technologies Limited. Digital Post-Production Hardware Systems for Additive 2.0 Manufacturing. https://amtechnologies.com/. Accessed 31 January 2023.
14. Tamburrino, F., Barone, S., Paoli, A., Razionale, A.V.: Post-processing treatments to enhance additively manufactured polymeric parts: a review **16**, 221 (2021)
15. Lorei, I.: Further Development of a process planning tool to automate the Control of hybrid additive-subtractive manufacturing process, Karlsruhe (2022)
16. Beckhoff Automation GmbH & Co. KG. Safety technology—TwinSAFE. https://www.beckhoff.com/en-en/products/automation/twinsafe/. Accessed 31 January 2023.

Dimensioning and Structural Optimization of Components Using a Distributed Simulation Approach Consisting of FEM and Multiphysics

Sebastian Anders, Eva Russwurm, Florian Faltus, and Jörg Franke

Abstract

One of several challenges while dimensioning components of handling systems is the consideration of force and torque influences acting during operation in the correct order of magnitude based on the process used. With existing design approaches dynamic loads in particular are usually not adequately taken into account and therefore high safety factors are selected. However, this also increases the weight and usually the costs. To reduce over-dimensioning, this paper presents a methodical approach for dimensioning and structural optimization of components. The method combines a multi-physical simulation and finite element analysis to improve the design of components, considering the process-related forces and moments applied to the system boundaries. These are calculated in the multi-physical simulation and transferred to an FEM simulation, where they are used to optimize the component's design and structure. The approach of distributed simulation helps to reduce the over-dimensioning of components and improve the development process of assembly system components.

Keywords

Structural optimization • FEM • Multiphysics • Distributed simulation

S. Anders (✉) · E. Russwurm · F. Faltus · J. Franke
Institute for Factory Automation and Production Systems (FAPS), Friedrich-Alexander-Universität Erlangen-Nürnberg, Erlangen, Germany
e-mail: sebastian.anders@faps.fau.de

159

S. Ihlenfeldt et al. (eds.), *Annals of Scientific Society for Assembly, Handling and Industrial Robotics 2023*, https://doi.org/10.1007/978-3-031-74010-7_14

1   Introduction

When dimensioning assembly systems, a large number of boundary conditions, such as product-related material and process-related load variations as well as requirements regarding the performance and kinematics of the handling system, must be taken into account. Based on the process used, one challenge is to apply the correct order of magnitude for all forces and moments acting during operation. The loads result from static and dynamic forces and moments, which arise from the interactions of the individual designed assembly system elements. Due to a high process diversity, the prevailing dynamic loads cannot be considered with existing design approaches. To avoid component failures, and thus possibly the entire assembly system, high safety factors are selected. However, this increases the weight of components and thus usually also the costs [1–3].

This paper presents a methodical approach to counteract this over-dimensioning. Using a multi-physical simulation, unknown loads are determined within a framework of a distributed simulation in an early phase of the product development process, which are then introduced into a Finite Element Analysis (FEA) in the domain of mechanics. With this method, a stress-appropriate design can be achieved.

2   State of the Art

This chapter will give an overview of fundamental tools and methods which are used in this paper.

2.1   Structural Optimization

During product development a variety of processes must be considered. In addition to requirements acquisition and the design of functions, this includes dimensioning and design of individual components. For this purpose, various process models and mechanical correlations with specific safety factors exist for component design. In many cases, the safety factors tend to be selected larger than necessary for the actual function. Topology optimization can be used to support this. Here, weight and stiffness are optimized on basis of defined boundary conditions and target values. This results in a proposal for the component design with optimized shape contour [4, 5]. One goal is the optimal distribution of wall thicknesses and cross-sectional dimensions with a constant, unchanging outer geometry of the construction space. Furthermore, material optimization supports the selection of the corresponding material properties such as modulus of elasticity, transverse contraction coefficient and density [4, 6].

2.2　(Multi-)Physics-Based Simulation

For the development of mechatronic systems, there is a multitude of discipline-specific as well as interdisciplinary process models, which are supported by simulation technology. Both interdisciplinary and discipline-specific simulation methods are used, which differ greatly in terms of the accuracy of their modelling. While interdisciplinary simulations are used in an early phase of the product development process, domain-specific simulations are used in a later phase of the product development process [7, 8].

In physics-based simulation, a physics engine is used to calculate the dynamic motion behaviour of simulated objects and their mutual interaction [9, 10]. A physics engine is a software library that enables the simulation of objects based on their geometric and physical properties [11]. Deformable bodies and particle systems can be represented for fluid simulations in addition to rigid bodies [12]. In this way, simulation objects are animated using physical laws and calculation methods from computer graphics, whereby a high level of detail is achieved through the mapping of physical properties and the consideration of further environmental parameters, such as gravity [9, 10]. In mechanical and plant engineering, physics simulation is predestined for virtual commissioning due to its real-time-capable response behaviour [9, 13].

2.3　Finite-Element-Analysis

The FEA can be used in many ways in the product development process, whereby the investigation objectives, and thus also the requirements for the model used, change continuously. If the dimensions and forces in the components are roughly determined for a quick estimate at the beginning of the design, detailed and meaningful finite element method (FEM) simulations ensure the strength of the entire product in a later phase of the product development process. This saves costs by avoiding design errors in conjunction with fewer physical prototypes. The basic idea of FEM is the decomposition of complex geometries into a finite number of discrete elements, which can be described via conventional systems of structural mechanical equations. The resulting total stress of a component can be determined by suitable equilibrium conditions in the connecting nodes between these elements [5, 14, 15]. A basic procedure for performing an FEA is described by Vajna et al. [5]. The starting point is the definition of the problem to be investigated, including all load cases, system properties, boundaries and its environment. In the subsequent step, the geometry of the real system is converted into an idealized model, which must have the system properties relevant for solving the problem [5].

2.4   Co-Simulation

Due to the increased complexity of mechatronic systems, it is necessary to consider the entire mechatronic system and the interactions between individual modules and models. Therefore, distributed simulation is in focus of research in recent years. Domain-specific sub-models are made accessible to a joint simulation to analyze the behavior of an entire system. Co-simulation makes it possible to link the discipline-specific simulation models that have already been created during the product development process, which are very precise in terms of mapping accuracy, and thus to consider all the interactions that occur between the models of the overall mechatronic system. A central challenge, however, is the unification of "models from different simulation disciplines into a multidisciplinary overall model" [1], since the model data relevant in the context of co-simulation must be exchanged via a uniform interface [1, 16, 17].

2.5   Distributed Simulation in Component Design

In Russwurm et al. distributed simulation is used to counteract the over-dimensioning of gripper systems. The concept includes a co-simulation consisting of a multi-physical simulation tool and a FEM simulation, a modelled development process and the real system itself. The unknown loads of the handling process are determined by means of a multi-physical simulation and introduced into an FEM simulation [1]. As Fig. 1 illustrates, the simulation-based design process begins within a multi-physics simulation tool, which, in addition to the handling device and the handling process, also depicts the gripper system itself and the handling object in a highly abstracted manner.

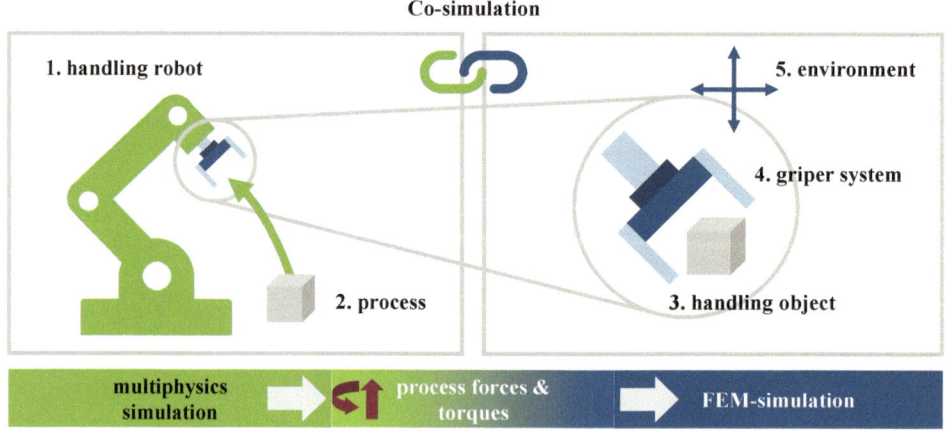

Fig. 1 Co-simulation for component design with reference to [1]

The process-related forces and moments are calculated within the multi-physical simulation and are transferred to the FEM-simulation. There the previously unknown, dynamic loads are used as boundary conditions for the system limits to design the gripper system. An important component of the methodology is the real system, which is used as a template for the system model and the system-specific data. This also enables the separate validation of individual multi-physical and FEM simulations, but also in combination within the framework of a co-simulation [1].

3   Research Objective

The state of the art has shown that simulation technology is widely used in the development of mechatronic systems. Due to the successively increasing complexity of mechatronic systems in recent years, it is no longer suitable to look at single domains, but to have multidomain distributed simulations [16].

FEA is widely used in mechanical design. An idealized model is transformed into finite elements by means of suitable element types, a specific material is assigned, boundary conditions are defined and loads are introduced [5, 18]. In the context of gripper systems, the latter include static, dynamic and process-related forces and moments that are applied at the system boundaries [3]. Due to different types of handling objects in connection with dynamic speeds and movements, a large number of different process forces and torques occur during the assembly process. As explained in the initial situation, these loads cannot be adequately considered with existing design approaches. As a result, high safety factors are usually selected, which, however, increase the weight of the components and thus usually also the costs.

To reduce the over-dimensioning of components, a methodical procedure for the design and structural-mechanical optimization based on the approach described in Sect. 2.5 is presented below. By considering the prevailing, process-related loads, the development process of assembly system components is improved.

4   Distributed Simulation for Dimensioning and Structural Optimization

In the following chapter, the concept and the approach of the distributed simulation will be discussed, in order to subsequently list the methodical procedure for the stress-appropriate optimization of mechanical components.

4.1   Approach

The concept of distributed simulation for component design and optimization shown in Fig. 2 is based on the approach of Russwurm et al. presented in Sect. 2.5. The process-related forces and moments applied to the system boundaries of an object under investigation are calculated within a multi-physical simulation and transferred to an FEM simulation via a neutral data format. There, the loads are introduced as boundary conditions at component level and are used for dimensioning and structural-mechanical optimization. The first, multi-physical simulation is carried out as part of the function definition and thus also forms the basis for interdisciplinary system development. In addition to determining the unknown loads at the system boundaries of the object under investigation, the developed solution concept can also be virtually validated.

Due to the reduced modelling effort with a simultaneously high mapping fidelity and the calculation of the forces and moments acting in joints between simulation objects at runtime, the physics-based simulation technology is predestined for an initial, interdisciplinary simulation of the entire mechatronic system. At the beginning of development, the simulation model comprises simplified 3D geometries, which are further concretized in the course of the development process. The simulation results are made available to the designers and simulation experts via a central database system. In addition, executable SysML process models enable a partially automated execution of the development process [1]. The loads are then subsequently introduced into a domain-specific FEM simulation. Based on the calculated analysis results, structural-mechanical optimizations are carried out. Improvements in topology, shape, cross-sectional dimensions or material usage are

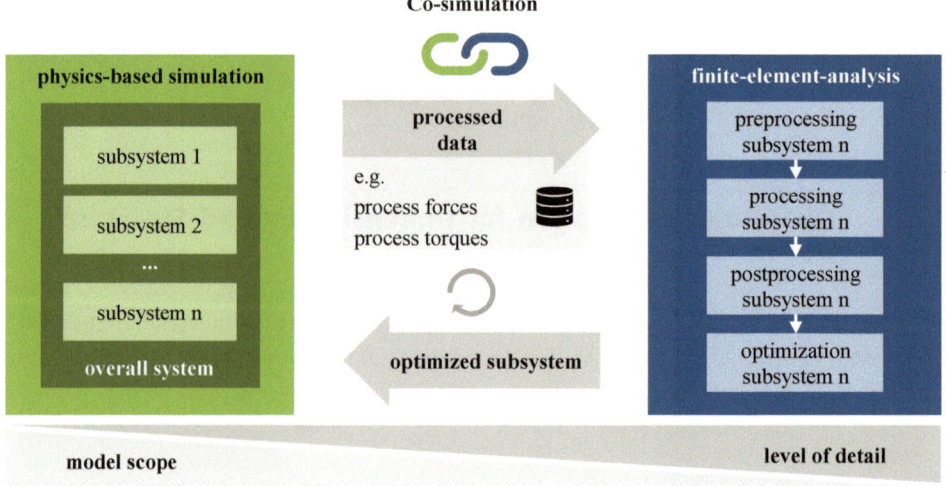

Fig. 2 Concept of a distributed simulation for component design and optimization

then evaluated in a further iteration before being transferred to the physical simulation and validated in the context of the entire mechatronic system. The integration of the mechanical component is further achieved through virtual tests on the digital model, which has been increasingly enriched along the development process.

4.2   Methodology

The conceived methodical procedure for component design and component optimization includes the simulation procedure model of [19], the procedure for model creation for physics-based simulation according to [10] as well as the procedural steps for carrying out an FEA according to [5]. The method aims to improve product design within the discipline of mechanics, in accordance with the concept from Sect. 4.1, and is divided into four successive phases as Fig. 3 illustrates.

Initial Phase. Starting with the definition of an overall objective the functionality of the system to be considered is described and the requirements for the optimization process

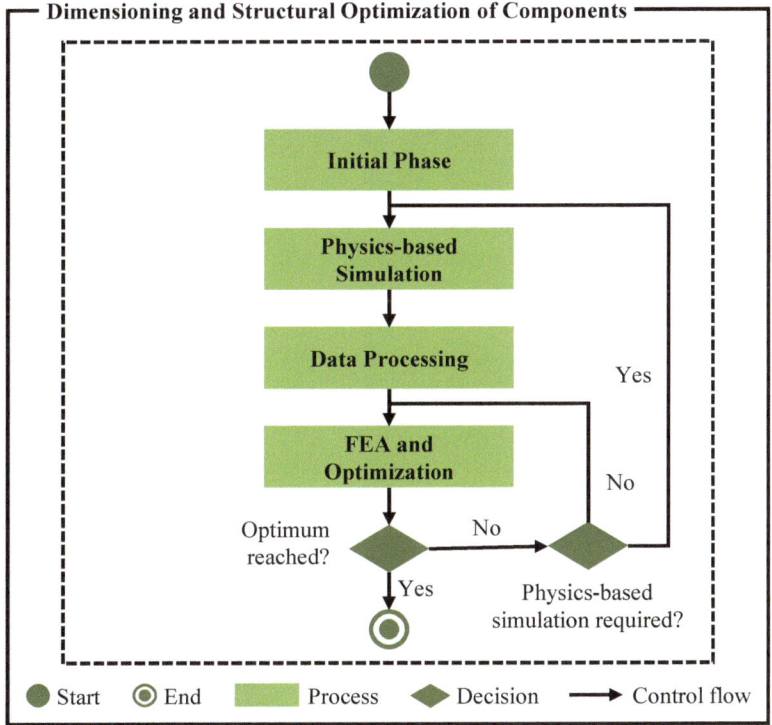

Fig. 3 Methodology for component design and optimization

are defined based on sub-objectives. A specification list includes the design variables to be varied, the requirements for the component and the load cases to be considered. In the subsequent task definition, the target description is further specified and fully described. The work contents include the definition of the required information and data as well as the assignment of the work packages to the departments involved. The system analysis is carried out based on the task specification, whereby the system under investigation is analyzed and abstracted according to the optimization objective. The results form the basis for the final system boundary and interface definition. In this process, the system boundaries and interfaces between the physics and FEM simulation required for the distributed simulation are defined.

Physics-based Simulation. In the physics simulation, a simulation model is first built according to the initial phase. In a second step, the unknown loads are determined in a simulation run and validated afterwards.

First, a collision model is derived, which is parameterized with physical parameters and expanded into a kinematic model by defining joints and other design metaphors. Joints of the kinematics model restrict the degrees of freedom of two model elements in relation to each other and thus enable the mapping of fixed connections, linear or rotational axes. In this modelling step, the object of investigation is cut free according to the system boundary and interface description based on joints to its system environment. The forces and torques acting in the joints between the simulated objects can thus be calculated by a physics engine at runtime. The kinematic model is provided with behavioral models and, in the last modelling step, a control strategy is implemented within the specific simulation environment according to the Model-in-the-Loop approach.

Following the modelling, the loads in the previously defined joints of the object under investigation are calculated in the context of a simulation run and validated afterwards. Suitable methods for validation include the limit value test and the sensitivity analysis, which are described in [20], or the calculation using rollover formulae or the comparison with measured data.

Data Processing. The validated simulation results are exported from the multiphysics simulation environment as comma-separated values and read into a database system. Database operations are used to filter the simulation results. Finally, the processed data is made available to the mechanical design for FEA and optimization.

FEA and Optimization. In this phase, a simulation model is built based on the initial phase, which is calculated and all necessary variables are evaluated. Based on the analysis results, the object under investigation is improved structurally.

In the first step of pre-processing, the free-cut geometry is converted into an idealized model by using component symmetries and substitute models. After the assignment of the specific material properties, a transfer into the finite elements takes place. The mesh is generated either manually or automatically using a mesh generator. The element shape and mesh density are to be adjusted to the analysis objective according to the task specification as well as the system analysis. Singularities in areas of point-related supports or loads are avoided by coupling elements, which connect the support or load point of the free-cut subsystem to the surrounding mesh. The object of investigation is then constrained by defining boundary conditions according to the system boundary and interface definition. Finally, the loads determined during the physical simulation are introduced into the FEM model.

After the calculation of the simulation model, the analysis results are visualized as well as assessed and validated. Especially in areas with high stress gradients or in areas of force application, mesh refinements and modifications may be necessary. The simulation results are then validated using suitable methods, such as comparison with an analytical substitute problem or with measurement results. Based on the analysis results, optimizations are made to the component geometry. For this purpose, the determined stresses and deformations are first used to verify the strength of the component. If the component safety corresponds to the required minimum safety and the specified requirements for the design are met, no optimizations are necessary. Otherwise, changes are made to the design variables according to the specification list. By varying the design variables, changes are made to the topology, shape, cross-sectional dimension or material. With the use of an optimization algorithm, the computer rather than the designer makes improvements to the components. The algorithm varies the design variables until the previously defined optimum is reached [4].

Finally, a new simulation run is performed to evaluate the improved component. If significant changes in component geometry or material properties are made during optimization, these are transferred to the physics-based simulation and validated in the context of the entire mechatronic system.

4.3 Validation Through an Exemplary Application

In the following section, the gripper jaws of a parallel gripper are optimized according to the presented method. For the physics-based simulation, Siemens' Mechatronic Concept Designer (MCD) and for the FEA, Simcenter Nastran are used. The gripper system handles objects with a mass of up to 100 g, and the objective function of the gripper jaws represents a maximum weight of 30 g. The constraints are the interface to the gripper system and the contact points to the object to be handled. The frictional connection between the gripper jaw and the base system absorbs the static weight, dynamic motion and acceleration, and process-related forces and moments, and represents the system boundary of the FEM simulation.

In the MCD, fixed joints were defined in the interface between the gripper base system and the gripper jaw, on the basis of which the forces and moments acting during a simulation run are calculated by the associated physics engine. The results of the physics-based simulation were then exported, validated by a check calculation, and introduced into the

Fig. 4 Object of investigation
with existing and optimized
gripper jaws

 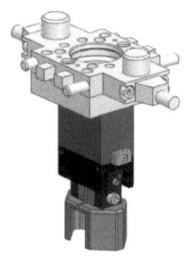

Exisiting gripper jaws
Material: EN AW 7075;
Weight: 152 grams

Optimized gripper jaws
Material: PA 2200;
Weight: 28 grams

FEM model as loads. The strain and stress distribution as well as the maximum deformation of the gripper jaw are the results of the FEM calculation. In order to achieve the
defined objective function, the design variables material, cross-sectional dimension and
shape of the gripper jaws were modified based on the FEM results, as shown in Fig. 4.

The optimized gripper jaws were then evaluated in a new simulation run and virtually
validated in the context of the entire gantry system in the MCD. The target function of
30 g was achieved within the given constraints, successfully completing the optimization
process. The example application showed that the concept and methodical approach for
component design and optimization can be successfully applied to the optimization of
components of a handling system, in this case gripper jaws.

5   Summary and Outlook

The challenge in planning assembly systems is to consider the prevailing, process-related
loads in the right order of magnitude. Due to a high process diversity, these loads cannot
be adequately considered with existing design approaches. As a result, high safety factors
are selected, which increases the weight of the system components and usually also the
cost. In this paper, a methodical approach was presented on how to counteract this over-
dimensioning. Using a physics-based simulation, unknown loads are determined in an
early phase of the product development process within the framework of a distributed
simulation which then are introduced into an FEA. Through the methodical procedure,
components of a mechatronic system are dimensioned and optimized structurally within
four phases in the domain of mechanics. The use of AI methods is a promising approach
for structural optimization. Using optimization methods such as evolutionary algorithms
or particle swarm optimization, design parameters are varied under restrictions until a
target value, such as the weight or the overall stiffness of the component, is minimized
or maximized [4, 5]. With the help of the presented method, the over-dimensioning of
plant components of mechatronic systems is counteracted by a stress-appropriate design.

A promising approach is the (partially) automated execution of the development process based on a modeled process [1].

References

1. Russwurm, E., Faltus, F., Laukotka, F., Schwan, L., Brossog, M., Krause, D., Franke, J.: Methodik zur frühzeitigen Integration prozessspezifischer Einflussfaktoren aus der automatisierten Montage in die Auslegung funktionskritischer Komponenten. In: Krause D, Paetzold K, Wartzack S (eds.) Proceedings of the 32nd Symposium Design for X (2021)
2. Haibach, E.: Betriebsfestigkeit. Verfahren und Daten zur Bauteilberechnung, 3rd edn. Springer, Berlin (2006)
3. Schmalz, J.K. B.: Rechnergestützte Auslegung und Auswahl von Greifersystemen. Dissertation, Technische Universität München (2018)
4. Schumacher, A.: Optimierung mechanischer Strukturen. Grundlagen und industrielle Anwendungen, 3rd edn. Springer Vieweg, Berlin, Heidelberg (2020)
5. Vajna S., Weber C., Zeman K., Hehenberger P., Gerhard D., Wartzack S. (eds.): CAx für Ingenieure. Eine praxisbezogene Einführung, 3rd edn. Springer Vieweg, Berlin (2018)
6. Seiler M. S.: Geometrische Restriktionen bei der geometriebasierten Strukturoptimierung von Maschinenbauteilen mit Freiformgeometrien. Dissertation, Rheinisch-Westfälische Technische Hochschule Aachen (2012)
7. Eigner M., Roubanov D., Zafirov R.: Modellbasierte virtuelle Produktentwicklung. Springer Vieweg, Berlin, Heidelberg (2014)
8. Bracht U., Geckler D., Wenzel S.: Digitale Fabrik. Methoden und Praxisbeispiele, 2nd edn. Springer Vieweg, Berlin, Heidelberg (2018)
9. Damrath F.: Physikbasierte mechanische Absicherung zur energieeffizienzorientierten Planung und Auslegung automatisierter Montageanlagen im Automobilbau. Dissertation, Universität des Saarlandes (2018)
10. Lacour F.-F.: Modellbildung für die physikbasierte virtuelle Inbetriebnahme materialflussintensiver Produktionsanlagen. Dissertation, Technische Universität München (2012)
11. Spitzweg M.: Methode und Konzept für den Einsatz eines physikalischen Modells in der Entwicklung von Produktionsanlagen. Dissertation, Technische Universität München (2009)
12. Drescher B., Stich P., Kiefer J., Reinhart G.: Physikbasierte Simulation im Anlagenentstehungsprozess–Einsatzpotentiale bei der Entwicklung automatisierter Montageanlagen im Automobilbau. In: Dangelmaier W, Laroque C, Klaas A (eds.) Simulation in Produktion und Logistik 2013. Heinz-Nixdorf-Inst., Paderborn (2013)
13. Röck, S.: Hardware in the loop simulation of production systems dynamics. Prod. Eng. Res. Devel. 5(3), 329–337 (2011)
14. Kestel P.: Assistenzsystem für den wissensbasierten Aufbau konstruktionsbegleitender Finite-Elemente-Analysen. Dissertation, Friedrich-Alexander-Universität Erlangen-Nürnberg
15. Rieg, F., Hackenschmidt, R., Alber-Laukant, B.: Finite Elemente Analyse für Ingenieure, 5th edn. Carl Hanser Verlag GmbH & Co. KG, München (2014)
16. Russwurm E., Faltus F., Franke J.: Systems Engineering als Basis für konstruktionsbegleitende Zusammenarbeit
17. Günther F. C.: Beitrag zur Co-Simulation in der Gesamtsystementwicklung des Kraftfahrzeugs. Dissertation (2016)
18. Fröhlich, P. (ed.): FEM-Anwendungspraxis, 1st edn. Vieweg, Wiesbaden (2005)

19. Verein Deutscher Ingenieure e.V. (2014): Simulation von Logistik-, Materialfluss- und Produktionssystemen. Grundlagen 03.100.10(3633-Blatt1). Beuth Verlag GmbH, Berlin
20. Rabe, M., Spieckermann, S., Wenzel, S.: Verifikation und Validierung für die Simulation in Produktion und Logistik, 1st edn. Springer, Berlin, Heidelberg (2008)

Visualization of Forces and Torques for Robot-Programming of In-Contact Tasks

Johannes Hartwig, Sabine Fischer and Dominik Henrich

Abstract

A variety of automation tasks require an accurate specification of forces and torques over time and space. Traditionally, only experts can provide such specifications, contrasting the need for intuitive robot programming in small and medium enterprises. Simulations and visualizations are typical approaches to increase the usability of robot programming frameworks. Thus, we extend our robot programming approach with force and torque visualizations, enabling non-experts to verify and adjust programs for in-contact tasks. In this paper, we contribute a comprehensive comparison of different force/torque visualization techniques and discuss their applications in robot programming systems. Furthermore, we evaluate visualizations of forces and torques generally as well as specifically adapted to our programming concept utilizing two user studies. Vector arrows for forces and curved arrows for torques showed promising results.

Keywords

Robotics · Non-expert robot programming · Programming system · Graphical user interface · Simulation

J. Hartwig (✉) · S. Fischer · D. Henrich
Lehrstuhl für Angewandte Informatik III (Robotik Und Eingebettete Systeme), Universität Bayreuth, Universitätsstraße 30, 95440 Bayreuth, Germany
e-mail: johannes.hartwig@uni-bayreuth.de

© The Author(s) 2025
S. Ihlenfeldt et al. (eds.), *Annals of Scientific Society for Assembly, Handling and Industrial Robotics 2023*, https://doi.org/10.1007/978-3-031-74010-7_15

1   Introduction

In industrial environments, experts usually program robot manipulators via a textual programming language. This leads to a time- and cost-consuming programming process, where
users must be highly experienced in programming and executing the task. These issues limit
the acceptance of robotic systems in small and medium enterprises (SMEs) [1]. Therefore,
a fast and easily re-programmable robot system is needed to be profitable for small batch
sizes and flexible production. We can solve this by using the programming by demonstration paradigm with kinesthetic teaching because the user only needs to guide the robot,
which can generally be considered intuitive [2]. With this approach, it is traditionally only
possible to enter trajectories, but it is also necessary to allow program structures and adaptions to the robot motions. For this purpose, extended playback programming can be used,
which enhances the concept of playback programming by a graphical programming interface allowing these operations [3]. After the programming process, feedback for the user
is also helpful. Here, the graphical user interface (GUI) with a robot simulation can help
non-experts verify whether the robot program will solve a task satisfyingly.

Some tasks require an accurate specification of forces and torques over time and space
(in-contact tasks). Such tasks are, for example, planing wood or tightening screws. To date,
many GUIs simulate only position-related robot motions. For easy use, the visualization of
in-contact robot tasks should extend the GUI to allow for verification before they are executed
on the real robot. The simulation is extended by visualization of forces and torques if a user
programs such motions. While some approaches exist to visualize forces and torques (see
Sect. 2), their usability and intuitiveness still need to be evaluated. Moreover, it needs to be
evaluated how and which visualization methods can be combined best to create an intuitive
representation of in-contact motions. In addition to visualizations, forces and torques can
be transmitted to the user by haptic signals or audio signals [4, 5]. Nevertheless, this work
is limited to visualizations since they can be directly integrated into the GUI.

In this paper, we evaluate supportive visualizations of forces and torques exerted by
the robot in our programming system through two user studies. The goal is to provide
the highest usability and intuitiveness for non-experts. Section 2 gives an overview of the
different visualization methods of forces and torques. We then describe the foundations and
assumptions for our approach and embed the possible visualization methods within them in
Sect. 3. Thereupon, Sect. 4 discusses the user studies conducted and their results. Section 5
summarizes the paper and discusses future work.

2   State of the Art

Forces and torques can be fully described by their magnitude, direction, and application point
or reference. However, different visualization methods are used in education and robotics.
The suitability of these representations in depicting the forces and torques of an in-contact

motion depends on various factors, including which of the aforementioned properties they aim to convey.

Forces are visualized in teaching various natural and engineering sciences in both school and university. Different effects of visualization types were investigated in direct comparison for educational applications. A study found that to illustrate Newton's law, displaying an animated depiction is better suited for force visualization than displaying the same as text [6]. However, the transferability of this finding to robot simulations is limited due to the different application areas and by the nature of the textual description, which does not occur in any other study from education or robotics. Also, it has been shown that arrows animated and displayed at runtime are better suited as visual force feedback when training surgeons than a time-force graph displayed after completing the task [7]. However, due to the different times of display and application domain, it cannot be concluded that arrows are also more suitable for force visualization in robot simulations. Unlike for forces, there are no studies evaluating different visualizations for torques. In terms of usability, both force visualizations and torque visualizations have yet to be quantitatively evaluated.

In robotic-related applications, forces and torques are nearly always considered over time. The visualizations are used here as a tool during the programming process and are rarely reviewed or evaluated. There, we find different programming systems with a variety of visualizations. They can be categorized by the way they represent time courses (single chosen time step, progressive animation, several time steps simultaneously), the symbols used (e.g., arrow, text, diagram), and the placement (e.g., on an object, in a separate GUI section). Figure 1 depicts such an overview for the reviewed force visualizations represented as a tree [5, 8–23]. The torque visualizations [10, 11, 13, 14, 20, 21, 24–26] are illustrated in the same classification in Fig. 2. In Sect. 3, the identified visualizations are reviewed and applied to our general programming approach. Subsequently, these are evaluated for their suitability for an intuitive robot programming system in Sect. 4, as there have been no findings in this area.

3   Visualization Methods

In order to determine which type of visualization is suitable for our approach, we first outline the requirements of the robot programming concept, which is based on playback programming [3]. This concept extends playback programming by enabling users to edit the trajectory, simulate the robot's motion, and add program structure. It shall be expanded to support in-contact tasks. This concept is intended to be usable in a general, task-agnostic manner. It does not use environment modeling, as it requires either a high level of online perception or previous offline modeling by an expert. Therefore, the included robot simulation models only the robot, not the environment. This circumstance eliminates all visualizations that require an environment representation.

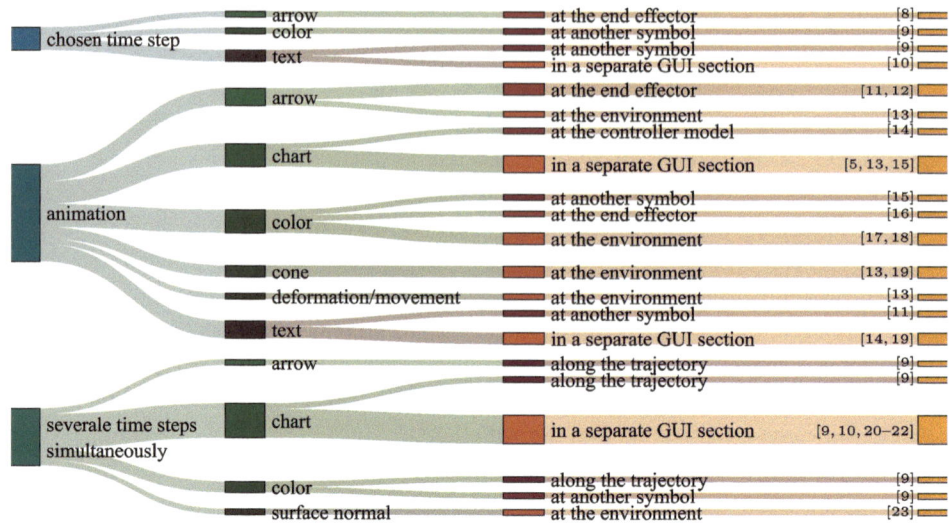

Fig. 1 Classification of reviewed visualizations of forces in robot simulations. The representations are split up by the categories (left to right): time courses, the symbols used and their placement

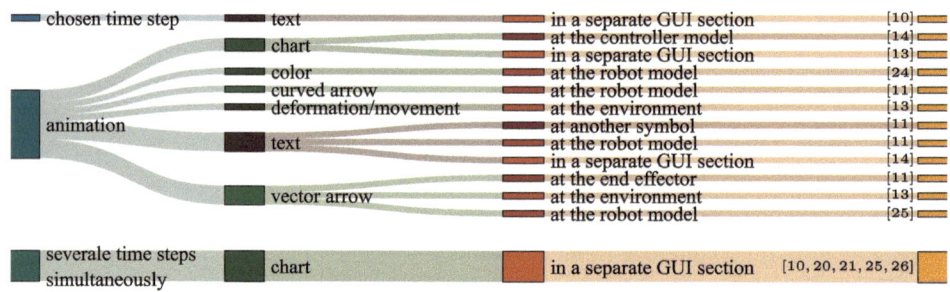

Fig. 2 Classification of reviewed visualizations of torques in robot simulations. The representations are split up by the categories (left to right): time courses, the symbols used and their placement

Additionally, all parameters of the in-contact motion are known as a user directly demonstrates it and modifies it using the GUI. Therefore, we can formulate these demonstrated in-contact motions as hybrid force-/motion-controlled motions [27]. For simplification, our formalization of in-contact motions omits the control part since we only want to visualize them. Using the 6D-vector representation of forces and torques, we get the wrench $\mathcal{F}(t) = (f_x, f_y, f_z, t_x, t_y, t_z)$ at a given time step t of our desired in-contact motion in the robots task frame C:

$$\mathcal{F}(t) = S(t)\mathcal{F}_{\text{position}}(t) + (\mathbb{I} - S(t))\mathcal{F}_{\text{force}}(t) \tag{1}$$

The wrench $\mathcal{F}_{\text{position}} \in \mathbb{R}^6$ generates the movement in the directions, which have no natural constraints by the environment to generate a force or torque (so they are not in-contact). Wrench $\mathcal{F}_{\text{force}}$ represents the desired forces and torques. Matrix $S(t)$ is the compliance selection matrix describing whether a dimension of the wrenches is position- or force-controlled. For the forces and torques, only $(\mathbb{I} - S(t))\mathcal{F}_{\text{force}}$ needs to be visualized. Since we have no modeled environment to use, e.g., for a physics simulation, we simulate all degrees of freedom for $\mathcal{F}_{\text{position}}$ of the motion in every time step t.

Combining the reviewed visualizations (see Sect. 2) and our requirements, possible visualization methods are: an arrow at the end effector, arrows along the trajectory of the end effector, a bar chart along the trajectory of the end effector, or three bar charts aligned in their GUI section, which plotted the forces along the world coordinate axes over time. Both curved and vector arrows are applicable for torques, while forces were only represented via vector arrows. The magnitude of the forces and torques is scaled using the maximum magnitude. The direction depends on the direction of the resulting force vector over all dimensions for the arrows, and for the curved arrows, the rotation axis of the resulting torque (using the right-hand rule). For the diagrams, we need no direction as the dimensions are represented separately. The application (or pivot) point is the origin of the task frame. We choose either a single time step or sampled points on the robot's trajectory for the temporal courses. We refrain from adding text and colors to the representations since they can always be added later. Moreover, an animation has been found to be more effective in conveying the information compared to a textual representation (see Sect. 2).

4   Evaluation

4.1   Pre-study

We used an online questionnaire based on the System Usability Scale (SUS) [28] to gauge the usability of different visualizations for forces and torques. To allow for an assessment of the chosen visualizations (see Sect. 3) without fully incorporating them within the GUI of a robot simulation, each of them was represented by two picture series. Thus, the questionnaire showed snapshots of a linear movement with a constant force or torque in one direction (see Fig. 3) and what attaching the second of four screws along a circular trajectory would look like.

Since we are specifically interested in the visualizations, we modified the statements of the SUS to refer to the visualization instead of the system throughout the questionnaire. Furthermore, "I found the various functions in this system were well integrated." was omitted. Finally, the questionnaire was translated into German. The overall SUS score is robust against these kinds of modifications [29].

In the online questionnaire, 32 participants rated the visualizations, out of whom 21 identified as male and the rest as female. Their age ranged between 19 and 79 years. Most

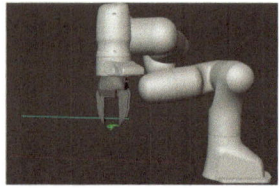 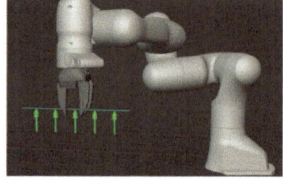 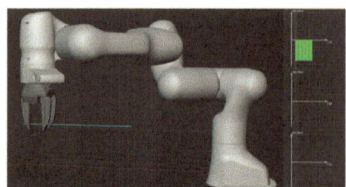

(a) curved arrow at the end effector (b) vector arrows along the trajectory (c) bar charts over time in a separate GUI section

Fig. 3 Visualization snapshots for a linear movement with constant torque

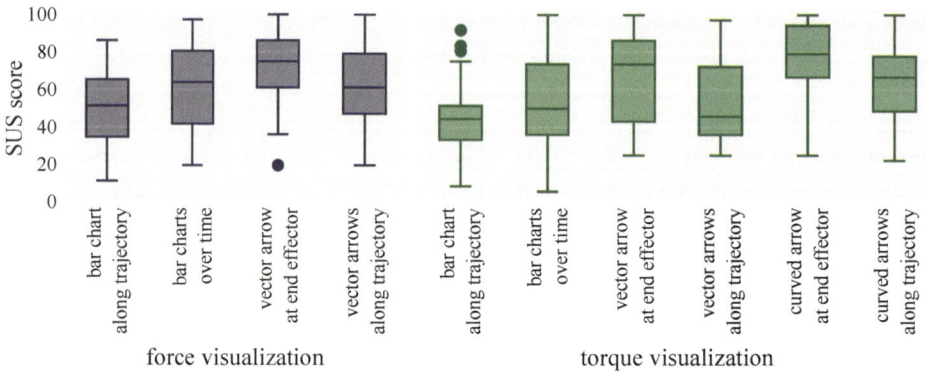

Fig. 4 Box-plot of SUS scores of the force/torque visualizations in the pre-study

participants were non-experts. About 40.5% of the participants reported having no experience with robots. Further 43.8% reported having used robots at least once but not frequently.

Figure 4 shows the resulting SUS scores of force and torque visualizations. Similar visualizations are ranked similarly based on their median scores for both quantities. Notably, the visualizations that reach the highest median scores use an arrow at the end effector. Based on a one-way repeated measures ANOVA the SUS scores of the force visualizations differ significantly ($p \approx 0.0001$). A Tukey HSD test shows that the SUS score of bar charts along the trajectory differ significantly($p_{adjusted} \approx 0.0002$) from the SUS score of vector arrows at the end effector. There is also weak evidence that the SUS score charts along the trajectory differ from those of charts over time ($p_{adjusted} \approx 0.07$) and from those of vector arrows along the trajectory ($p_{adjusted} \approx 0.08$). Similarly, the SUS scores of the torque visualizations differ significantly based on a one-way repeated measures ANOVA ($p \approx 0.0000$). A Tukey HSD test shows that the SUS scores of curved arrows at the end effector differ significantly from the SUS scores of bar charts along the trajectory ($p_{adjusted} = 0.0$), vector arrows along the trajectory ($p_{adjusted} \approx 0.0001$) and bar charts over time ($p_{adjusted} \approx 0.0002$). Furthermore, there are statistically significant differences between the SUS scores of bar charts along the trajectory and of vector arrows at the end effector ($p_{adjusted} \approx 0.002$) as well as between the SUS scores of bar charts along the trajectory and curved arrows along the trajectory ($p_{adjusted} \approx 0.002$).

Other than the described significant results, only tendencies can be derived based on the surveys for the individual aspects, the representation of the temporal courses, the symbols used, and their placement. This is also due to the sample size (N = 32). Overall, the survey showed a clear tendency to prefer arrows over diagrams as symbols for both forces and torques. Concerning the symbol used, the visualization selected for forces via vector arrows at the end effector is consistent. This observation also fits with the results of the study from Sect. 2 on visualization of forces in surgical procedures [7]. For the curved arrows, which should be used to represent torques based on this survey, only one proof of feasibility has been provided so far, in which these arrows were localized differently [11]. The two visualizations rated highest by participants on the SUS, vector arrows at the end effector for forces and curved arrows at the end effector for torques, were based on progressing animations. They received significantly higher scores than visualizations that also use arrows but visualize the complete time course at a glance. These results suggest that animation is best suited for visualizing temporal sequences for robot programming systems.

4.2 User-Study

We conducted a second user study on a real robot system to further investigate the usability of force and torque visualizations to represent in-contact motions. To this end, we integrated arrows at the end effector into the GUI of our robot simulation (see Fig. 5). We used vector arrows for forces and curved arrows for torques as our findings from Sect. 4.1. As these represent only a single time step in the animation, we added continuous line charts over time for both forces and torques in a different section of the GUI to evaluate if a simultaneous overview of the whole movement helps the users. We chose to use line charts over bar charts after receiving feedback from the online questionnaire that suggested they could provide better performance. Thus, we evaluated three visualization combinations using only arrows in animation, line charts, or both. All methods display all forces and torques exerted by the end effector simultaneously.

The user study consisted of three sections, one per combined force/torque visualization. First, participants were asked to recreate two in-contact motions displayed in the GUI by hand guiding a Franka Emika Panda robot. Together these two motions consisted of four parts with different force/torque components, namely no forces and torques, a force along the x-axis, a force with components along both the x- and the z-axis, and a force along the z-axis in combination with a torque around the z-axis, all with respect to the world coordinates. The order of these parts and the sense of direction of the forces/torques varied between sections. For each replication attempt, we recorded the poses of the robot end effector and the forces and torques measured between the flange and end effector using an additional

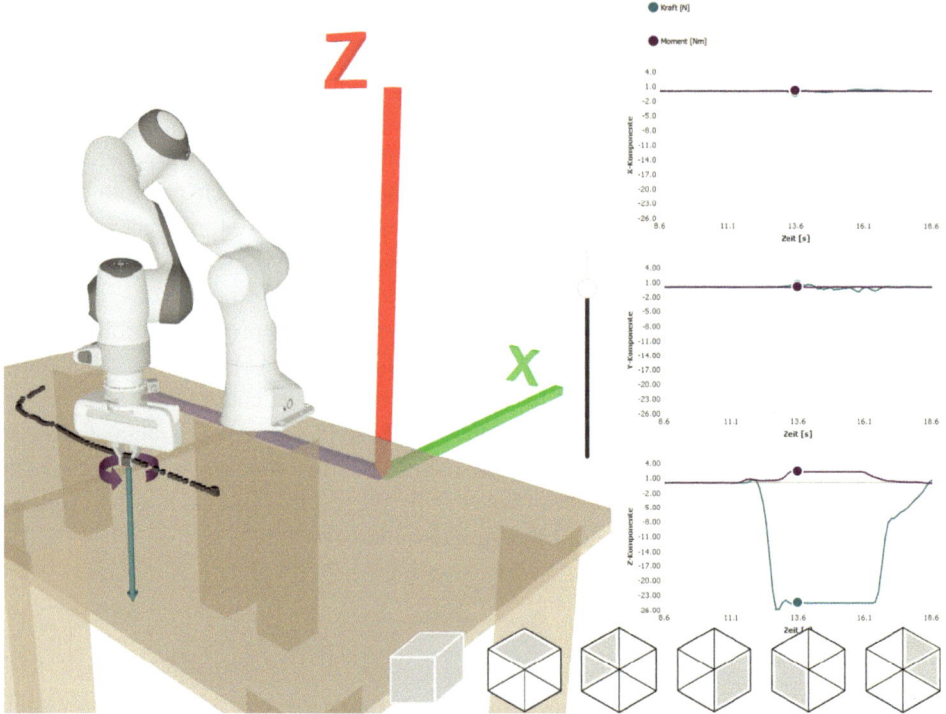

Fig. 5 Force/torque visualization in our robot simulation with arrows and line charts

force/torque sensor. At the end of each section, participants answered the SUS questionnaire [28] translated to German but otherwise unmodified.

Half of the 12 participants identified as female, and the other half as male. Their age ranged from 20 to 28 years. Most participants were non-experts. Seven participants reported having no experience with robots. Another three participants reported having used robots at least once but not frequently.

The recorded forces and torques revealed that participants only applied forces and torques correctly in 60.6% of the in-contact sections where forces/torques were displayed. At about 45.7%, the correct reproductions rate is also relatively low for the torques around the z-axis. In some cases, participants applied forces and torques along the correct axis but in the wrong direction. As shown in Fig. 7 this happened for all combinations of forces and visualizations but was also more prevalent for torques than for forces. The rate of correctly recreated forces and torques differed between visualization methods. Only displaying arrows at the end effector produced the highest rate overall and in terms of the single forces/torques tested. This rate is also higher for tasks displaying charts and arrows than for only line charts. This observation indicates that arrows are better suited to visualize the forces and torques of in-contact motion than charts.

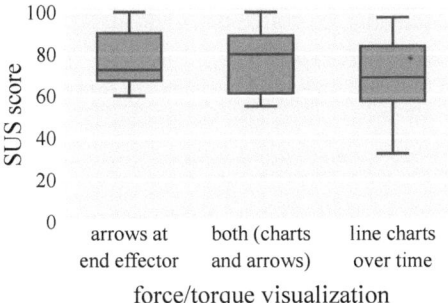

Fig. 6 Box-plot if SUS scores of force/torque visualizations. Arrows consist of vector arrows for forces and curved arrows for torques

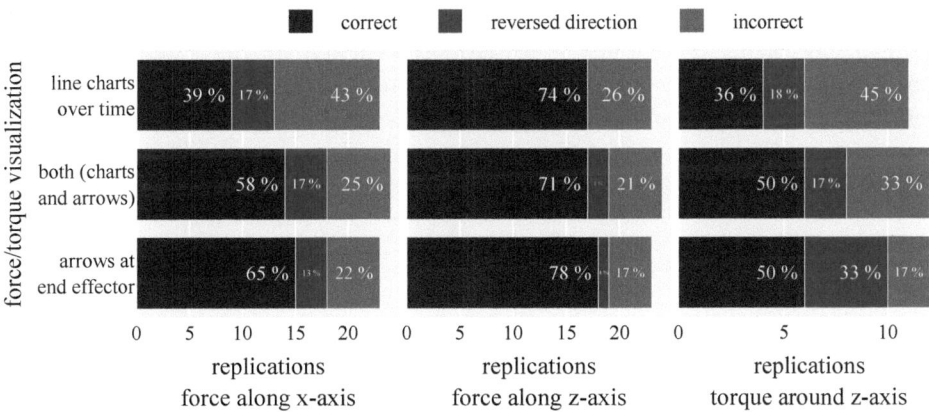

Fig. 7 Correctness of the forces and torques applied by hand guiding the robot based on the visualized in-contact motions excluding replications where the transmission of measured forces/torques was interrupted

Considering the absolute SUS score (see Fig. 6), it is high for arrows and can be rated as acceptable [30]. Thus, the perceived usability is to be rated high. On the other hand, the connection between actual usability and intuitiveness needs to be further investigated because of the comparatively low correct execution rate, especially for torques (see Fig. 7). Here, it should be noted that the participants had no feedback on task completion. In addition, the experiment showed that the execution of the task is also motorically challenging.

An analysis of the SUS scores of the visualization methods confirmed this trend. A one-way repeated ANOVA test showed significant differences between the SUS scores of the three force/torque visualizations ($p \approx 0.036$). However, a pairwise comparison of the SUS scores of these visualizations using a Tukey HSD test did not reveal statistically significant differences. Nonetheless, as shown in Fig. 6 there is a clear trend that visualizations including arrows perform better than the visualization only using charts. Interestingly, less information (arrow only, single time step only vs. both (charts and arrows)) yields to similar results. This leads us to the design decision for our programming concept: For the visualization, we use arrows in the simulation and add an expert mode with line graphs.

5   Conclusion

Through two user studies, we evaluated supportive visualizations of forces and torques for robot programming systems. As a result, we provide a visual method with the highest usability and intuitiveness for non-experts shown by our studies. Users perceive the visualization as highly usable, but the task performance shows that intuitiveness can still be improved. Future work may include integrating these visualizations in a skill-based visual programming framework (e.g. [31]) to validate our conclusions. Furthermore, we could investigate the gap between perceived usability and task completion if we repeat the experiment and provide the participants feedback on whether the task was successful. Finally, utilizing a less mathematical representation of forces and torques, e.g., using intuitive words (up-down, left-right) instead of axis labels, the intuitiveness of our approach could be increased even further.

References

1. Kildal, J., et al.: Potential users' key concerns and expectations for the adoption of cobots. Proc. CIRP **72** (2018)
2. Villani, V., et al.: Survey on human-robot collaboration in industrial settings: safety, intuitive interfaces and applications. Mechatronics **55** (2018)
3. Riedl, M., Henrich, D.: A fast robot playback programming system using video editing concepts. In: Tagungsband des 4. Kongresses Montage Handhabung Industrieroboter. Springer, Berlin (2019)
4. Ebrahimi, A., et al.: Real-time sclera force feedback for enabling safe robot-assisted vitreoretinal surgery. In: 2018 40th Annual International Conference of the IEEE Engineering in Medicine and Biology Society (EMBC) (2018)
5. Yusof, A.A., Kawamura, T., Yamada, H.: Evaluation of construction robot telegrasping force perception using visual, auditory and force feedback integration. J. Robot. Mechatron. **24**, 6 (2012)
6. Rieber, L.P., Tzeng, S.-C., Tribble, K.: Discovery learning, representation, and explanation within a computer-based simulation: finding the right mix. Learn. Instruct. **14** (2004)
7. Rodrigues, S.P., et al.: Influence of visual force feedback on tissue handling in minimally invasive surgery. British J. Surg. **101** (2014)
8. Liu, H., et al.: Interactive robot knowledge patching using augmented reality. In: 2018 IEEE International Conference on Robotics and Automation (ICRA) (2018)
9. Leutert, F., Schilling, K.: Augmented reality for telemaintenance and -inspection in force-sensitive industrial robot applications. IFAC **48** (2015)
10. Toz, M., Kucuk, S.: Dynamics simulation toolbox for industrial robot manipulators. Comput. Appl. Engin. Educat. **18** (2010)

11. Hulin, T., Hertkorn, K., Preusche, C.: Interactive features for robot viewers. In: Intelligent Robotics and Applications. Springer, Berlin (2012)
12. Ortmaier, T., et al.: Robot assisted force feedback surgery. In: Advances in Telerobotics. Springer, Berlin (2007)
13. Miller, A.T., Allen, P.K.: Graspit!. A versatile simulator for robotic grasping. IEEE Robot. Automat. Mag. 11 (2004)
14. Williams, L.E.P., et al.: Kinesthetic and visual force display for telerobotics. In: Proceedings 2002 IEEE International Conference on Robotics and Automation (2002)
15. Talasaz, A., Trejos, A.L., Patel, R.V.: The role of direct and visual force feedback in suturing using a 7-DOF dual-arm teleoperated system. IEEE Trans. Hapt. **10** (2017)
16. Akinbiyi, T., et al.: Dynamic augmented reality for sensory substitution in robot-assisted surgical systems. In: 2006 International Conference of the IEEE Engineering in Medicine and Biology Society (2006)
17. Erickson, Z., et al.: What does the person feel? Learning to infer applied forces during robot-assisted dressing. In: 2017 IEEE ICRA (2017)
18. Haouchine, N., et al.: Vision-based force feedback estimation for robot-assisted surgery using instrument-constrained biomechanical three-dimensional maps. IEEE Robot. Automat. Lett. **3** (2018)
19. Vogl, W., Ma, B.K.L., Sitti, M.: Augmented reality user interface for an atomic force microscope-based nanorobotic system. IEEE TNANO **5**, 4 (2006)
20. Fonseca Ferreira, N.M., Tenreiro Machado, J.A.: ROBLIB: an educational program for robotics. In: IFAC Proceedings Volumes, vol. 33 (2000)
21. Gil, A., et al.: Development and deployment of a new robotics toolbox for education. Comput. Appl. Engin. Educat. 23 (2015)
22. Tenreiro Machado, J.A., Galhano, A.M.S.F.: A program for teaching the fundamentals of robot modelling and control. In: IFAC Proceedings Volumes. Elsevier (1994)
23. Quintero, C.P., et al.: Robot programming through augmented trajectories in augmented reality. In: 2018 IEEE/RSJ IROS (2018)
24. Manring, L., et al.: Augmented reality for interactive robot control. In: Dervilis, N. (ed.) Special Topics in Structural Dynamics and Experimental Techniques, vol. 5. Springer, Cham (2020)
25. Rossmann, J., Jung, T.J., Rast, M.: Developing virtual testbeds for mobile robotic applications in the woods and on the moon. In: 2010 IEEE/RSJ International Conference on Intelligent Robots and Systems (2010)
26. Gupta, V., Chittawadigi, R.G., Saha, S.K.: RoboAnalyzer: robot visualization software for robot technicians. In: Proceedings of the Advances in Robotics. Association for Computing Machinery, New York (2017)
27. Raibert, M.H., Craig, J.J.: Hybrid position/force control of manipulators. J. Dyn. Syst., Measur., Control **103** (1981)
28. Brooke, J.: SUS-A quick and dirty usability scale. Usabil. Evaluat. Indust. **189**, 194 (1996)
29. Lewis, J.R.: The system usability scale: past, present, and future. Int. J. Human-Comput. Interact. **34** (2018)
30. Bangor, A., Kortum, P.T., Miller, J.T.: An empirical evaluation of the system usability scale. Int. J. Human-Comput. Interact. **24** (2008)
31. Riedelbauch, D., Sucker, S.: Visual programming of robot tasks with product and process variety. In: Annals of Scientific Society for Assembly, Handling and Industrial Robotics 2022 (to appear) (2022)

A Layered Pipeline for Natural Language Robot Programming with Control Structures

Sascha Sucker and Dominik Henrich

Abstract

Natural language is an intuitive interface to supplement programming in modern automation settings. However, most natural language frameworks are specialized and not universally applicable. We contribute a novel layered pipeline that transforms the instructions of laypersons into robot programs with non-linear control flow and that facilitates reuse. The instructions are analyzed regarding grammatical features. From this, the syntactical analysis derives programs with nested control structures and references to physical parts within a scene to be manipulated. These programs are semantically interpreted during online execution in a concrete scene—i.e., the control structures are evaluated, and part specifications are grounded to physical parts. With that, a fully specified skill is created and executed by a robot system. Since only the input/output interface of each pipeline stage is defined, they are adjustable independently of each other. Our experiments demonstrate how industrial robots in diverse domains can be verbally programmed using the pipeline.

Keywords

Dependency grammar • Modular architecture • Intelligent robots • End-user programming • Syntax • Semantics

S. Sucker (✉) · D. Henrich
Chair for Robotics and Embedded Systems, University of Bayreuth, Universitätsstraße 30, 95447 Bayreuth, Germany
e-mail: sascha.sucker@uni-bayreuth.de
URL: https://robotics.uni-bayreuth.de

D. Henrich
e-mail: dominik.henrich@uni-bayreuth.de

S. Ihlenfeldt et al. (eds.), *Annals of Scientific Society for Assembly, Handling and Industrial Robotics 2023*, https://doi.org/10.1007/978-3-031-74010-7_16

1   Introduction

Shorter innovation cycles and small-batch production pose significant challenges for automation [4]. One response is to increase the accessibility of robot programming [17, 18, 24]. We envision future programming to feel equally natural as instructing human co-workers. Inspired by the main human interaction method [23], we follow the approach of *natural language programming* (NLP). However, the few currently existing NLP frameworks in robotics are limited to specific applications and are thus not universally applicable.

Therefore, we aim to advance NLP for industrial robots (Fig. 1). Compared to mere robot commanding our notion of programming differs in the utilization of control structures, thus, allowing the control flow of the program to be adapted online. We distinguish between the typically used control structures: Sequences, selections, and loops. The latter is further subdivided into number-, condition-, and set-controlled loops. Other typical control structures are not explicitly considered, as they often can be composed of the control structures above (e.g., switches or post-test loops). To facilitate a high degree of flexibility and quick adaptations to the desired domain, the approach must be modular. Thus, the programming system should be divided into independent sub-components connected with clearly defined interfaces. This can be achieved using a pipeline architecture. The transformation of the spoken instruction into robot movements traverses different layers of abstraction—hence, resulting in a layered pipeline.

We present a layered pipeline for natural language robot programming (Fig. 3). Our contribution is twofold: (i) We transfer the concepts of natural language programming using control structures to the context of industrial robots. (ii) Owing to the modular design, individual pipeline stages can be adjusted with little effort, facilitating reuse and quick adaptation to the desired domain.

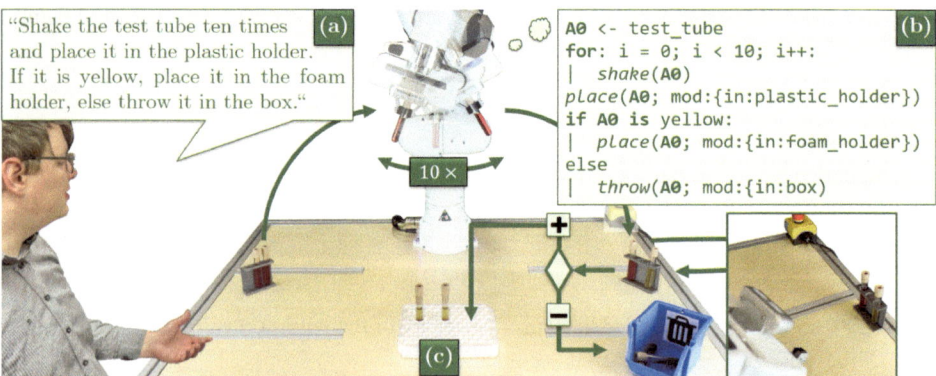

Fig. 1 Natural language (**a**) is transformed into programs with control structures (**b**). Consecutively, the robot applies operations to the current scene (**c**)

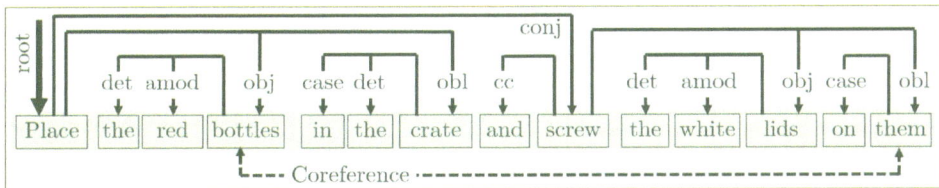

Fig. 2 The dependency tree encodes grammatical connections between words in a sentence. These dependencies include determiners (det), adjective modifiers (amod), objects (obj), obliques (obl), and conjunctions (conj). Coreferences specify groups of words referring to the same thing

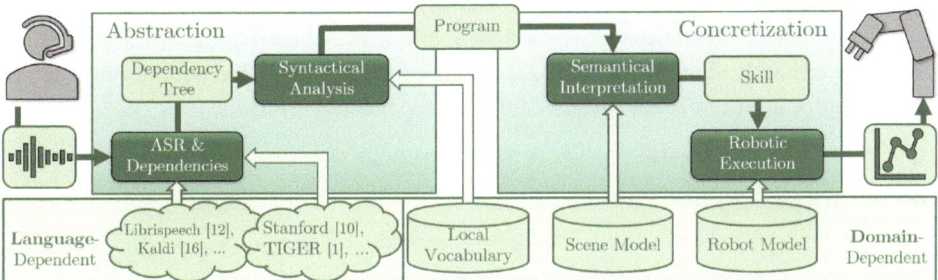

Fig. 3 The layered pipeline transforms spoken commands into fitting robot movements: Such commands are converted to text, whose grammatical dependencies are analyzed. These are parsed into a program with control structures. The program is interpreted for a scene to create skills, which are executed by the robot

2   Related Work

A common method of robot commanding is through the usage of skills [14], e.g., with graphical [18] or natural language programming. Regarding NLP, the instruction is interpreted by determining the skill from the verb and deriving parameters from its valences, i.e., modifiers of the verb [6, 15, 21, 24]. Since the execution's control flow is not adapted, this approach alone does not allow the creation of complex programs. Nevertheless, we build upon this approach in order to define the operations embedded within the control structures (Sect. 3.1).

The morphology and grammatical dependencies of a naturally instructed sentence can be analyzed to derive control structures: For example, conjunctions of main clauses can be converted into sequences with the help of feasibility calculations [9]; selections are derived from the conditional member clauses [20]; condition- and amount-controlled loops can be generated from function words like "while" or from mentioning the number of repetitions [8, 11]. Set-supervised loops can be programmed by defining the set of parts to be manipulated [5]. However, these approaches only focus on a few distinct control structures.

In contrast, analyzing the dependency tree of the instruction is a promising approach to derive all relevant control structures (as shown in [7, 22]). Dependency trees represent the syntactical make-up of an instruction by connecting words based on their grammatical relation (cf. Fig. 2) [10]. Starting from a root word, dependencies (e.g., subject, adjective, conjunction) are established to other sentence constituents. Through this, grammatical composition can be analyzed uniformly. Coreferences identify references between mentions of the same identity [3], e.g., in Fig. 2 'them' refers to 'bottles'.

In this paper, we expand on the work of Landhäußer et al. [7] and Weigelt et al. [22] in the context of flexible robot programming. So far, their approach has only been exemplarily applied to two specific robotic systems (humanoid/mobile robot). It neither includes a conceptual pipeline for natural language robot programming nor provides an adequate interface to cover diverse robot domains. Additionally, the grounding of underspecified part specifications to concrete parts (as discussed in our previous work [18]) is implemented prototypically only. This work addresses these problems to enable natural language programming for arbitrary robot systems utilizing dependency analysis.

3   Approach

The goal of our layered pipeline is to transform spoken instructions into robot movements which perform the desired manipulation of parts within the scene (Fig. 3). This pipeline subdivides into abstraction and concretization stages. The former converts the spoken instructions into a program. Consequently, the abstraction includes Automatic Speech Recognition (ASR), Dependency Parsing, and Syntactical Analysis. We regard those stages as abstraction since information is primarily stripped from the instruction—ASR removes the explicit octave of the speech, and syntactical analysis may extract equal program statements from different sentence structures. The concretization stages interpret the program, thus transforming it into the desired robot motions for a concrete scene. To further increase modularity, the concretization has two stages: During Semantical Interpretation, the program is gradually interpreted online yielding robot skills, which the Robot Execution stage converts into concrete trajectories. Hence, the program is transformed to increasingly more explicit representations.

In more detail, the instructions are transformed by this pipeline as follows: The ASR stage transcribes the spoken audio signal into text. This text then is parsed with regard to its grammatical dependencies (Sect. 2). Program code is synthesized by exploring these dependencies and matching keywords to the local vocabulary during the Syntactical Analysis (Sect. 3.1). The instruction provides an implicit execution flow that must be transformed into concrete control structures with associated robot operations and part specifications. The resulting program is scene-independent, allowing it to be used with different task variants, e.g., by not having to precisely define the position of the individual parts in advance. Accordingly, the program is executed by providing a scene state during the Semantical Interpretation (Sect. 3.2), wherein the underspecified part specifications and conditionals are grounded to physical parts. The specific robot skill is then determined based on the grounding and the previous program state. Utilizing the skill definition of Pedersen et al. [14], hardware independence is guaranteed during this stage. Finally, the skills are transformed into concrete motion sequences that achieve the desired manipulation of the scene.

Both ASR and Dependency Parsing are studied extensively within computational linguistics. Thus, optimized methods are readily available. Robot Execution mostly follows the approach of executing skills [14]. In this paper, we therefore mainly cover the Syntactical Analysis and Semantical Interpretation.

3.1 Syntactical Analysis

The Syntactical Analysis converts dependency trees into programs (Fig. 4). We require such programs to represent the instructed parts, operations, and control structures. Based on this, the program is divided into declaration and procedure sections. The *declaration* lists all named part specifications $\tilde{P}$ and assigns unique labels to each of them. We call each declaration list entry a variable, distinguishing between atomics and compounds. *Atomic variables* always refer to one part specification, whereas *compounds* link several atomic or other compound variables with one common operator ('and', 'or'). Depending on the domain, part specifications $\tilde{p} \in \tilde{P}$ can include features such as the type p_t, number p_n, color p_c, or location p_l (see Sect. 3.2), resulting in the tuple

$$\tilde{p} = (p_n, p_t, p_c, p_l). \tag{1}$$

The *procedure section* contains the control flow and operations that access the declared part variables. We group the operations and the control structures into the superordinate term *statement*. Each operation is defined by its type, part variable to be manipulated, and other instructed modifications that describe the process in more detail. Programs are created by analyzing the dependency tree. Starting from the root node, the dependencies linking nodes within the tree are transformed into program constituents based on their

"Place three blue bottles in the crate and screw the white lids on them."
------------- Declaration -------------
A0 <- bottle(num: 3, col: blue)
A1 <- lid (num: *, col: white)
------------- Procedure -------------
place(A0; mod: {1:field, to:right})
screw(A1; mod: {on:A0})

Depend.	Mark	Control Structure
conj	And	Sequence
sconj	If	Selection – Positive
parataxis	Instead	Selection – Negative
sconj	While	Conditional-loop
sconj	Until	Conditional-loop (Neg.)
obl	Times	Count-loop
...	...	...

(a) Example program.   (b) Excerpt of the control structure table.

Fig. 4 Programs are synthesized from instructions utilizing their grammatical dependencies (**a**). The control structures connecting statements are derived from a table of grammatical dependencies and function words (**b**)

mark (e.g., 'if' within a subordinated conjunction). The operation is identified by the verb node and its dependencies [21]. These include the parts to be moved (accusative object) and other specifying parameters (e.g., adverbs or obliques). A part specification is created if a dependency relates to a nominal node whose identifier is assigned to a part type in the language model. By analyzing its connected dependencies, additional part features are extracted (e.g., adjectives, numerals, or compounds).

A table of dependencies contains the control structures between statements with associated function word markings (Fig. 4). This method is reasonable since only few distinct function words exist (e.g. 'if', 'while', 'until') and they are rarely altered [2], resulting in a table with few entries. Several statements are linked to a sequence employing chained conjunctions of main sentences with the marker 'and' (coordinating conjunction dependency). If the user instructs several sentences, the determined partial programs are likewise strung together as one sequence. However, humans rarely structure instructions concisely without ambiguities or contradictions so that the desired instruction sequence does not necessarily correspond to the one spoken [13]. Therefore, the feasibility of the possible orderings must be considered [9]. Selections are identified by the dependency subordinated conjunction (sconj) and the associated marker 'if'. The condition is derived from the subordinate sentence and its positive case from the main sentence. If there is additionally a parataxis dependency on the main clause (e.g., with an 'instead' marker), the subordinate clause is considered the negative case. Conditional loops are determined analogously to the selections using, e.g., the marking 'while', whereby the main clause corresponds to the contents of the loop body. Repetitions of an operation ("Press it five times") are identified using an oblique dependency resulting in an amount loop. Set-loops are not derived directly from this table since they are inherently encoded in the part specifications by defining the required number of parts ("Move four cubes"). The operation performs similarly for all specified parts resembling a set-loop.

Occasionally, a part should be manipulated multiple times in succession. This can be taken into account by analyzing the coreferences as follows: Only references between nominals related to parts are considered, though one compound may also combine multiple references to such nominals. We obtain this set of part-related nominals $\mathcal{N}$ by analyzing the dependency tree and the local vocabulary, where each id refers to the word at the corresponding position in the instruction. We use the set of coreference chains $\mathcal{K}$, where one coreference chain $k = (k_{\mathrm{msm}}, k_{\mathrm{ref}})$ contains two word id sets: k_{msm} (the most specific mentions) and k_{ref} (all references to k_{msm}). We define the word id set of all atomic mentions

$$\mathcal{A} = \{w \in \mathcal{N} \mid (\forall k \in \mathcal{K} : w \notin k.k_{\mathrm{ref}})\} \tag{2}$$

as a subset of $\mathcal{N}$, whose ids are not referenced in any coreference ($k.k_{\mathrm{ref}}$). Every atomic mention is transformed into an atomic variable with corresponding part specification (Eq. 1). Compounds link atomics or other compounds via one conjunction. Accordingly, $\forall k_c \in \mathcal{K}_c \subseteq \mathcal{K}$ with

$$\mathcal{K}_c = \{k \in \mathcal{K} \mid |k.k_{\mathrm{msm}}| > 1\} \tag{3}$$

one compound must be formed by linking the appropriate atomics or compounds of the $k.k_{\mathrm{msm}}$. The remaining conjunctions of atomics and compounds without coreferences should also be usable as contiguous variable in the operation. For example, a sentence 'Move the bottle and lid to the left' should result in two atomic ('bottle', 'lid') and one compound variables. Therefore, such conjunctions must also be linked within a compound variable. Following the outlined stages, we can transform natural instructions into programs with nested control flow.

3.2   Semantical Interpretation

In this stage, the program is interpreted within a scene resulting in robot skills with concrete parameters. For this purpose, the program is processed sequentially, and the part specifications in the current program stage are grounded (Fig. 5). Operations are converted to concrete skills utilizing grounding as introduced in our previous work [18]. Grounding refers to assigning physical parts to part specifications that are associated to atomic variables. If a grounding was already found for one variable and is referenced, the existing grounding result is reused, allowing repeated manipulations. The parts are grounded dynamically during the execution, i.e., only a subset of parts in the scene are grounded at one execution stage. This is for two reasons: (i) The scene state and, thus, future assignments may change; and (ii) the parts specified in the body of a selection must only be present if its condition resolves.

The operation is converted to a concrete skill utilizing grounding [18]. The current world state $\hat{P} = \{\hat{p}_0, \hat{p}_1, ...\}$ is the set of the part states in a given scene. One state encompasses the relevant and identified features of the part (e.g., geometrical shape, color, or position). In contrast, part specifications $\tilde{P}$ define boundary conditions that part states must satisfy in the context of an operation execution. To use the concept from [18] we convert every specification into p_n equal part templates with the same properties. Such part templates $P = \{p_0, p_1, ...\}$ are defined equal to the specifications without the amount p_n. Therefore,

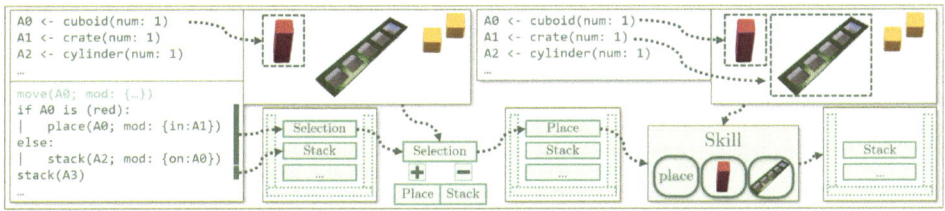

Fig. 5 The program is interpreted and consecutively transformed into precise skills, partially grounding the part specification on demand. Here, after the first operation (move) the grounding for 'A0' is reused to interpret the selection

we utilize the function $\gamma : P \to \hat{P}$, which maps all templates injectively to a part state, where each mapping must suffice the function $\sigma : \hat{P} \times P \to \{ \text{TRUE}, \text{FALSE} \}$. Function σ return whether a part state $\hat{p}$ satisfies the boundaries defined by part template p. Thus, for each part template p, exactly one state $\hat{p}$ must be found that satisfies the boundaries of p. This general construction allows grounding to be performed on arbitrary domains. Since a local grounding is performed in this case, the states can be assigned to the templates in a greedy manner. Grounding allows skills to be filled with concrete parameters.

For example, a template $p = (p_t, p_c, p_l)$ might contain information about the geometric type p_t, the color p_c, and location p_l (based on Eq. 1). A corresponding state $\hat{p} = (\hat{p}_t, \hat{p}_c, \hat{p}_l)$ is analogous in structure. Hereby, named features (such as type or color) can be described as an entry within a taxonomy. Such a taxonomy captures "is-a"-relations between a set of nodes T. Leaf nodes $\hat{T} \subset T$ denominate *concrete features* which part state may exhibit. When ascending from leaf nodes upwards towards the root node, encountered inner nodes encode increasingly abstract descriptions. Thus, such inner nodes may occur exclusively in similarly abstract part templates. An example of a taxonomy of part types in the palletizing domain might include the term 'box', which covers products ranging from small tea boxes to large packages. The set of part locations $L = L_s \cup L_u$ includes well-specified affine transformations (L_s) and underspecified constraints on the part transformation (L_u). For example, an underspecified location may describe an area in scene space in which a part should be present. Each location $l \in L$ is associated with a *location function* is_at $: \hat{p} \times l \to \{\text{TRUE}, \text{FALSE}\}$, which outputs whether the pose of $\hat{p}$ matches the location l. Analogously to named features, part states may only exhibit locations $l_s \in L_s$, while templates underlie no such restriction. In this domain, the satisfies function σ would correspond to

$$\sigma(\hat{p}, p) = \text{is\_a}^{\text{type}}(\hat{p}_t, p_t) \wedge \text{is\_a}^{\text{color}}(\hat{p}_c, p_c) \wedge \text{is\_at}(\hat{p}, p_l). \tag{4}$$

The program is executed utilizing a call stack analogous to high-level programming languages. Statements are pulled successively, where operations are transformed into skills and control structures manipulate the call stack depending on their type. Sequences put their statements in reverse order on the stack, making the first statement in the sequence the next pulled from the stack. Selections push either the positive or negative case based on their condition evaluation. If the head of a loop resolves, the loop is pushed again along with its body. Thereby, the body is pulled next, and the loop head is rechecked afterward. The resulting grounded skills can be executed on any suitable robot system.

4   Experimental Validation

We designed and modeled four benchmark tasks to highlight specific aspects of natural language programming using our proposed architecture (Fig. 6): In *Task T1*, 'labeled' is an additional binary feature of part states corresponding to a taxonomy with an agnostic root node and two leaf nodes. The subordinated conjunction with 'if'-marking is transformed into a selection with the 'labeled' condition and positive/negative branches. This task shows that the definition of parts can be adapted by adding or removing features. With *Task T2*, an electrical connection is completed within an assembly benchmark toolkit [19]. One arbitrary conductor and one resistance conductor must be inserted in a pre-assembled electrical circuit. The general concept of our part types allows part states with unique naming. Therefore, the goal locations of the conductors can be uniquely identified by directly addressing the plates (e.g., 'P0' and 'Q0'), bypassing the possibly ambiguous grounding of part states. However, this task also highlights the drawback of local grounding: Based on the type taxonomy, 'resistance' is the child of the 'conductor' type. For example, the resistance may be mistakenly grounded for the first operation, leaving no matching part for the second operation. *Task T3* shows the instruction of a for-loop within a laboratory domain. Due to our general concept, skills more complex than pick-and-place can be defined (e.g., 'shaking' and 'emptying' the measuring cylinder). This relates both to the required robot movements and the applied manipulation to the parts (e.g., a 'mixed'-feature may be increased during shaking). *Task T4* displays a condition-loop in a service domain by instructing the cleaning of a whiteboard and the extension to operations with force-control. Our prototype could transform the benchmark instructions into correct programs that were successfully executed on our robot system (Fig. 6). Each task is situated within its unique domain and utilizes different control structures—thus, demonstrating the flexibility and adaptability of our approach.

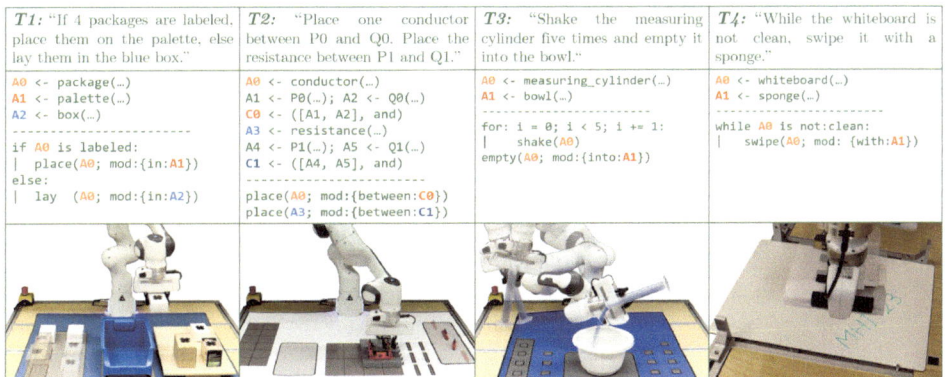

Fig. 6 Four benchmark tasks highlight the system's capabilities. The first row shows the instruction(s), which are transformed into programs (second row). The third row displays a robot executing the instruction in the context of a corresponding domain

5 Conclusion and Future Work

In this paper, we contributed a modular pipeline for natural language robot programming by laypersons. We achieved this by analyzing the grammatical speech patterns within the implicit instructions and transforming them into programs with explicit control structures (Sect. 3.1). Thus, we synthesized operations and part specifications embedded within control structures from language. By grounding these specifications, we can parametrize the operations to be suitable for industrial robots (Sect. 3.2). We showed the high adaptability of our pipeline by customizing it for diverse domains (Sect. 4).

Our approach may be extended in future work: (i) Currently, interpretation errors are propagated through the pipeline without mechanisms to seek user feedback. A dialog component may resolve this error propagation. (ii) Additionally, we defined the input/output within the pipeline to be human-readable. Thus, intermediate results may help by error detection and correction. (iii) The local grounding may lead to additional errors (Sect. 4) and must therefore be extended to a global grounding approach [18]. (iv) Furthermore, behavioral programming can be incorporated into natural language such that the programming more closely resembles the training of human co-workers. (v) Finally, we plan to compare the usability of our prototype with other programming systems.

Acknowledgements This work was partly funded by the Deutsche Forschungsgemeinschaft (DFG) under grant agreements He2696/18 (VerbBot). The authors would like to thank Carsten Scholle (B.Sc.), Philipp Jahn (B.Sc.), and Siegfried Köhler (M.Sc.) for their valuable assistance in implementing and testing the demonstrator.

References

1. Brants, S., et al.: TIGER: linguistic interpretation of a German corpus. J. Lang, Comput (2004)
2. Carnap, R.: The Logical Syntax of Language. Routledge & Kegan Paul (1937)
3. Clark, K., Manning, C.D.: Deep reinforcement learning for mention-ranking coreference models. In: EMNLP, Association for Computational Linguistics (2016)
4. Dietz, T., et al.: Programming system for efficient use of industrial robots for deburring in SME environments. In: German Conference on Robotics, Munich, DEU (2012)
5. Han, J., et al.: Structuring human-robot interactions via interaction conventions. In: IEEE International Conference on Robot and Human Interactive Communication (RO-MAN) (2020)
6. Knoll, A., et al.: Instructing cooperating assembly robots through situated dialogues in natural language. In: IEEE International Conference on Robotics and Automation (ICRA) (1997)
7. Landhäußer, M., Hug, R.: Text understanding for programming in natural language: control structures. IEEE/ACM International Workshop on Realizing Artificial Intelligence Synergies in Software Engineering, Florence, ITA (2015)
8. Lauria, S., et al.: Mobile robot programming using natural language. Robot. Autonom, Syst (2002)
9. Liu, R., Zhang, X.: Generating machine-executable plans from end-user's natural-language instructions. Knowl.-Based Syst. (2018)

10. De Marneffe, M.C., et al.: Universal Stanford dependencies: a cross-linguistic typology. In: International Conference on Language Resources and Evaluation, Reykjavik, ISL (2014)
11. Matuszek, C., et al.: Learning to parse natural language commands to a robot control system. In: Experimental Robotics, Springer, Heidelberg, DEU (2013)
12. Panayotov, V., et al.: Librispeech: an asr corpus based on public domain audio books. In: IEEE International Conference on Acoustics, Speech and Signal Processing (ICASSP) (2015)
13. Pane, J.F., et al.: Studying the language and structure in non-programmers' solutions to programming problems. Int. J. Human-Comput, Stud (2001)
14. Pedersen, M., et al.: Robot skills for manufacturing: from concept to industrial deployment. Robot. Comput.-Integrat, Manufact (2016)
15. Pires, J.N.: Robot-by-voice: experiments on commanding an industrial robot using the human voice. Int. J. Indust, Robot (2004)
16. Povey, D., et al.: The Kaldi speech recognition toolkit. In: IEEE Workshop on Automatic Speech Recognition and Understanding (2011)
17. Riedelbauch, D., Hartwig, J., Henrich, D.: Enabling end-users to deploy flexible human-robot teams to factories of the future. In: IROS 2019 Workshop (2019)
18. Riedelbauch, D., Sucker, S.: Visual programming of robot tasks with product and process variety. In: Annals of Scientific Society for Assembly, Handling and Industrial Robotics (to appear) (2022)
19. Riedelbauch, D., Hümmer, J.: A benchmark toolkit for collaborative human-robot interaction. In: IEEE Interenational Conference on Robot and Human Interection Commmunication (RO-MAN) (2022)
20. Rybski, P. et al.: Interactive robot task training through dialog and demonstration. In: ACM/IEEE International Conference on Human-Robot Interaction. New York (2007)
21. Spangenberg, M., Henrich, D.: Grounding of actions based on verbalized physical effects and manipulation primitives. In: IEEE/RSJ International Conference on Intelligent Robots and Systems (IROS) (2015)
22. Weigelt, S., Hey, T., Steurer, V.: Detection of conditionals in spoken utterances. In: IEEE International Conference on Semantic Computing (ICSC). Laguna Hills, USA (2018)
23. White, L.: THE SYMBOL: the origin and basis of human behavior. In: A Review of General Semantics, ETC (1940)
24. Wölfel, K., Henrich, D.: Grounding verbs for tool-dependent, sensor-based robot tasks. In: IEEE International Symposium on Robot and Human Interaction Communication (RO-MAN) (2018)

Experimentable Digital Twins in the Loop

Jorge Luis Jiménez Aparicio and Jürgen Roßmann

Abstract

Digital twins (DTs) have become prominent in the digitalization era due to the advantages they bring to fields of application like Industry 4.0. In this contribution, we briefly go over the role of experimentable DTs in a V-model-based development cycle of a system, which includes XiL methods. We also provide a proof of concept (in the form of two application examples) of their use in the model-in-the-loop development stage, where the controller model of the system is implemented in real-time capable hardware.

Keywords

Digital twins • Simulation • Model-in-the-loop

1 Introduction

Digital twins (DTs) offer several advantages in the Industry 4.0 era, where a DT follows the life cycle of its physical twin to provide better support and greater insight for tasks like analysis, optimization, maintenance, upgrading, and updating, among others. Here, we use the definition introduced in [16] for the so-called "Experimentable Digital Twins" (EDT) interchangeably with a DT. Virtual Testbeds (VTBs) [14] are advanced 3D simulation

J. L. Jiménez Aparicio (✉) · J. Roßmann
Institute for Man-Machine-Interaction, RWTH Aachen University, Ahornstrasse 55, 52074 Aachen, Germany
e-mail: jimenez@mmi.rwth-aachen.de
URL: https://www.mmi.rwth-aachen.de

J. Roßmann
e-mail: rossmann@mmi.rwth-aachen.de

© The Author(s) 2025
S. Ihlenfeldt et al. (eds.), *Annals of Scientific Society for Assembly, Handling and Industrial Robotics 2023*, https://doi.org/10.1007/978-3-031-74010-7_17

environments that represent all required parameters of the application to simulate a realistic behavior. These VTBs are enhanced with the use of DTs, as it allows an even more accurate description of the agents in the simulated environment.

Before having a functional physical twin, and its corresponding DT, all products or systems have to go through a development cycle. A popular model-based development cycle is the V-model, which was first introduced for software development in [4], was later adopted for mechatronic systems in the VDI 2206 guideline [21], and has continued to evolve to be applied in diverse sectors. One iteration of this model includes development steps where elements of the system are tested in a loop that includes parts of a greater system, or components of the planned target system itself, to have a faster and cheaper overall development process. These are commonly known as X-in-the-loop (XiL), where normally X stands for model, software, or hardware.

DTs and VTBs could be implemented in this development cycle to further increase the benefits inherent to it. This applies generally to many types of development cycles, in which an improved design, early detection of issues, increased efficiency, and better collaboration can be achieved [15, 19] when compared to development cycles that rely solely on physical prototypes.

In this contribution, we aim to analyze the feasibility of integrating DTs into the V-model-based development cycle of a product, which also includes XiL methods. To achieve this, we provide a brief overview of every XiL iteration with DTs and present two application examples as the proof of concept for the use of DTs with model-in-the-loop (MiL) development, along with the use of real-time capable hardware as part of the life data exchange schema.

2 State of the Art

The are several approaches of modeling DTs, most commonly including physics-based, data-driven, hybrid, and reduced-order modeling [2, 9, 12]. However, the majority of these are not based on the V-model, and only a few take into account XiL methods. For instance, the proposed big data-driven model in [17] follows a cyclical circle pattern.

The work in [10] focuses on creating a high fidelity twin of hardware/software interaction in a MiL setup. However, the framework is limited to only MiL from the XiL methods and it is not based on the V-model for development. Furthermore, the work does not involve real-time capable hardware, which is typically seen in in hardware-in-the-loop (HiL) setups, such as the one in [7].

Another study, proposed by the authors in [18], introduces a twin-driven product design framework that employs DTs in the V-model. However, the stages of the V-Model in this framework are more oriented towards overall concepts such as market analysis, customer requirements, and sales among many others; rather than product development with the use of XiL methods.

In the publication [22], DTs are defined as a combination of models and data that creates a virtual copy of a system. The authors explore the use of DTs in engineering dynamic applications and provide an overview of integrating them into the V-model development cycle. They propose a "W-model" for product design, where the initial stage is done virtually with a virtual model, and the DT is only introduced in the later stage of development. From a virtual prototype, they start building a DT. In contrast, our perspective advocates the use of DTs from the outset of the development cycle. We believe that developing DTs from the beginning provides increased benefits such as improved collaboration, early issue detection, and more efficient designs.

The work presented in [20] introduces the Hybrid Testbed as a novel holistic approach to integrate DTs and VTBs into a XiL-based development cycle. The authors propose a modular and scalable framework that includes real data to improve the accuracy and effectiveness of the system. Although our work differs in the sense that we use the V-model as the basis for the development cycle and propose the use of DTs at a more granular level, we share the same goal of achieving a seamless transition from the design stage to the final product. Consequently, our work could serve as a proof of concept for the use of DTs and VTBs in a model-in-the-loop (MiL) setup. Overall, the concepts proposed in [20] align with our approach and provide valuable insights into the integration of DTs in the XiL-based development cycle.

3   Overview of the Model

The V-model is a well-known model-based development cycle used for developing products and systems. It consists of two parallel branches, one for design and one for testing. As shown in Fig. 1, the DT can be integrated into several stages of the V-model as a central component. This means that the DT should be included in these development stages of the model. There are mainly two types of development that can be applied in this context. The first type is for the development of a final product, which is considered as a whole once it is created. Here, the DT should be created before or, at the latest, together with the physical twin. The second type is for the development of a product that is part of a larger system, for example, designing the steering wheel of a vehicle. In this case, DTs of the larger system or its components can already be employed during the creation of the product's DT. In both cases, VTBs with DTs can be used to enhance the development and testing process.

In the following, we give an overview of the various XiL methods defined in [1, 3, 5, 8, 11] and their alignment with our approach of using DTs and VTBS.

Model-in-the-loop (MiL): All components involved are executed in the form of simulation models in development computers. These components are usually not identical to the target system. If a new element of a larger system is being developed, using the DT to represent said system can already provide more accurate integration results of the model and speed up its development. If a new system is being developed, this would be the first step to create

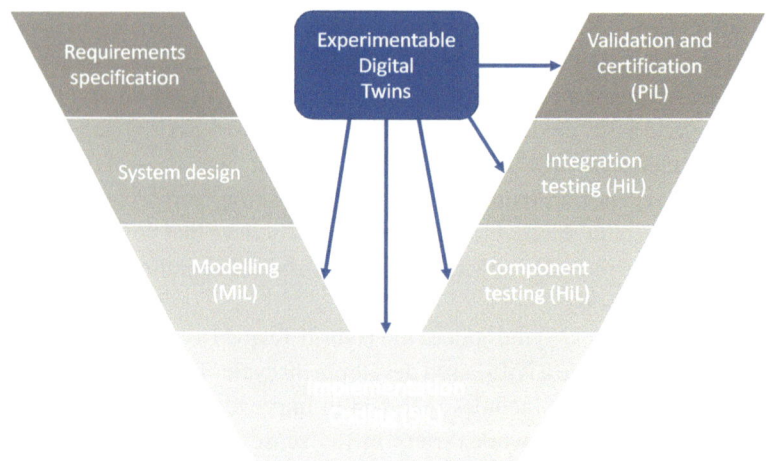

Fig. 1 V-model diagram of a product with DTs integration. DTs are integrated into all XiL methods for development, together with VTBs for testing

a DT of the target system. Our example applications shown in Sect. 5 are based on this part of the development cycle. *Software-in-the-loop (SiL):* Software (e.g. a control algorithm) is designed and developed with the target hardware in mind. It is usually executed on a development computer and connected to the real process or parts of it. Here, the software can be implemented in a DT of the target hardware (e.g. an embedded system) for a more accurate behavior. This arrangement could be connected to a DT of the larger target system or a VTB representing the real process. *Hardware-in-the-loop (MiL):* This improves on SiL by implementing the developed software on the target hardware and testing it with a simulation model of the real process running on a development computer. Here too, the larger system should be represented by its DT or the real process can simulated in a VTB. *Product-in-the-loop (PiL):* A more generalized analogy of the specific term used in the automotive industry Vehicle-in-the-loop (ViL). Here, The entire target system is tested and validated. The DT of the developed product is now part of the larger system's DT and they are tested as a whole. Testing can be done with the help of VTBs, which in turn can have more DTs. If the final product is not part of a larger system, the DT is finalized here and can keep being updated together with the physical twin during its life cycle.

4   Methodology for the MiL Implementation with Real-Time Capable Hardware

In an MiL setup, the DT is used to represent the behavior of a physical system or component in a virtual environment. This virtual environment typically includes other DTs, components, and systems with which the component or system being modeled interacts. By replacing

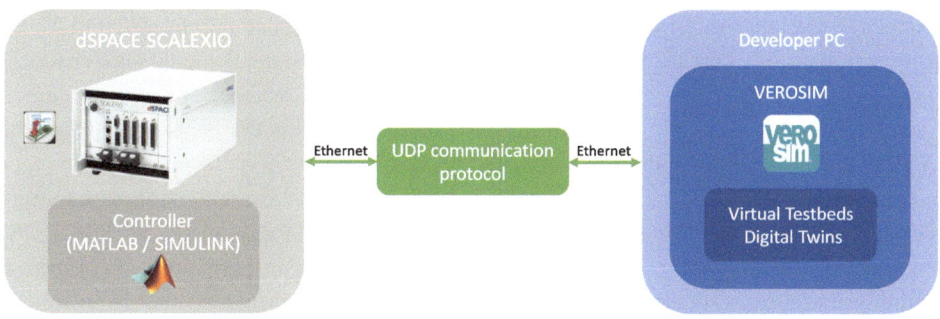

Fig. 2 Concept for the architecture implemented to enable MiL with DTs and VTBs and real-time capable hardware. Examples of the simulation software (VEROSIM) and hardware (Scalexio) are given. Communication between them is handled by an UDP protocol to support real-time communication

the physical system or component with its virtual representation, the DT can be integrated into the MiL setup, enabling simulation and testing of the system or component's behavior in a controlled environment. This methodology allows for a more efficient and accelerated development process, reducing the need for physical prototypes and field testing.

In this section, we describe the concept used to implement the application examples in Sect. 5. Our goal was to develop a modular and reusable architecture for MiL simulations in VEROSIM, our physically accurate 3D simulation software, by incorporating real-time capable hardware. As shown in Fig. 2, this architecture utilizes specific real-time hardware with its standard connection port, but it can be adapted to work with other real-time hardware. The DT in development is simulated within VEROSIM together with other DTs and the VTB.

To enable communication between the DT and the hardware, we have implemented an interface module in VEROSIM, which can be customized to handle different I/O data structures and port connections. We chose to utilize the UDP transport protocol due to its ability to provide the necessary speed for real-time applications, despite its lower reliability compared to the TCP protocol [13]. While there may be a risk of data loss during transfer, this can be mitigated by implementing a more robust controller for the system.

5   Application Examples

This section presents two application examples that demonstrate the use of DTs in MiL. The first one involves a turntable system where we control the angular velocity based on user input. The second involves the Gemini VIII spacecraft, where we control the rendezvous and docking with the Agena target.

Following the architecture shown in Fig. 2, we used the real-time capable hardware Scalexio from dSPACE to run the controller of the target system. The controller model is created in Matlab/Simulink and loaded into the Scalexio. As stated before, we used the UDP connection protocol for the real-time communication between the Scalexio and VEROSIM. Finally, a DT of the target system was running in VEROSIM, whose controlling parameters were given by the Scalexio. Depending on the application, this DT is also deployed within a VTB for testing and validation.

Before setting up the connection with the Scalexio for the application examples, we conducted tests with a simple plant and controller in Matlab/Simulink, with VEROSIM serving as an echo server. The tests revealed no difference in performance compared to running the system within Simulink. However, when we performed the same test with the Scalexio as the controller, we measured a significant behavior change, likely due to information loss inherent to the UDP protocol and minor communication delays. This can be seen in Fig. 3. As a result, we had to make adjustments to the controller parameters when implementing the application examples to compensate for this. We achieved this after some iterations of the parameters, which we rapidly changed and tested within the explained setup thanks to the flexibility provided by the use of DTs instead of physical prototyping.

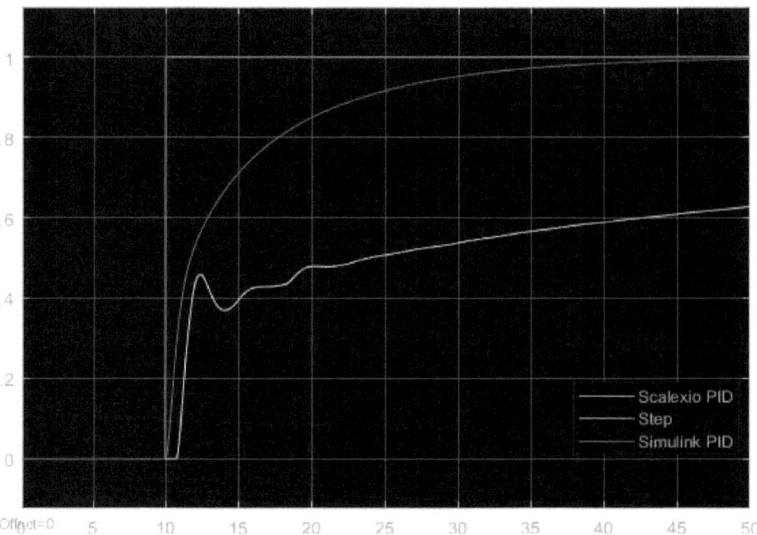

Fig. 3 Step response of a PID controller with a plant in Matlab/Simulink. Reference step response (blue), response with the Simulink controller (orange), and response with the Scalexio controller (yellow)

Fig. 4 Turntable in the 3D
simulation software VEROSIM

5.1   Turntable

The turntable system consists of a turning plate mounted on a square mount (see Fig. 4). Our model includes a laser sensor that measures the presence of an object on the plate. This allows the calculation of an angular velocity error, which is the input of the controller. The Scalexio has a PID controller that will output the torque force to be applied by the turntable's motor to reduce the angular velocity error to zero. The target angular velocity is set by the user.

The turntable setup is shown in Fig. 5. To the left, the behavior of the target's angular speed (cyan), the object detection (orange), and the torque response (yellow) are shown. A small delay in the response of the torque can be observed, likely due to delays in the data transmission. Nevertheless, the PID can successfully control the angular velocity of the turntable. This setup still had room for improvement in performance but served as a first step to test the combination of the Scalexio and a MiL development stage.

5.2   NASA's Gemini Program

As part of NASA's Gemini program [6], the first rendezvous, docking, and extravehicular activity mission was conducted. This mission planned to launch, rendezvous, and dock the Gemini VIII spacecraft with the Gemini-Agena target vehicle (GATV). Figure 6 shows the DT of the Gemini VIII used for this test, together with the DT of the GATV, present in the VTB. Components for sensor and rigid body simulation as well as for orbital dynamics are used within the VTB to simulate a close-to-reality scenario for the DT under development.

The external guidance, navigation, and control (GNC) system was provided by one of our industry partners, Jena Optronik. It was implemented in Matlab/Simulink and simulated the space environment and Lidar measurements for a standard rendezvous and docking

Fig. 5 Simulation of the Turntable system in VEROSIM. Left is the user interface where the target angular velocity can be set and some system variables are shown: target's angular speed (cyan), object detection (orange), and torque response (yellow). Right are the properties of the interface module where the IP, port, and I/O values of the UDP connection are displayed

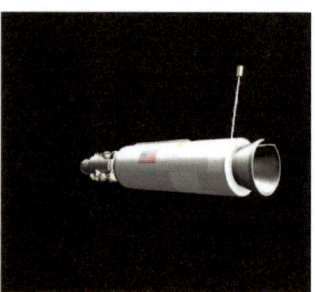
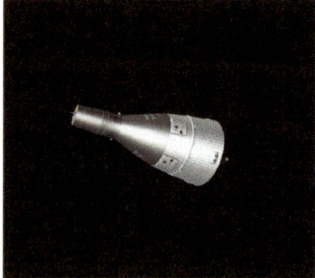

Fig. 6 DTs of the GATV (left) and GEMINI VIII (right)

mission. From this GNC, we were able to extract the necessary control system to implement in the Scalexio, which is depicted in Fig. 7. The control system is composed of the UPD communication blocks, a PD controller, a saturation block, a PWM generator, and low pass filters. The only differences between this version and the one loaded in the Scalexio, are the UDP communication blocks from Simulink, which are replaced by UDP blocks provided by dSPACE for compatibility with the Scalexio.

As the controller was initially designed for a chaser spacecraft docking to the International Space Station (ISS), it required adjustments in the hyperparameters, primarily scaling, to control the thrusters of the Gemini VIII properly. Further modifications were necessary to

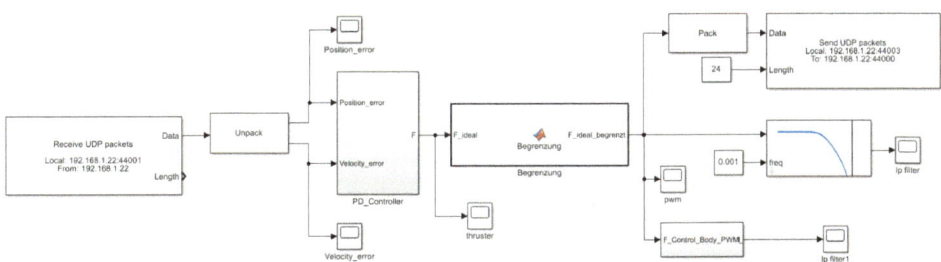

Fig. 7 Simulink model of the controller extracted from the GNC system provided by Jena Optronik. The controller was complemented with UDP communication blocks. The main component is contained in a black box named "Begrenzung"

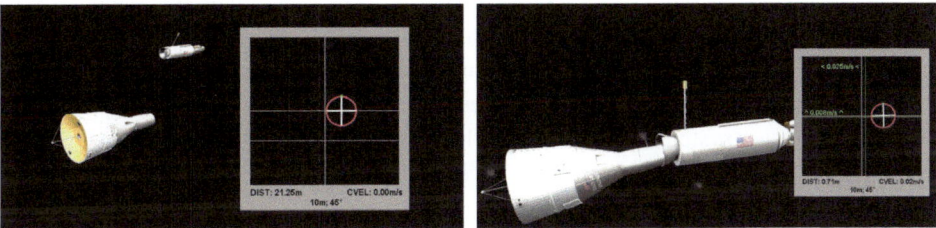

Fig. 8 Initial position of the GATV and GEMINI VIII's DTs in the VTB in VEROSIM (left). Final position of the rendezvous and docking maneuver (right)

account for the time delays during communication. However, due to the constant communication of the DT with the control's model, the modification process was done with increased efficiency.

In the left section of Fig. 8, the initial position of the Gemini VIII and the GATV is shown, with a distance of 21.25 m between them. The controller receives the current position and velocity error of the Gemini VIII with respect to the target and outputs the x,y,z thrust force vector. This is received by the Gemini VIII's DT, which operates the thrusters accordingly. Although the controller was not thoroughly optimized, the spacecraft successfully reaches the target, as observed on the right section of Fig. 8, where the distance has closed to 0.71 m. This helps demonstrate the successful use of DTs in an MiL setup with the use of real-time capable hardware. Finally, this particular experiment highlights the benefits of using DTs in product development instead of physical prototyping. Running these experiments with physical prototypes would be extremely expensive and dangerous, if not impossible.

An additional approach of a mix between MiL and HiL was also conceptualized. Here, an embedded system specialized in space application (Jena Optronik's DevBoard) handles the movement of the Gemini VIII and gives its pose information to VEROSIM, where the VTB with the DT of the GATV is implemented. The pose is given to the GNC inside the Scalexio and a Thrust vector is returned to VEROSIM and passed on to the embedded system. Although this configuration is left for future work, it is mentioned here as an additional benefit to using DTs in MiL/HiL. This approach is useful for developers to test their GNC models running in the Scalexio together with the DevBoard in a HiL setup.

6 Conclusions

This contribution has shown a proof of concept for the integration of DTs at the MiL development stage of a system, together with the use of real-time capable hardware to run the controller of said system. This could be extended to other stages of the XiL development, where other parts of the system can be tested with the help of DTs and VTBs. The integration of DTs increases the efficiency, behavior accuracy, and predictability of the performance of the final product increasing the inherent benefits of development with the V-model. This is especially important in fields of application where running real-life development, testing, and validation can be very expensive and dangerous (e.g. in space).

A limitation to this approach is the necessity of DTs of other VTB's components (e.g. sensors) or a larger system (e.g. vehicles) to have closer-to-reality results. The compatibility of such DTs could be a problem if the development does not follow a scalable, modular approach like the one presented in [20].

This study is in its initial stages and only examines the use of DTs in the MiL stage, as demonstrated by the two examples presented. Further research is required to assess the potential benefits of integrating DTs and VTBs into other stages of the V-model and XiL development cycles. This future work should also determine the applicability of DTs in each stage and identify which stages stand to benefit the most from their use.

Acknowledgements This work is part of the project "ViTOS-3", supported by the German Aerospace Center (DLR) with funds of the German Federal Ministry of Economics and Technology (BMWi), support code 50 RA 2120.

References

1. Bacic, M.: On hardware-in-the-loop simulation. In: Proceedings of the 44th IEEE Conference on Decision and Control, pp. 3194–3198 (2005)
2. Bao, J., Guo, D., Li, J., Zhang, J.: The modelling and operations for the digital twin in the context of manufacturing. Enterprise Inf. Syst. **13**(4), 534–556 (2019)
3. Berg, G., Nitsch, V., Färber, B.: Vehicle in the loop. In: Handbook of Driver Assistance Systems, pp. 1–9. Springer International Publishing (2015)
4. Adolf, P.: Börhl and Wolfgang Dröschel. Der Standard für die Softwareentwicklung mit Praxisleitfaden. Oldenbourg Wissenschaftsverlag, Das V-Modell (1993)
5. Demers, S., Gopalakrishnan, P., Kant, L.: A generic solution to software-in-the-loop. In: MILCOM 2007 - IEEE Military Communications Conference, pp. 1–6 (2007)
6. Hacker, B.C., Grimwood, J.M.: On the Shoulders of Titans: A History of Project Gemini

7. Han, J., Hong, Q., Feng, Z., Syed, M.H., Burt, G.M., Booth, C.D.: Design and implementation of a real-time hardware-in-the-loop platform for prototyping and testing digital twins of distributed energy resources. Energies **15**(18) (2022)
8. Isermann, R., Schaffnit, J., Sinsel, S.: Hardware-in-the-loop simulation for the design and testing of engine-control systems. Control. Eng. Pract. **7**(5), 643–653 (1999)
9. Kapteyn, M.G., Knezevic, D.J., Huynh, D.B.P., Tran, M., Willcox, K.E.: Data-driven physics-based digital twins via a library of component-based reduced-order models. Int. J. Numer. Meth. Eng. **123**, 2986–3003 (2020)
10. Morris, B., Cheng, T., Koontz, C.: A model-in-the-loop technique for digital twin simulations (MIL-DT): a generalized approach for simulating embedded software, 2022-11
11. Plummer, A.R.: Model-in-the-loop testing. Proc. Inst. Mech. Engin., Part I: J. Syst. Control Engin. **220**(3), 183–199 (2006)
12. Qi, Q., Tao, F.: Digital twin and big data towards smart manufacturing and industry 4.0: 360 degree comparison. IEEE Access **6**, 3585–3593 (2018)
13. Rind, A.R., Shahzad, K., Abdul Qadir, M.: Evaluation and comparison of tcp and udp over wired-cum-wireless lan. In: 2006 IEEE International Multitopic Conference, pp. 337–342 (2006)
14. Rossmann, J., Schluse, M., Rast, M., Atorf, L.: E-robotics combining electronic media and simulation technology to develop (not only) robotics applications. In: E-Systems for the 21st Century. Apple Academic Press (2016)
15. Schleich, B., Anwer, N., Mathieu, L., Wartzack, S.: Shaping the digital twin for design and production engineering. CIRP Ann. Manufact. Technol. **66**, 141–144 (2017)
16. Schluse, M., Priggemeyer, M., Atorf, L., Rossmann, J.: Experimentable digital twins—streamlining simulation-based systems engineering for industry 4.0. IEEE Trans. Indust. Inf. **14**(4), 1722–1731 (2018)
17. Tao, F., Cheng, J., Qi, Q., Zhang, M., Zhang, H., Sui, F.: Digital twin-driven product design, manufacturing and service with big data. Int. J. Adv. Manufact. Technol. **94**, 3563–3576 (2018)
18. Tao, F., Sui, F., Liu, A., Qi, Q., Zhang, M., Song, B., Guo, Z., Lu, S.C.Y., Nee, A.Y.: Digital twin-driven product design framework. Int. J. Product. Res. **57**(12), 3935–3953 (2019)
19. Tao, F., Zhang, H., Liu, A., Nee, A.Y.: Digital twin in industry: State-of-the-art. IEEE Trans. Indust. Inf. **15**(4), 2405–2415 (2019)
20. Thieling, J., Rossmann, J.: Scalable sensor models and simulation methods for seamless transitions within system development: From first digital prototype to final real system. IEEE Syst. J. **15**(3), 3273–3282 (2021)
21. VDI. 2206: Entwicklungsmethodik für mechatronische systeme. Verein Deutscher Ingenieure 2 (2004)
22. Wagg, D.J., Worden, K., Barthorpe, R.J., Gardner, P.: Digital twins: state-of-the-art and future directions for modeling and simulation in engineering dynamics applications. ASCE-ASME J. Risk Uncert. Engin. Syst. Part B Mech. Engin. **6**(3) (2020)

Analysis of Point Set Registration Algorithms for Industrial Parts

Oguz Kedilioglu, Yanming Zhao, Martin Landesberger,
Christian Hofmann, Michael Hofmann, Jörg Franke
and Sebastian Reitelshöfer

Abstract

In this paper, we investigate the suitability of existing point set registration algorithms for the task of pose estimation of industrial parts. Numerous algorithms with different characteristics have been developed already. However, no comparative analysis regarding the performance in the context of industrial parts have been conducted yet. We implement a comparative study by applying various point set registration algorithms to an industrial part and varying the point cloud density. The results of our study provide insights into the performance of different algorithms and shed light on the best practices for point set registration in the context of industrial parts.

Keywords

Point set registration · Pose estimation · Computer vision

1 Introduction

The advancement of automation technology in the industrial sector has led to a growing demand for high-precision 6-degree-of-freedom (6DoF) pose estimation. The accurate determination of the pose of objects, such as industrial parts, is crucial for a variety of appli-

O. Kedilioglu (✉) · Y. Zhao · C. Hofmann · J. Franke · S. Reitelshöfer
Institute for Factory Automation and Production Systems, Friedrich-Alexander-Universität
Erlangen-Nürnberg (FAU), 91058 Erlangen, Germany
e-mail: oguz.kedilioglu@faps.fau.de

M. Landesberger · M. Hofmann
Heinz Maier-Leibnitz Zentrum (MLZ), Technical University of Munich, Garching 85748, Germany

© The Author(s) 2025
S. Ihlenfeldt et al. (eds.), *Annals of Scientific Society for Assembly, Handling
and Industrial Robotics 2023*, https://doi.org/10.1007/978-3-031-74010-7_18

cations, ranging from quality control and inspection to robotic manipulation and autonomous navigation.

6DoF pose estimation [1] is a complex process that involves several critical components, including feature detection, image-to-image correspondence computation using epipolar geometry, 3D point triangulation, and point set registration. Of these, point set registration plays a vital role in determining the pose of the object.

Point set registration is a fundamental task in computer vision, which aligns two or more point sets by minimizing the distances between their corresponding points. In the context of 6DoF pose estimation, the point sets represent the 3D model and the scene, and the registration process is used to determine the transformation that maps the model to the scene. This transformation provides the 6DoF pose of the object in the scene [2].

In addition to its importance in 6DoF pose estimation, point set registration is also relevant for other tasks, such as simultaneous localization and mapping (SLAM). In SLAM, a mobile robot or device builds a map of its environment while simultaneously determining its position within that map. Point set registration plays a crucial role in this process by aligning the model of the environment with the observed data from the sensors [3].

As a result of the growing importance of point set registration, there are numerous algorithms available with different strengths and weaknesses. For example, some algorithms may provide accurate results in real-time, while others may have better accuracy but are computationally expensive. So far, there has been no comprehensive comparative study that analyses the performance of different point set registration algorithms for industrial parts.

This paper aims to fill this gap by providing a thorough analysis of the strengths and weaknesses of various point set registration algorithms. The study focuses on the use of these algorithms for industrial parts, considering factors such as accuracy and computational complexity. Examined algorithms include Coherent Point Drift (CPD) [4], FilterReg [5], TEASER++ [6] and RANSAC [7]. The paper is structured with sections dedicated to a review of related works, the presentation of the proposed methodology, an evaluation of the results, and a conclusion summarizing the key findings and insights.

2   Related Works

Point set registration is a fundamental problem in computer vision and robotics, and its solution is crucial for many industrial applications such as high-precision 6DoF pose estimation. In recent years, many algorithms have been proposed to solve the problem of point set registration. In this section, we will briefly summarize the most commonly used algorithms, including the algorithms Iterative Closest Point (ICP), the Coherent Point Drift (CPD), FilterReg, TEASER++, and the RANSAC.

The ICP algorithm [8] is a popular iterative method for point set registration. It consists of two main steps: finding the closest points between two point sets and updating the transformation between them. The method iteratively repeats these steps until convergence, producing an accurate registration result. It has a high requirement for the initial position of the two point sets and is therefore often used for fine-tuning after the application of coarse registration algorithms.

The CPD algorithm [4] is a non-rigid point set registration method that is well-suited for dealing with deformable shapes. It employs Gaussian mixture models to model the probability density functions of the point sets, allowing it to handle non-linear transformations. The optimization process involves finding the parameters of the Gaussian basis functions that minimize the distance between corresponding points in the two point clouds.

The FilterReg algorithm [5] is a novel point set registration method that utilizes a filter-based optimization framework, which iteratively refines the correspondences between points in the two point clouds by taking into account both the similarity between points and their spatial arrangements in the two clouds.

The TEASER++ algorithm [6] is a recent point set registration algorithm that is designed to work well for RGB-D scenes. It leverages a hierarchical optimization framework to produce fast and accurate registration results. It is a global registration algorithm, i.e., it relies on ICP for fine registration.

The RANSAC algorithm [7] is a popular algorithm for robust estimation in computer vision. In the context of point set registration, RANSAC can be used to identify inliers in a noisy point set, allowing for robust registration even in the presence of outliers. It is also used in combination with ICP.

In the field of point set registration for industrial parts, various benchmark datasets have been introduced to evaluate the performance of different algorithms. These benchmark datasets contain 3D point clouds obtained from real-world industrial parts and corresponding ground truth poses.

Bellekens et al. [9] provide a comprehensive survey of existing algorithms for rigid 3D point cloud registration. The authors evaluate and compare different algorithms based on accuracy and computational efficiency, using several benchmark datasets. But they don't analyze the effect of the point cloud density specifically.

Another benchmark is the FAUST dataset [10], which provides a benchmark for evaluating point set registration algorithms on human body scans. This dataset has been used to evaluate the performance of different algorithms in handling complex and large point clouds.

Finally, the ETH dataset [11] is another benchmark used to evaluate point set registration algorithms in the context of mobile robotics. The ETH dataset includes point clouds of indoor and outdoor environments. But it doesn't focus on industrial parts which are characterized by feature-poor metallic surfaces and lead to sparser point clouds.

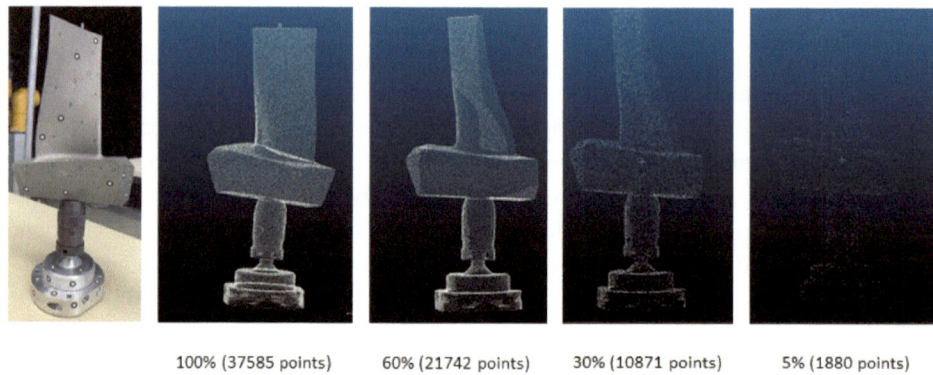

100% (37585 points)   60% (21742 points)   30% (10871 points)   5% (1880 points)

Fig. 1 Original reference point cloud with different density levels

3 Methodology

Point set registration is a fundamental problem in computer vision and robotics that involves aligning two or more point sets to a common reference frame. The objective of point set registration is to find the optimal transformation that maps one point set to another, so that they become geometrically aligned. The transformation could be a rigid transformation that involves rotation, translation, and scaling, or it could be a non-rigid transformation that involves non-linear deformations. Here, we focus on rigid transformations without scaling represented by the homogeneous transformation $T = (R, t)$ with $R \in \mathbb{R}^{3 \times 3}$ and $t \in \mathbb{R}^3$.

The original point cloud was generated using a high-precision 3D scanner ATOS Compact Scan from GOM Metrology and contains 37585 points. To examine the effect of point cloud density on the registration quality of different algorithms, we decrease the density of the point cloud to 60, 30 and 5% of the original amount of points (Fig. 1). The density is decreased by uniformly removing points from the total point set across the whole geometry of the part. The created point subset is then fitted into the original point set by the registration algorithms for analyzing their performance behavior.

In this paper, we focus on the performance of four well-known point set registration algorithms, CPD, FilterReg, TEASER++ and RANSAC, applied to a metallic 3D-printed turbine blade.

4 Evaluation

The results of the evaluation of the four point set registration algorithms, namely CPD, FilterReg, TEASER++, and RANSAC, are shown in Figs. 2 and 3. The point cloud from the GOM scanner was matched with a point subset created from it, and the registration results were analyzed for different densities of the point cloud, including 5, 30, 60, and 100%.

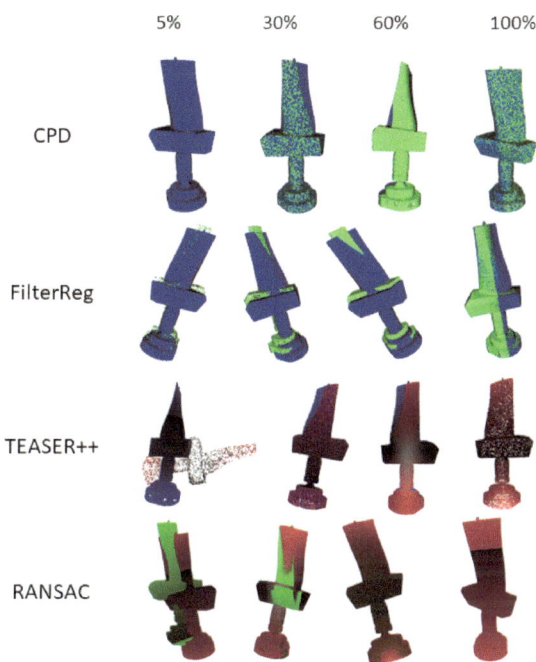

Fig. 2 Registration results of CPD, FilterReg, TEASER++ and RANSAC with 5, 30, 60 and 100% of the original points. In the CPD, FilterReg and TEASER++ plots the blue point cloud represents the complete reference point set. In the RANSAC plot it is represented by the red point set

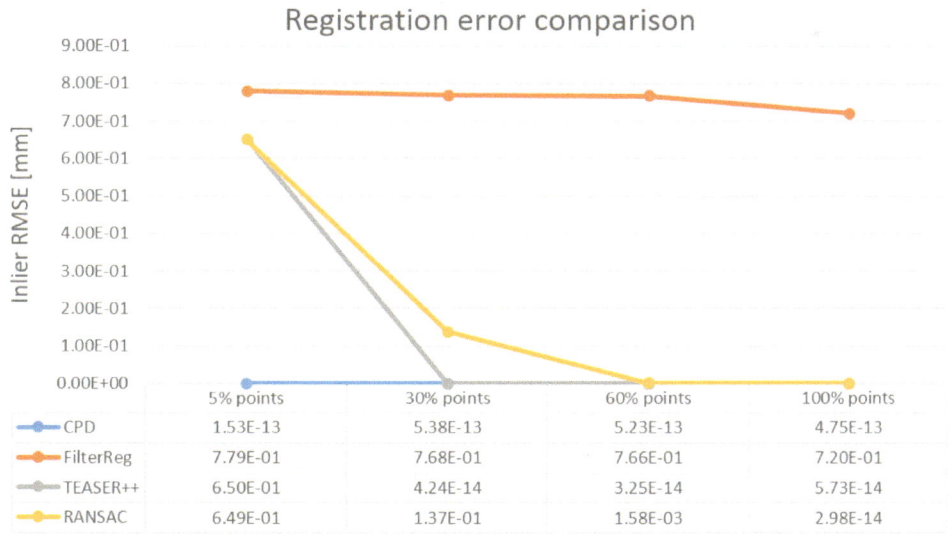

	5% points	30% points	60% points	100% points
CPD	1.53E-13	5.38E-13	5.23E-13	4.75E-13
FilterReg	7.79E-01	7.68E-01	7.66E-01	7.20E-01
TEASER++	6.50E-01	4.24E-14	3.25E-14	5.73E-14
RANSAC	6.49E-01	1.37E-01	1.58E-03	2.98E-14

Fig. 3 Inlier RMSE of CPD, FilterReg, TEASER++ and RANSAC for different synthetic point cloud densities. The results for the density levels at 5, 30, 60 and 100% are shown

Table 1 Runtime of CPD, FilterReg, TEASER++ and RANSAC with 100% of the point set

	CPD	FilterReg	TEASER++	RANSAC
Runtime (s)	40173.4	61.8	522.0	204.7

The Root Mean Square Error (RMSE) is used as a metric (Eq. 1),

$$RMSE = \sqrt{\frac{\sum_{i=1}^{n} \left(x_i - \hat{x}_i\right)^2}{n}} \tag{1}$$

where n is the number of corresponding points and $(x_i - \hat{x}_i)$ is the Euclidean distance between corresponding points after registration. In the optimal solution, the distance between the corresponding points is 0 after complete registration.

The results show that CPD achieves very low RMSE values for all densities of the point cloud, making it the most robust algorithm among the four (Fig. 3). However, its runtime of 40173.4 s is significantly slower compared to the other algorithms, making it less suitable for real-time applications (Table 1).

FilterReg, on the other hand, was the fastest algorithm with a runtime of only 61.8 s. However, its registration quality was the worst among the examined algorithms, which is evident from its relatively high RMSE values.

TEASER++ was very fragile when the density of the point cloud was low, but it improved rapidly when more points were available. With 100% point cloud density, TEASER++ was already able to outperform CPD and FilterReg.

RANSAC was even more sensitive regarding sparse point clouds and needed the largest number of points to return good registration results. It even outperformed TEASER++ by a small margin at 100% density.

In addition to the simulated data, the evaluation was also conducted with real data, where the target point set was not acquired by downsampling the original set from the scanner, but instead was retrieved via reconstruction with the tool CloudCompare [12] and images from 20 megapixel grayscale cameras (Fig. 4). The observations from the simulated data were confirmed by experiments with the real data, where CPD again showed its robustness against sparse point clouds, while TEASER++ and RANSAC struggled with it (Fig. 5).

5 Conclusions

In this study, four well-known point set registration algorithms, CPD, FilterReg, TEASER++ and RANSAC, applied to a metallic 3D-printed turbine blade have been analyzed by varying density levels from 100 to 5%. The results show that CPD has the lowest RMSE values but a significantly slower runtime compared to the other algorithms. FilterReg is the fastest

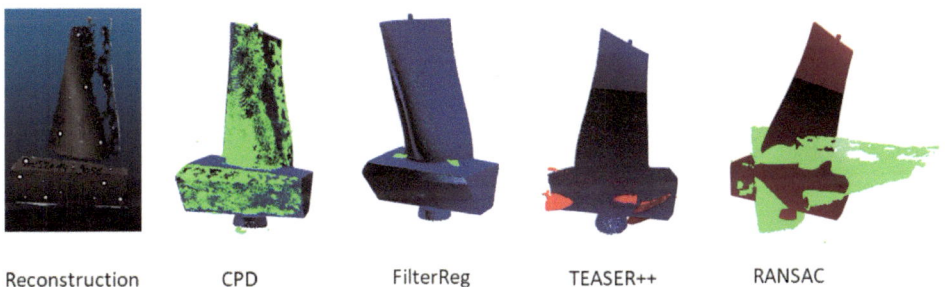

Reconstruction   CPD   FilterReg   TEASER++   RANSAC

Fig. 4 Registration results of CPD, FilterReg, TEASER++ and RANSAC with real data. In the CPD, FilterReg and TEASER++ plots the blue point cloud represents the complete reference point set. In the RANSAC plot it is represented by the red point set

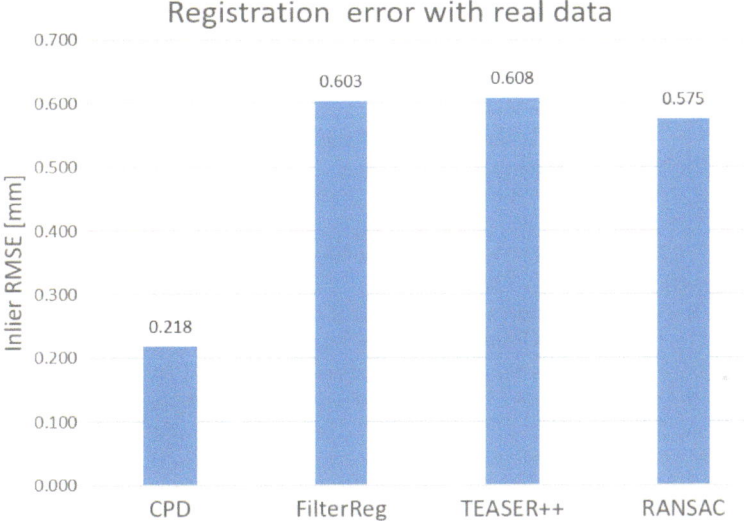

Fig. 5 Inlier RMSE of CPD, FilterReg, TEASER++ and RANSAC with real data. The real data point clouds are provided by reconstructions from camera images. CPD outperforms the other three algorithms clearly

but has the worst registration quality. TEASER++ improves rapidly when more points are available but struggles with sparse point clouds. RANSAC is even more sensitive to sparse point clouds and needs the largest number of points to return good results. The evaluation was conducted with simulated and real data, and the results confirmed that CPD was the most robust against sparse point clouds, while TEASER++ and RANSAC struggled with it. Future work can examine other characteristics of registration algorithms by creating data sets with very specific manipulation operations on them. This targeted analysis of datasets will allow an isolated view of relevant properties of the algorithms.

Acknowledgements The project receives funding from the German Federal Ministry of Education and Research under grant agreement 05K19WEA (project RAPtOr).

References

1. He, Z., Feng, W., Zhao, X., Lv, Y.: 6d pose estimation of objects: Recent technologies and challenges. Appl. Sci. **11**(1), 228 (2020)
2. Maiseli, B., Yanfeng, G., Gao, H.: Recent developments and trends in point set registration methods. J. Vis. Commun. Image Represent. **46**, 95–106 (2017)
3. Khairuddin, A.R., Talib, M.S., Haron, H.: Review on simultaneous localization and mapping (SLAM). In: 2015 IEEE International Conference on Control System, Computing and Engineering (ICCSCE), pp. 85–90. IEEE (2015)
4. Myronenko, A., Song, X.: Point set registration: coherent point drift. IEEE Trans. Pattern Anal. Mach. Intell. **32**(12), 2262–2275 (2010)
5. Gao, W., Tedrake, R.: Filterreg: robust and efficient probabilistic point-set registration using Gaussian filter and twist parameterization. In: Proceedings of the IEEE/CVF Conference on Computer Vision and Pattern Recognition, pp. 11095–11104 (2019)
6. Yang, H., Shi, J., Carlone, L.: Teaser: fast and certifiable point cloud registration. IEEE Trans. Robot. **37**(2), 314–333 (2021)
7. Derpanis, K.G:. Overview of the RANSAC Algorithm, 4, no. 1, pp. 2–3. Image Rochester, NY (2010)
8. Li, P., Wang, R., Wang, Y., Tao, W.: Evaluation of the ICP algorithm in 3d point cloud registration. IEEE Access **8**, 68030–68048 (2020)
9. Bellekens, B., Spruyt, V., Berkvens, R., Penne, R., Weyn, M.: A benchmark survey of rigid 3d point cloud registration algorithms. Int. J. Adv. Intell. Syst. **8**, 118–127 (2015)
10. Bogo, F., Romero, J., Loper, M., Black, M.J.: Faust: dataset and evaluation for 3d mesh registration. In: Proceedings of the IEEE Conference on Computer Vision and Pattern Recognition (CVPR), June 2014
11. Strasdat, H., Montiel, J., Davison, A.J.: Scale drift-aware large scale monocular slam. Robot.: Sci. Syst. VI, **2**(3), 7 (2010)
12. Girardeau-Montaut, D.: Cloudcompare. EDF R&D Telecom ParisTech, France (2016)

A New AI-Based Approach for Contextualization and Prediction of Human Activities in Industrial Robot Applications

Sebastian Krusche⬥, Ibrahim Al Naser⬥, Mohamad Bdiwi⬥, and Steffen Ihlenfeldt⬥

Abstract

Human activity prediction in industrial environments with content semantics using 3D point clouds can ensure human safety and process efficiency. This work aims to train and optimize context-based AI models to develop a framework for anticipating human group activities and estimating human motion. Main Contributions of this work: 1. Design of multi-layer structure of various DNN Classifiers to segment, detect and track dynamic agents (e.g., humans, robots, or AGVs) 2. Training of context-based sequential relational anticipation model to predict human activities and positions in the early stage 3. empirical experiment with 60 subjects for collecting datasets of human actions and activities in on industrial setting. All these procedures are merged in the proposed framework and evaluated in six scenarios of one industrial Use-Case.

Keywords

Early action prediction • Intention anticipation • Motion trajectory estimation

1 Introductions

In order to satisfy the future challenges of industrial production, flexible production lines are required that can react situationally to changes in the process chain and product diversity. Depending on the order situation, agile machine systems have to be changed to new

S. Krusche (✉) · I. Al Naser · M. Bdiwi · S. Ihlenfeldt
Fraunhofer Institute for Machine Tools and Forming Technology, 09126 Chemnitz, Germany
e-mail: sebastian.krusche@iwu.fraunhofer.de

S. Ihlenfeldt et al. (eds.), *Annals of Scientific Society for Assembly, Handling and Industrial Robotics 2023*, https://doi.org/10.1007/978-3-031-74010-7_19

production processes in the shortest possible time, which poses a significant challenge in terms of safety and control technology. One solution to ensure high efficiency and adaptability is to design a production line with fenceless workstations that allow collaboration between humans and robots [1]. This approach to workplace design places high demands on safety and control technology. The control system must safely recognize the human in the robot's workspace and predict the human's future intentions. Fulfilling this condition requires methods that can reliably predict human activities and movements at a very early stage. Activity prediction of this kind would allow the human to work very closely with the robot during certain product steps without the risk of collision. Conversely, there is no loss of machine utilization, as the system can increase speed again when an allowable safety distance is reached [2]. By ensuring safety, interactions such as teaching a heavy-duty robot can be enabled by gestures in close proximity to the robot without the need for complex operator releases. The no-code approach allows the human to teach the robot directly by guiding, pointing, or demonstrating without requiring knowledge of complex programming languages [3, 4].

This paper proposes a new AI-based approach for contextualizing and predicting human activities in industrial robot applications. Our approach aims to predict complex sequential actions based on partially observed images by observing the relationships and interactions between dynamical agents such as production workers, robots, or AGVs. To ensure that the action sequences are captured across the entire surveillance space, we use up to four sensors that capture the scene from various perspectives. Our contributions can be summarized as follows:

1. Design of multi-layer structure of various DNN Classifiers to segment, detect and track dynamic agents (e.g., humans, robots or AGVs).
2. Training of context-based sequential relational anticipation model to predict human activities and positions in the early stage.
3. Conducting and Evaluation of empirical experiments with 60 subjects for collecting datasets of human actions and activities in on industrial setting.

All these procedures are merged in the proposed framework and evaluated in six scenarios of one industrial Use-Case with robots.

2   Related Work

Most publications on human action prediction refer to the detection of temporal or spatial changes in 2D color images using neural networks [5–7], which usually include the detection or classification of the action in the early stage. The paper [7] presents a method that predicts human interaction using deep temporal information from videos. Flow coding is used to represent low-level motion information, and a deep convolutional neural network is used to extract deep temporal features. To ensure that the two-stream network also anticipates hard-to-predict actions in the early stage, an additional memory module that records the sequence and provides the prediction is used in the approach [8]. Unlike traditional two-stream architectures, the two-stream model in the paper [9] leverages both modalities to train both input gates jointly and both forget gates in the network rather than treating the two streams as separate entities with no information about the other.

The work [10] improves the training of deep temporal models to better learn activity progression for activity detection and early detection tasks. Conventionally, when training a Recurrent Neural Network, specifically a Long Short-Term Memory (LSTM) model, the training loss only considers classification error. The recognition value of the correct activity category or the recognition value distance between the correct and incorrect category should not decrease monotonically if the model is to observe more activities. For this, a novel ranking loss is designed that directly penalizes the model when such monotonicities are violated and is used together with classification loss in training LSTM models. While many works are concerned with predicting raw RGB pixel values, the approach [11] focuses on predicting motion evolution in future video images. To this end, dynamic images (DIs) are constructed by grouping moving pixels into a sequence of future images. Training a convolution-based LSTM will predict the next DIs based on an unsupervised learning process. The associated activity becomes predictable with forecasted DI. In some approaches, the environment is also included. To this end, a multi-stage LSTM architecture was developed to leverage context-aware and action-aware features [12]. It also introduces a novel loss function that promotes the model to predict the correct class as early as possible. To infer the appropriate action or activity from the static scene, the developers train a particular classifier in the paper [13], which predicts the action based on unique environment structures. The weakness of this approach is that the scene structures should significantly differ to get an accurate action prediction. An extension of these approaches is the additional detection of object interactions during action [14], including the complete environment for predicting walking paths and activities [15]. In contrast, a unique approach considers temporal and spatial changes of random events in the image [16], which are not directly related to the action of the human. No specific people or objects are tracked, and the observation includes the whole scene of interaction. Very often, the approaches refer to pure recognition of a completed activity. However, in some cases, the recognition is also used to predict future actions that are directly related to the current action. By hierarchically dividing the activity into two levels, future activity can be predicted based on the combination of simple actions [17]. Another long-term prediction approach involves observing sports group activities to predict future positions and plays [18]. The work [19] describes a joint prediction approach where both activity features (motion-based) from previous activities and object features present in the scene are used. The system is able to infer the corresponding action and derive its start time.In addition to the above drawbacks, detecting and tracking multiple people in a video is a significant challenge in crowded and cluttered scenes. None of the approaches uses a 3D multi-modal sensor system approach to predict group activities. Furthermore, the approaches focus on non-industrial areas where individuals' actions are mainly predicted without explicitly considering the interaction with the environment or other dynamic objects.

3   Approach

Industrial environments are very complex, especially where humans and robots collaborate on a task in a confined workspace. The sensor system must be able to infer future actions and activities based on the movements of various dynamic agents such as humans, robots, or AGVs. Our approach is to train a Sequential Relational Anticipation Model (SRAM) based on the partial observation of the interaction and activities of several active agents from various sensor perspectives. As a motivation, we use a model for predicting volley-ball moves in the context of single-player movements [20]. Our work aims to apply this group activity prediction approach to human–robot cooperation scenarios in an industrial context.

Figure 1 shows the general architecture of our approach. A multi-modal sensor system is provided for the action sequence acquisition, which observes the scene from different perspectives in real-time. Based on color images and point clouds, proper segmentation and tracking of humans and dynamic objects is performed based on a multi-layer structure of various DNN classifications [21, 22]. The observed tracklets of the single dynamic agents are passed to the Sequential Relational Anticipation Model and used to predict the group activity. By partially observing the interaction between the single dynamic agent, the future group behavior in activity and position is predicted.

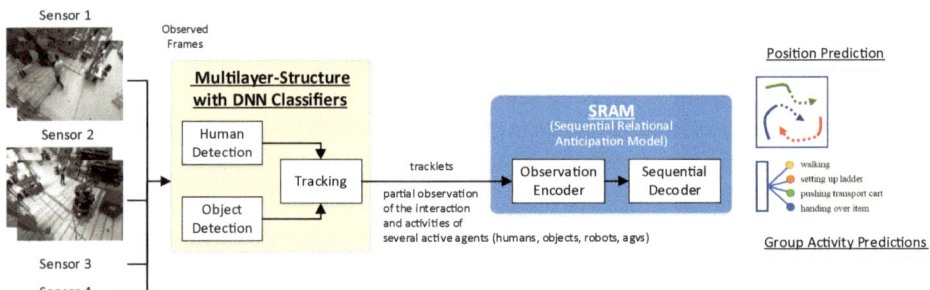

Fig. 1 Overall architecture of AI-based method for contextualization and prediction of human activities in industrial robot applications

3.1   Data Preparation

A prerequisite for training context-based AI classifiers is that a sufficiently large number of contextual datasets containing industrial actions and activities are available. In our work, an automated annotation tool [23] is used to automatically label human actions in multimodal datasets consisting of color images and 3D point clouds with low time and resource requirements. The core of the tool consists of a multi-layer structure of various DNN classifiers that ensure that human motions and actions are reliably detected in industrial environments. In addition, it is possible to process synchronous data from a multi-sensor system that views the scene from multiple perspectives. The fusion of the single sensor data increases the confidence of the hypotheses. It ensures that the human's action is tracked over a larger workspace with more complex structure.

The results of the automatic annotation process can be seen in Fig. 2. The annotation tool can locate, track and classify human and dynamic objects in the sensor data. The reliable classification of the human is done using the image data from the camera sensors (Right image), which view the scene from multiple angles. Various AI-based human pose classifiers [21, 22] are used, whose hypotheses are tested using a plausibility filter. Subsequently, in the 3D workspace (Left image), the humans or objects are segmented in the form 3D bounding box of environment structure based on the 3D point clouds of 4 sensors. The results of the annotation tool included human action labels in addition to the 3D and 2D bounding boxes.

The objective is to use the automatic annotation tool to create a new data set with a structure similar to [24]. It is essential for this that the action sequence is completely captured from all sensor perspectives. Each annotated action frame of a perspective consists of 41 images, where 20 images are captured before the target frame, and 20 images are

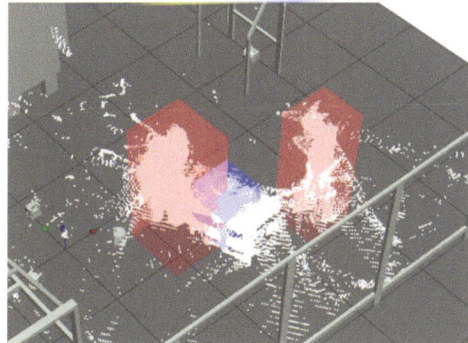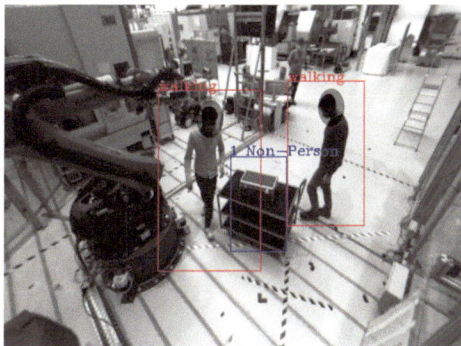

Fig. 2 Automatic annotation tool results—(Left image) 3D workspace with 3D point clouds and 3D bounding boxes for human (red) and object (blue); (Right image) 2D sensor view with 2D bounding boxes for human (red) including action marker and object (blue)

captured after the target frame. In contrast to [25], dynamic objects such as robots and AGV's are included in addition to the acting humans to represent the group activity.

3.2   Group Activity Prediction (SRAM)

The SRAM approach captures the interactions of multiple individuals with observational encoders and anticipates their future relationships through a sequential decoder (shown in Fig. 3). The approach differs from existing action prediction methods [26, 27] that can only predict the action of a single person. Compared to group activity detection methods [26, 28, 29], the SRAM method performs sequential prediction of group activities concerning future positions and activity representations. Activity prediction is also facilitated by explicitly predicting the future positions of individuals. A sequential decoder allows the prediction of group representations across multiple roll-off phases, even with very few observed frames, which is a significant performance improvement over other approach. This decoder is guided by a sequential reconstruction loss that mimics how a partial observation gradually approaches its complete observation. An adversarial loss that ensures that the generated full observation features can no longer be distinguished from the correct full observation features.

The selected sequential Relational Anticipation Model for predicting group activity of employees and dynamic objects is to be learned based on the annotated action sequences. The prerequisite is that a sufficiently large number of action frames of the respective action class are available. To ensure the large number of data, extensive subject tests were carried out, described in the experiments section.

In our work, we generated a large number of valid annotated 2D and 3D datasets of industrial actions and activities using the Automatic Annotation Framework and a

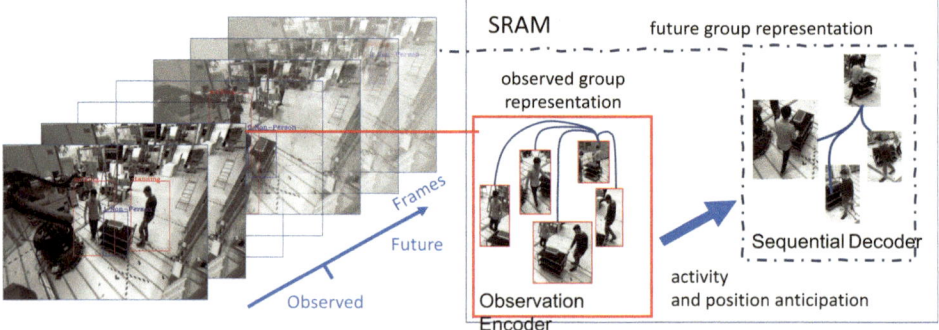

Fig. 3 Based on the initial frames, the SRAM method models the relational dynamics of a group. It predicts a group activity by predicting the future group activity representation and their positions for the unobserved frames (Illustration adapted Fig. 1 from [20])

multi-sensor system, which were used for training the Sequential Relational Anticipation Model. After completion of the training, the final model was integrated and tested in our processing pipeline, as shown in Fig. 1, for 3D workspace monitoring.

4 Experiments

A large amount of data with various actions and activities in an industrial context is needed to train the Sequential Relational Anticipation Model in our work. For this purpose, a large number of subject experiments have been conducted in the past in which humans performed various actions in an industrial environment in an uncontrolled manner. The scene was captured by a multimodal 3D sensor system from several perspectives during the experiments. The color images and point clouds of the 3D sensors were acquired synchronously in time and combined into a data set. The experimental scenarios range from very simple to very complex and include various actions related to the environment structure. Table 1 gives an overview of the test scenarios with corresponding action types and subject number.

4.1 Experiment Setup

The HRC cell at Fraunhofer IWU, corresponding to the standard of a robot cell without a protective fence from industrial production, was selected for recording the sensor data. The front of the cell is open and allows barrier-free access for humans and objects (shown in Fig. 5—Left). The large open area inside the cell ensures that the four sensors are optimally aligned with the action scene and are not obscured by machine parts and plant

Table 1 Test scenarios with corresponding action types and subject number

Nr	Title	Action types	Subjects
1	Person walks into robot cell	Standing (static), walking	1
2	Person walks with item	Standing (static), walking, setting up ladder	1
3	Person pushes a transport cart	Standing (static), walking, pushing transport cart	1
4	Person place a object	Standing (static), walking, place object	1
5	2 persons walk into robot cell	Standing (static), walking	2
6	2 persons hand over an item	Standing (static), walking, handing over item	2
7	2 persons with a transport cart	Standing (static), walking, Pushing transport cart	2

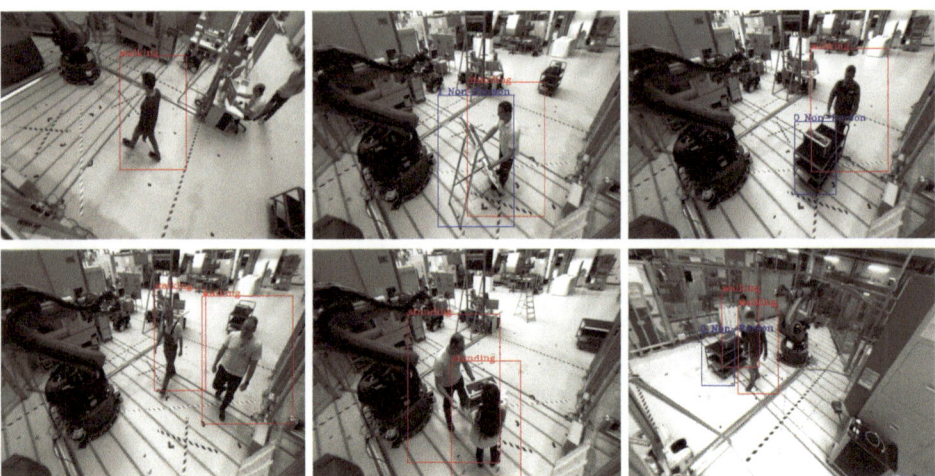

Fig. 4 Activity Overview—First Row: (Left) single Person walking, (Center) setting up a ladder, (Right) single Person pushing transport cart; Second Row: (Left) two Person walking, (Center) handing over an item, (Right) two person pushing transport cart; Object (blue bounding box) and Human (red bounding box)

structures. The sensor layout is shown in Fig. 5 (Right) and ensures that the acting human is recorded from several perspectives.

4.2   Prediction and Annotations Results

With the help of the automated annotation framework, it is possible to generate many datasets for our work to train the Sequential Relational Anticipation Model. The advantage is that the scene is captured from multiple sensor perspectives, increasing the number of valid sequences. Annotation of actions and activities showed that simple actions such as walking and standing are the most common, as opposed to complex activities such as setting up a ladder or pushing a transport cart. The prediction results of the trained model in Fig. 4 show that a SRAM can be used to predict actions and activities related to the group relationship of humans and objects.

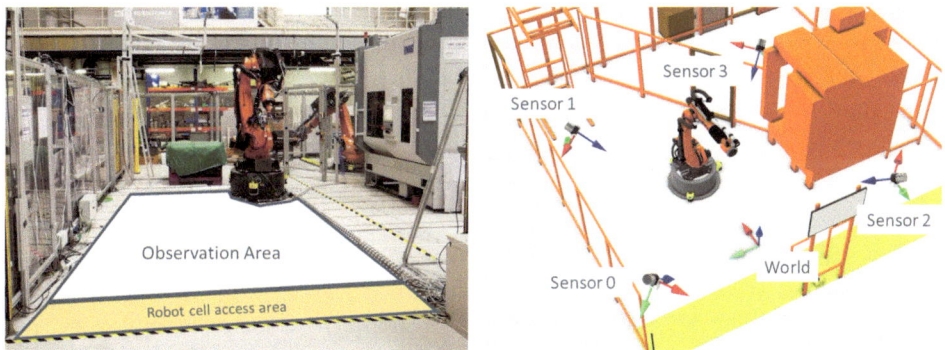

Fig. 5 (Left) Front view of HRC-Cell at Fraunhofer IWU, (Right) Sensor layout with 4 sensors

5 Conclusion

The proposed AI-based approach for contextualizing and predicting human activities in an industrial context is intended to help optimally control processes in production facilities in a predictive manner and ensure human safety during human–robot interactions. To implement and evaluate this approach, the following contributions were made to our work:

1. By designing and implementing a multi-layer structure of multiple DNN classifiers, it is possible to reliably segment, recognize and track humans and other dynamic objects in the sensor data. Result correlation in 3D and 2D space ensures that the group action sequence is captured over the entire workspace and duration.
2. Our work focused on training the Sequential Relational Anticipation Model to predict group activity in an industrial context. This approach includes the movements and states of other dynamic agents, such as robots and AGVs, in addition to human interactions. The advantage is that the prediction incorporates the interaction of all dynamic agents, and no other hybrid prediction is required.
3. The empirical experiments with more than 60 subjects to collect data sets of human actions and activities in an industrial environment allowed us to guarantee a reasonable basis for training the Sequential Relational Anticipation Model and to ensure the verification of the framework. The different complex scenarios allowed us to provide a valid number of action sequences.

A weak point of the SRAM is that the group activity must always have a fixed location reference. Transferring the learned model to other scenarios is challenging due to the position reference of the single dynamic agents. To learn new group activities, new data sets must be recorded and annotated. This paper aims to stimulate the development of new methods of activity prediction in an industrial context and lead to innovations in data-driven computer vision in the coming years. By exploring such for the industrial application domain, new possibilities for autonomous control of machines and robots in production processes will be opened. In addition to reducing floor space by designing hybrid workstations of human–robot cooperation, action prediction approaches also reduce cycle time interruptions and enable higher machine utilization.

References

1. Vogel, C., Fritzsche, M., Elkmann, N.: Safe human-robot cooperation with high-payload robots in industrial applications, In: 2016 11th ACM/IEEE International Conference on Human-Robot Interaction (HRI), pp. 529–530 (2016)
2. Rashid, A., Peesapati, K., Bdiwi, M., Krusche, S., Hardt, W., Putz, M.: Local and global sensors for collision avoidance, In: 2020 IEEE International Conference on Multisensor Fusion and Integration for Intelligent Systems (MFI), Karlsruhe, Germany, pp. 354–359 (2020)
3. Halim, J., Eichler, P., Krusche, S., Bdiwi, M., Ihlenfeldt, S.: No-code robotic programming for agile production: A new markerless-approach for multimodal natural interaction in a human-robot collaboration context. Frontiers in robotics and AI **9**, 1001955 (2022). https://doi.org/10.3389/frobt.2022.1001955
4. Bdiwi, M., Rashid, A., Putz, M.: Autonomous disassembly of electric vehicle motors based on robot cognition. In: 2016 IEEE International Conference on Robotics and Automation (ICRA), Stockholm, Sweden, pp. 2500–2505 (2016)
5. Ruiz, A.H., Gall, J., Moreno-Noguer, F.: Human Motion Prediction via Spatio-Temporal Inpainting. (2018). Available: http://arxiv.org/pdf/1812.05478v2
6. Feichtenhofer, C., Pinz, A., Zisserman, A.: Convolutional Two-Stream network fusion for video action recognition. (2016). Available: http://arxiv.org/pdf/1604.06573v2
7. Hua, G., Jégou, H.: Human interaction prediction using deep temporal features. Springer International Publishing, Cham (2016)
8. Kong Y, Gao S, Sun B, Fu Y.: Action prediction from videos via memorizing hard-to-predict samples. p. 8, (2018)
9. Lin Sun, Kui Jia, Kevin Chen, Dit-Yan Yeung, Bertram E. Shi, Silvio Savarese.: Lattice long short-term memory for human action recognition. (2017)
10. Ma, S., Sigal, L., Sclaroff, S.: Learning activity progression in LSTMs for activity detection and early detection. In: 2016 IEEE Conference on Computer Vision and Pattern Recognition (CVPR), pp. 1942–1950, (2016). Accessed: May 19 2021. Available: http://ieeexplore.ieee.org/document/7780583/
11. Rezazadegan, F., Shirazi, S., Baktashmotlagh, M., Davis, L.S.: On encoding temporal evolution for real-time action prediction. p. 7, (2017)
12. Mohammad Sadegh Aliakbarian, Fatemeh Sadat Saleh, Mathieu Salzmann, Basura Fernando, Lars Petersson, Lars Andersson.: Encouraging LSTMs to anticipate actions very early
13. Tuan-Hung Vu, Catherine Olsson, Ivan Laptev, Aude Oliva, Josef Sivic.: Predicting actions from static scenes. (2014)
14. Gkioxari, G., Girshick, R., Dollár, P., He, K.: Detecting and recognizing human-object interactions. (2017). Available: http://arxiv.org/pdf/1704.07333v3
15. Liang, J., Jiang, L., Niebles, J.C., Hauptmann, A., Fei-Fei, L.: Peeking into the future: predicting future person activities and locations in videos. (2019). Available: http://arxiv.org/pdf/1902.03748v3

16. Baradel, F., Wolf, C., Mille, J., Taylor, G.W.: Glimpse clouds: human activity recognition from unstructured feature points. Available: http://arxiv.org/pdf/1802.07898v4
17. Morais, R., Le Vuong, Tran, T., Venkatesh, S.: Learning to abstract and predict human actions. (2020). Available: http://arxiv.org/pdf/2008.09234v1
18. Chen, J., Bao, W., Kong, Y.: Group activity prediction with sequential relational anticipation model. (2020)
19. Tahmida Mahmud, Mahmudul Hasan, Amit K. Roy-Chowdhury.: Joint prediction of activity labels and starting times in untrimmed videos. (2017)
20. Chen, J., Bao, W., Kong, Y.: Group activity prediction with sequential relational anticipation model. (2020). Available: http://arxiv.org/pdf/2008.02441v1
21. Zhe Cao, Gines Hidalgo, Tomas Simon, Shih-En Wei, Yaser Sheikh.: OpenPose: Realtime Multi-Person 2D pose estimation using part affinity fields. *CoRR*, abs/1812.08008, (2018)
22. Li, J., Wang, C., Zhu, H., Mao, Y., Fang, H.-S., Lu, C.: CrowdPose: Efficient crowded scenes pose estimation and a new benchmark: arXiv
23. Krusche, S., Al Naser, I., Bdiwi, M., Ihlenfeldt, S.: A novel approach for automatic annotation of human actions in 3D Point Clouds for flexible collaborative tasks with industrial robots. (2023). Accessed: Feb. 7 2023. Available: https://www.frontiersin.org/articles/10.3389/frobt.2023.102 8329/abstract
24. Ibrahim, M., Muralidharan, S., Deng, Z., Vahdat, A., Mori, G.: A hierarchical deep temporal model for group activity recognition. (2015). Available: http://arxiv.org/pdf/1511.06040v2
25. Ibrahim, M.S., Muralidharan, S., Deng, Z., Vahdat, A., Mori, G.: Hierarchical deep temporal models for group activity recognition. (2016). Available: http://arxiv.org/pdf/1607.02643v1
26. Kong, Y., Tao, Z., Fu, Y.: Adversarial action prediction networks. IEEE Trans. Pattern Anal. Mach. Intell. **42**(3), 539–553 (2020). https://doi.org/10.1109/TPAMI.2018.2882805
27. Mengshi Qi, Jie Qin, Annan Li, Yunhong Wang, Jiebo Luo, Luc Van Gool.: StagNet: An Attentive Semantic RNN for Group Activity Recognition
28. Mostafa Ibrahim, Greg Mori.: Hierarchical relational networks for group activity recognition and retrieval
29. Wu, J., Wang, L., Wang, L., Guo, J., Wu, G.: Learning actor relation graphs for group activity recognition. (2019). Available: http://arxiv.org/pdf/1904.10117v1

Cost Modeling as Decision Support in Early Planning Phases of Agile Assembly Systems

Hannah Lickert, Paul Wiesener, Pinar Bilge, and Franz Dietrich

Abstract

Production environments and systems are transforming, driven by changing economic concepts, consumer demands and ecological guidelines. Linked to this, concepts for small-scale production systems in urban areas are evolving. These systems attempt to incorporate flexibility and individuality in both processes and products in range, variety and amount, enabled by growing technical possibilities. Therefore, small scale urban production systems are affected by other framework conditions than large-scale mass producer, in particular shortened planning phases due to frequent replanning, redesign, and reconfiguration. This requires constant adaptation to changing situations and basic planning tools. As such a cost model is proposed to support decision-making in early planning phases. The functionality of the model is investigated to achieve sufficient output accuracy with a reasonable amount of input. The input contains general information about the product, the processes, and the assembly equipment and is collected exemplarily in a case study while the output suggests different options, e.g., on assembly layout arrangements, machine selection, or depth of performance.

Keywords

Cost estimation • Agile assembly system • Urban production

H. Lickert (✉) · P. Wiesener · P. Bilge · F. Dietrich
Institute of Machine Tools and Factory Management, Chair of Handling and Assembly Technology Research, Technische Universität Berlin, Pascalstraße 8-9, 10587 Berlin, Germany
e-mail: h.lickert@tu-berlin.de

S. Ihlenfeldt et al. (eds.), *Annals of Scientific Society for Assembly, Handling and Industrial Robotics 2023*, https://doi.org/10.1007/978-3-031-74010-7_20

1   Introduction

Alike the requirements for assembly, the technological possibilities are constantly evolving. For realization in practice, planning tools have to adapt as well, to make economic representation of processes feasible. Cost modeling is used to estimate whether a project is economically viable. At the same time, with the goal of reducing costs, it can provide a basis for decision-making. Many existing approaches to model cost use a broad and accurate data basis, for example the precise determination of process times, to achieve accurate results for detailed planning. With shortened planning times in versatile systems, this effort is no longer economically viable and more flexible planning methods are advantageous. Therefore, this paper examines the relevant factors for cost estimation in flexible and agile as well as partially automated assembly environments.

Section 2 introduces the aspects the cost model tries to incorporate, by addressing the technological and structural developments in assembly and the influences on cost modeling. Section 3 is a short outline of methods for production cost estimation, documenting which approaches are researched and used for modeling. For comprehensibility, Sect. 4 presents the proposed model including the mathematical basis and cost factors. Section 5 applies the model in a case study and discusses the results. The case study is selected based on available research equipment and is therefore a pre-consideration for practical experiments. Finally, Sect. 6 elaborates the results and limitations of the model and how this deducts the need for further research.

2   Urban and Agile Assembly System

Current technological developments and the production change driven by sustainable transformation affect a shift of paradigm in assembly systems [1, 2]. In this section, the framework conditions for small scale urban assembly systems are presented as influencing factors leading to the paradigm shift. Concluding their effect on cost-estimation as one planning tool and the need for new approaches in planning flexible, changeable and adaptable systems in the context of urban and agile assembly activities.

Consumer demand and requirements. The enormous pressure for sustainable transformation in the face of the climate crisis requires a rethinking of the widespread method of mass production. Apart from the legislative means that force companies to take responsibility, the increased interest of consumers focus on ecological sustain products and production methods. This is connected with the consumers want for products that are better tailored to their individual needs and are durable in the sense of operability but also usefulness. [X].

The result is the increasing demand for individuality, while production volumes are decreasing, targeting small-scale production. Furthermore, the variant diversity of new products is decreasing because of less rapid technical innovations and the aim for better repairability, and spare part availability is increasing. Finally, this requires more versatile systems that offer a broader range of processes, for example disassembly.

Processes and value creation. According to the sustainable transformation, the concept of circular economy is inevitable to reduce, reuse and recycle. Converting the linear "take-make-dispose" economy into a closed loop includes new processes and organization in manufacturing [4]. Consequently, processes like disassembly are gaining importance and structures like reverse logistics need to be established, thus creating new value streams. Since used products are afflicted with many uncertainties about product variant, quantity, usage, or defect this requires flexibility. More flexible systems can deal with fluctuating production volumes or continually varying processes in depth and time, enabling more individual treatment and adaptability.

Planning and organization. Agile manufacturing is an innovative methodology, addressing the need for more flexibility and responsiveness toward uncertain and turbulent market requirements. It is often associated with lean or flexible manufacturing, whereas agile manufacturing is a proactive approach to deal with uncertain situations. Overall, it is assumed that the current challenges of the manufacturing sector can be overcome by introducing modular product structures, digital transformation, and by transforming systems and organization according to agile methods to remain competitive and increase productivity [5].

Technology capabilities. The recent development in manufacturing towards "Industry 4.0" is associated with digitization and technologies such as the internet of things, cloud computing, cyber physical systems, or big-data analytics. This is supposed to enable real-time connection between systems, machines, tools, workers, customers, and products, increasing production flexibility with reconfigurable machines and allowing profitable personalization and flexible automation [6]. Due to digitalization, it is possible to create collaborative manufacturing systems that can react in real time to changing conditions at the site, in the supply network, in customer requirements, and thus working flexibly and adaptively [7].

Urban Production. Urban production is a redefinition of the production site due to increasing urbanization and the coexistence of production site and the city, following current trends towards sustainable living and digital working [8]. Companies therefore focus on designing customer- or market-oriented processes to add more value to products or services. This implies a stronger involvement of customers in the development process of the product also called co-creation or co-design [1, 9]. Decentralizing companies or manufacturing sites offers scope for rethinking and developing organizational structures and processes, for example, to improve innovation cycles for new products. In the past, the price of the product and productivity were decisive for the development of a basic idea of a company. Today, also the product quality, flexibility, the ability to change and learn, as well as knowledge and services are decisive. Processes and structures must be consistently aligned with the market and adapted constantly. [10]

Paradigm shift towards agile assembly systems. Since industrialization, production methods and paradigms have changed. From craft production, in which products were produced locally at high cost and at customers request, to mass production, in which large quantities of products were produced at low cost, but with limited variants, to the current trend of mass customization, which follows the demand for higher product diversity. Through product family planning and modular products, different assembly combinations enable economic efficiency via an economy of scale at the component level. Currently, emerging technologies are creating a new paradigm the personalization of products tailored to the individual needs and preferences of consumers [1].

In line with production methods, production and thus assembly systems have adapted. In mass production, dedicated assembly systems were designed for a single task to achieve the highest possible throughput rates. For mass customization, systems need to be more responsive to changes in structure, resources, and product capacity and therefore reconfigurable assembly systems (RAS). As an extension the flexible assembly system (FAS) consists of machines capable of performing operations on a random sequence of parts of different types with little or no changeover time or effort. In summary, a RAS is a manufacturing system with custom flexibility and a FAS is a manufacturing system with general flexibility. The on-demand assembly system extends even further, being able to respond quickly to personalized consumer needs, and therefore requires configuration and reconfiguration in a cost-effective manner, which is possible through cyber-physical systems and agile manufacturing methods [1, 2].

Effect on cost estimation. The described changes in assembly systems and framework conditions show, that the requirements for flexibility, adaptability and fast reactivity are increasing, as well as the inclusion of more complex and variable processes. Resulting in shorter planning phases due to frequent and continuous adaptation of processes and operating resources. The reduction of planning times directly affects the functionality of planning tools. Considering the time effort, rough but reliable indicators gain significance compared to the prediction accuracy. In early planning phases, the precise prediction of process times as an indicator of efficiency and driver of cost per unit becomes less relevant especially in relation to preparation and setup times.

Therefore, cost estimation as a planning tool should not only depend on process times. The estimation has to incorporate a range of influential factors and provide transparency regarding cost origin to support decision-making. Furthermore, the amount of input must be limited, so that the model can easily be adjusted, remaining responsive to changes.

3 Methods of Cost Estimation

Cost estimation is particularly interesting as pre-calculation in early planning phases, providing a basis for decision-making and identifying a scope for actions and targets [11]. Cost estimation procedures are often methodologically systematized into qualitative and quantitative procedures [13]. A overview of relevant procedures is presented and assessed regarding their suitability for the framework described in Chap. 2.

Qualitative procedures. Qualitative procedures for cost-estimation indicate whether a solution is better or worse but without absolute values. These procedures rely on heuristic rules, good-bad examples or cost-structures based on expert knowledge or about previous productions. In general, they are rarely used or only in combination with quantitative methods, due to the fact, that they do not allow specific or transparent derivation of costs, hence thy are not discussed further [13].

Quantitative procedures. Quantitative procedures, on the other hand, represent a correlation between manufacturing processes and materials used and the expected costs [12]. These procedures can be classified into statistical/parametric (1), analogous (2), generative-analytical approach (3).

Parametric approaches (1) are based on statistical methods by determining parameters with significant influence on the production costs. These methods are characterized by a fast calculation of results. However, the accuracy corelates with the number of parameters and the work effort. Analog estimation methods (2) approximate costs using known costs for similar and comparable products. Therefore, such methods require comparable processes and products and are mostly used for minor product adjustments, as the estimation accuracy is related to similarity. For analytical procedures (3) products and processes are divided into units or features, thus also referred to as feature based. From these units or features, individual costs are calculated and summed up to total costs. Due to that, detailed knowledge about the processes or products is required, which can be difficult in early planning phases and some methods only determine a part of the total costs. Especially for the parametric, but also for the other approaches, machine learning methods are already applied, for example neuronal networks. Since these methods require an extensive database, they are not considered for the beginning, but for further research and refinement of the cost model [12].

None of the described approaches fulfills the requirements entirely for cost estimation in early planning phases in the described assembly framework. The analog methods are unsuitable at the product level, since they cannot represent the degree of flexibility. On the process level, it should be tested whether these methods are suitable for drawing conclusions from similar assembly operations, for example for recurring operations such as automated screwing, factoring in the number of screws. The analytical and feature-based approaches are suitable because they draw conclusions about the cost origin and thus support decision-making, while parametric models enable analysis of cost relationships and dynamic adjustment of fidelity.

Thus, to model the cost, logical cost blocks according to influential factors are specified. The separate costs are determined using parameterized methods, analytical methods and existing methods that are tailored for the individual considerations, for example the process costs according to Swift and Booker [14].

4   Description of the Model

The proposed cost model aims to resolve the contrast between accuracy of cost estimation and effort of data acquisition, as an economical solution for the assembly of small or single series. The goal is to achieve sufficient accuracy with relatively small effort. Furthermore, the model strives to fully consider the mentioned framework conditions, thus it should provide a basis for decision-making in small scale assembly systems as well as for extensions, such as the representation of disassembly processes.

The costs are estimated by using a mathematical optimization model. Following the principal equations illustrate comprehensibly the approach and the cost drivers but not the complete model. The indices and quantities are explained with the respective equation. Furthermore, variables are written in lower case and constants in upper case.

Total cost. The total cost (c_{tot}) equation contains 10 cost blocks, which are described individually below. The result is the cost per unit.

$$c_{tot} = c_{prep} + c_{mat} + c_{train} + c_{en} + c_{spc} + c_{proc} + c_{trans} + c_{buy} + c_{disp} + c_{serv} \quad (1)$$

Target function. Further, the objective for the optimization is formulated. Since economically sensible decisions should be found, the aim is to minimize costs.

$$\min c = c_{tot} + \sum_{g=1}^{G} \sum_{k=1}^{K} loa_{kkg} \quad (2)$$

Prerequisite and assumption. With a predefined precedence of operations, an operation or process sequence is deduced. Each operation i is assigned a unique position in this sequence and a workstation k. The order in which the resources are used also determines whether a station change occurs between operations and whether setup times are required. The first station always requires a setup time. The consideration of setup times is relevant because it is assumed that the operating stations are universally applicable. In addition, each station k is assigned to a specific level of automation (loa_k), based on the feasible types of processing, for example hybrid assembly with collaborative robot. The approach to determine the level of automation or mode of processing by costs was introduced by Salmi et al. [14].

Preparation. The preparation costs (c_{prep}) result from the costs for the modification of the assembly layout (c_{layout}) and for the setup costs at the individual stations (c_{setup}), both calculations take the lot size into account. The number of stations and their characteristics as well as their spatial arrangement in different layouts can be predefined in the model. The layout primarily influences the transport between the stations.

$$c_{prep} = c_{layout} + c_{setup} \tag{3}$$

Material. Material costs (c_{mat}) arise for each operation i at the corresponding station k. This equation is simplified, it contain a binary decision variable that is crucial to assign each operation to exactly one resource, which is relevant for the logic, but not for the understanding of cost determination.

$$c_{mat} = \sum_{k=1}^{K} \sum_{i=1}^{I} C_{mati} \tag{4}$$

Training and Programming. Training or programming costs (c_{train}) are incurring at each station k for an operation i. These costs are particularly difficult to determine, since the effort involved depends heavily on the experience of the employees, both in manual and automated assembly. The ratio of training costs decreases with the batch size N.

$$c_{train} = \frac{1}{N} \times \sum_{i=1}^{I} \sum_{k=1}^{K} C_{traink} \tag{5}$$

Processes. The process costs (c_{proc}) are calculated, according to Swift and Booker, from the process times (T_{proc}) and costs per time incurred at a station k (c_k) for manual and machine operation [14]. The division into manual or machine work, according to Salmi et al., depends on the stations level of automation (loa_k) [15]. The operating equipment costs (c_k) include maintenance, personnel, operating life, working hours and employees experience. The process times (T_{proc}) are comprised of joining and handling times, including additional factors for difficulties and special features of the operation [14].

$$c_{proc} = \sum_{i=1}^{I} \sum_{k=1}^{K} c_{kik} \times T_{procik} \tag{6}$$

Energy. The energy costs (c_{en}) are calculated from the electricity costs (c_E), the electricity (E_{con}) consumption of the particular station k and the process time (T_{proc}).

$$c_{en} = \sum_{k=1}^{K} \frac{E_{conk} \times C_E}{3600} \times \sum_{i=1}^{I} T_{procik} \qquad (7)$$

Space. The space costs (c_A) are calculated from the space costs per time as well as the space taken (A_{con}) and the process time (T_{proc}) of the particular station k.

$$c_{spc} = C_A \times (\sum_{i=1}^{I} \sum_{k=1}^{K} A_{conk} \times T_{procik}) \qquad (8)$$

Transport. The transport costs (c_{trans}) are calculated from the costs for transport (C_{trans}) and the transport times (T_{trans}) from a station m (predecessor) to a station k in a certain layout l. To consider the two options of outsourcing and disposal during transport, the K stations are extended by $K + 1$ outsourcing and $K + 2$ disposal.

$$c_{trans} = C_{trans} \times \sum_{k=1}^{K+2} \sum_{m=1}^{I} \sum_{l=1}^{L} (T_{transkml} \times \sum_{r=2}^{I} x_{translkmrl}) \qquad (9)$$

Outsourcing. Outsourcing costs (c_{buy}) are incurred for each operation i that is handed over to an external service provider.

$$c_{buy} = \sum_{i=1}^{I} C_{bi} \qquad (10)$$

Disposal. Disposal costs (c_{disp}) are incurred for each operation i that is disposed of. This alternative becomes attractive when the other costs of the processes, preparations exceed the value of the order (V). If a product is disposed of in operation i, the model also forces all subsequent operations to station $K + 2$ disposal. This option is only practical for disassembly and related processes.

$$c_{disp} = (C_D + V) \qquad (12)$$

Services. Service costs (c_{serv}) are calculated in the same way as process costs. The only difference is that it was specified for all services d that they have to be executed manually. Due to the complexity of the integration of such services, they were not represented in the previous expressions, but they offer the first approach to represent processes of disassembly in the model.

$$c_{serv} = \sum_{d=1}^{D} \sum_{k=1}^{K} c_{kdk} \times T_{procdk} \qquad (13)$$

5 Case Study

To apply the model, it was implemented in the optimization software Jump. Parallel a data set was formulated for an exemplary assembly environment, oriented to existing operating equipment at the TU Berlin. The exemplary assembly environment at the TU Berlin is established for research and experiments in the field of urban circular cloud assembly and therefore possesses a majority of the attributes mentioned in Sect. 2 [16]. In preparation tests on manual assembly of smoke detectors were conducted.

The case study is intended to test the functionality of the model, the feasibility of collecting relevant input data, and the comprehensibility and usefulness of the results. Furthermore, it was examined whether the model makes different decisions based on the input data regarding the degree of automation and process depth.

Assembly environment design. The data for the model regarding the environment are: 5 stations [incoming and outgoing goods, manual assembly, manual assembly with cobot, miniature robot, robot arm] in 2 different arrangements [layout 1, layout 2] with the automation levels [manual, hybrid, machine] and defined additional factors for [sensitivity and controls]. The observed process is the assembly of a smoke detector consisting of 5 parts [housing and housing cover, printed circuit board (PCB) with photodiode, battery, smoke collector] with the operations [input and output of goods, connecting PCB with smoke collector, connecting PCB with battery, inserting into housing, closing with housing cover]. A batch-size of [100] pieces was assumed.

Data set. In addition, data and costs specific to the assembly environment or process must be defined—cost and useful life of equipment, space and energy costs, duration of process steps, material costs. Within the case study, realistic values were researched for the location of Berlin. For example, the energy costs (c_E) a price of 0,2664 €/kWh was expected. Assuming that the qm costs 30€ per month and operating times are 1760 h per year, space costs of 0,0,000,568 €/m^2*sec was expected. The space taken (A_{con}) **for each station was calculated from the machine dimensions adding areas for accessibility and the** electricity (E_{con}) consumption was captured during machine operation. This data is known or captured in real environments, thus it is trivial to describe them in detail.

Result. The results are the total costs broken down by the described cost blocks as well as the determined order of operations and assignment to a station. This includes the decision which operations should be performed with which level of automation and about the possibility to outsource the operation. The possibility of disposal is to be considered reasonably only for disassembly (Fig. 1).

However, the determination of process times is not yet implemented in the model and was therefore performed in advance. Also, the selection of stations according to the degree of automation is only implemented exemplarily and does not reflect the decisions complexity. It is based on the factors of the programming and setup times as well as the batch size, since these times are only approximated, the quality of the statement is to be

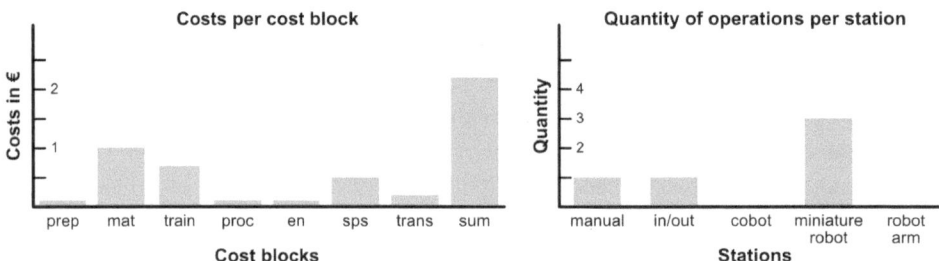

Fig. 1 Visual output of the model

doubted strongly. Nevertheless, it can be shown that such a decision can be made on the basis of more accurate data.

6   Conclusion

The model represents a first attempt for a suitable planning tool for agile, automated, and urban assembly environments. The logic blocks within the model, allow to trace the costs origins and alter them during planning. Furthermore, the optimization goal to minimize costs provides the opportunity to decide between different options. Recent options are the assembly layout, process depth and degree of automation.

The assembly layout is based on the relation between the rearrangement effort and the work piece transport in the respective layout. The case study included 2 layouts, which can be extended. The decision about the process depth is based on the possibility to cancel, dispose, or outsource processes. The decision about the degree of automation is based on the preliminary work of Salmi et al. [15], which has to be modified for the assumed input and according to recent technological developments in automation preparation. The other limitations and the further planned research are addressed separately.

The usefulness of the data input has to be evaluated in practice. The experimental data set is based on research, so it must be clarified whether the data used are also available in practical applications and can be generated with an appropriate effort.

The determination of process times has to be integrated. Process times, as previously mentioned, tend to be less important in terms of their accuracy, as they have a lower influence on the total costs in small-scale assembly systems. Nevertheless, the simplified (non-simulation based) determination of operating times of robots is a challenge as well as the determination of joint processing times of humans and robots in collaborative assembly environments [17]. While methods-time measurement is an established method to predict times for manual processes, its equivalent for robots has not been applied in practice and additionally requires quite accurate information about the individual process steps. Therefore, the estimation of process times with little information offers an important field of research and needs to be further addressed.

Further aspects that are not jet considered have to be integrated. This includes the holistic consideration of all assembly-related processes—assembly, disassembly, reassembly and recycling processes in terms of circular economy as well as the determination of the recycling path based on the cost model. In addition, ecological costs could be considered, e.g., through life cycle assessment of CO_2 emissions through production and crediting them in the total costs. Furthermore, the question arises how social aspects can be taken into account and whether further risk factors such as defect rates and scrap are necessary.

Further research needs to conducted. First, experiments are planned to check the accuracy of the cost estimation. The assembly environment described in the case study as well as the individual parts and products are available and preliminary tested for manual assembly and disassembly. Next, these are prepared for automated processes. After that, it is planned to expand the considered range of processes to be able to map important decisions for circular economy business models.

References

1. Hu, S.J.: Evolving paradigms of manufacturing: from mass production to mass customization and personalization. Procedia CIRP **7**, 3–8 (2013)
2. ElMaraghy, H.A.: Flexible and reconfigurable manufacturing systems paradigms. Int. J. Flex. Manuf. Syst. **17**, 261–276 (2005)
3. Balderjahn, I.: Nachhaltiges management und konsumentenverhalten, 2. vollständig überarbeitete Auflage. UVK Verlag, Stuttgart, (2021)
4. Jawahir, I.S., Bradley, R.: Technological elements of circular economy and the principles of 6R-Based closed-loop material flow in sustainable manufacturing. Procedia CIRP **40**, 103–108 (2016)
5. Ali, A., Wasim, A.: Innovative framework for assessing the impact of agile manufacturing in small and medium enterprises. Sustainability **14**, 11503 (2022)
6. Bastidas-Cruz, A., Heimann, O., Haninger, K., Krüger, J.: Information requirements and interfaces for the programming of robots in flexible manufacturing. In: Schüppstuhl, T., Tracht, K., Henrich, D. (eds.) Annals of Scientific Society for Assembly, Handling and Industrial Robotics. Springer Vieweg, Heidelberg (2020)
7. Cohen, Y., Faccio, M., Pilati, F. et al.: Design and management of digital manufacturing and assembly systems in the Industry 4.0 era. Int J Adv Manuf Technol **105**, 3565–3577 (2019)
8. Brandt, M., Gärtner, S., Meyer, K.: Urbane Produktion: Ein Versuch einer Begriffsdefinition. In: Forschung Aktuell (08), pp. 1–14 (2017)
9. Binner, H. F.: Organisation 4.0: MITO-Konfigurationsmanagement. Masterplan zur prozessorientierten Organisation. Springer, Wiesbaden (2018)
10. Buck, H.:Aktuelle Unternehmenskonzepte und die Entwicklung der Arbeitsorganisation. Visionen und Leitbilder. In: Spath, D., Westkämper, E., Bullinger, H. et al. (eds.): Neue Entwicklungen in der Unternehmensorganisation, pp. 57–77. Springer, Heidelberg (2017)

11. Schwabe, O., Bilge, P., Hoessler, A., Tunc, T.; Gaspar, D., Price, N., Sharir, L., Pasher, E., Erkoyuncu, J., Dietrich, F., et. al.: A maturity model for rapid diffusion of innovation in high value manufacturing. procedia CIRP 96, pp. 195–200 (2020)
12. Coenenberg, A.G., Fischer T.M., Günther T.: Kostenrechnung und Kostenanalyse. 9. end. Schäffer-Poeschel Verlag, Stuttgart (2016).
13. Layer, A., Brine, E.T., Van Houten, F., Kals et al.: Recent and future trends in cost estimation. Int. J. Computer Integrated Manufacturing, 15(6), pp. 499–510. (2002)
14. Swift, K.G., Booker, J.D.: Assembly costing. In: Manufacturing process selection handbook, pp. 393–409. Butterworth-Heinemann, Oxford (2013)
15. Salmi, A., David, P., Blanco, E., Briant, O., Summers, J.: A cost estimation model to support automation decision in assembly systems design. Int. J. Prod. Res. 56(24), 7426–7443 (2018)
16. D. Schröter, D., Kuhlang, P., Finsterbusch, T., Kuhrke, B., Verl, A.: Introducing process building blocks for designing human robot interaction work systems and calculating accurate cycle times. Procedia CIRP, 44, pp. 216–221, (2016)
17. Oguz, A., Bilge, P., Glodde, A., Rahlfs, S., Dietrich, F.: Assembly and through life servides in the context of urban cloud manufacturing. In: Future Automotive Production Conference Wolfsburg (2022)

Towards Myoelectric Control for Industrial Exoskeletons

Oliver Ott and Robert Weidner

Abstract

Physical support systems such as exoskeletons are gaining importance to support workers in smart and connected factories by reducing physical strain. In order for the exoskeleton to be integrated seamlessly with humans and tasks within the work environment, it must be able to detect and adapt to human intentions. Compared to conventional control, myoelectric control, i.e., control based on measured muscle activity, enables improved human–robot interaction and more intuitive interaction with humans. This paper explores how myoelectric control can improve the interaction of industrial exoskeletons. For this purpose, the anatomical basics of electromyography (EMG) for detecting muscle activity are presented, followed by existing myoelectric control strategies. The insights gained are applied to the implementation of myoelectric control for industrial exoskeletons. It is shown that EMG-based control benefits from human adaptivity, making it particularly suitable in the case of variant movements and changing conditions. Even though using electromyographic control for adaptation of the exoskeleton includes efforts in terms of setup and calibration, the insights gained into the physiological state of the human (e.g., muscle activity, fatigue) are crucial for the adaptation to the user.

O. Ott (✉) · R. Weidner
Chair of Production Technology, Institute of Mechatronics, Universität Innsbruck, Technikerstraße 13, 6020 Innsbruck, Austria
e-mail: oliver.ott@uibk.ac.at

R. Weidner
Laboratory of Manufacturing Technology, Helmut-Schmidt-University/University of the Federal Armed Forces Hamburg, Holstenhofweg 85, 22043 Hamburg, Germany

S. Ihlenfeldt et al. (eds.), *Annals of Scientific Society for Assembly, Handling and Industrial Robotics 2023*, https://doi.org/10.1007/978-3-031-74010-7_21

Keywords

Situation-dependent system behavior • Industrial exoskeleton • Electromyographic (EMG) control

1   Introduction

Digitalization and automation have led to a more intelligent and connected factory [1]. Consequently, the work environment is increasingly characterized by intensive human–machine collaboration and interaction [1, 2]. There are, however, still a substantial number of manual tasks involved in industrial processes [3], such as heavy lifting in logistics or repetitive assembly tasks in unergonomic postures in manufacturing processes. Especially the resulting physical strains on operators increase the risk of musculoskeletal disorders [4]. For this reason, more and more attempts are made to provide physical support to the human through technical aids [1], which, if used appropriately and in a targeted manner, can contribute to preserving their health [5]. Exoskeletons have become increasingly important as physical support systems for operators in industrial workplaces [1, 6, 7] to reduce physical strain.

Exoskeletons are technical support systems worn outside the body and support the human musculoskeletal system [8]. For implementation, diverse system approaches that differ in functionality and morphological structure [3] are pursued and intended to reinforce, simplify, stabilize, or complement movements [8]. In the industrial application, the provided support should be adapted to the work context, characterized by the human, the task, and the exoskeleton within the work environment [8]. However, it can be very versatile according to the variability of these individual factors. At the same time, it is crucial to support core activities (e.g., lifting and carrying, working at and above head level) without interfering with secondary activities (e.g., walking, sitting down, bending) [8].

Additionally, workers of the future factory must be highly flexible and adaptable due to the variety of work tasks and the high degree of individualization of products [1]. This means that support systems and their interaction with the operator must also be able to adapt to these changing conditions [9]. Ideally, the exoskeleton integrates seamlessly with humans and is controlled intuitively like a natural extension of their body [10]. To accomplish seamless integration, detecting the user's intentions is essential for the exoskeleton's control [11].

During the use of exoskeletons, the user's intention is approximated by measuring their physical interaction by observing executed movements, resulting forces, or moments [12]. Alternatively, by decoding the human body's biomechanical signals of skeletal muscles, i.e., electromyography (EMG), human–robot interaction (HRI) can be taken to a neuromuscular level [13]. A temporal delay in human neuromusculoskeletal physiology allows the determination of the exoskeleton forces and moments required for support before the motion [12]. Consequently, this kind of HRI paves the way for an almost seamless integration of exoskeleton and human [14]. Despite the temporal advantage, using EMG signals as a control input, i.e., myoelectric control, is not very common in industrial applications [13].

For this reason, this paper aims to investigate ways to improve the interaction of industrial exoskeletons through the use of myoelectric control. In this respect, existing myoelectric controls are examined. Since myoelectric controls are mainly used in rehabilitation and assistance [13, 15], thorough research includes these areas. The existing myoelectric control methods will be presented, and the general advantages and disadvantages of the application of myoelectric control will be shown. The insights gained on how myoelectric control can improve the interaction and adaptivity of exoskeletons will be illustrated using the example of an existing industrial exoskeleton.

2   Myoelectric Control Strategies

2.1   Biomechanical Foundations

The human body consists of more than 400 skeletal muscles, which account for about 40% of its weight [16]. By contracting, these muscles generate forces exerted on the musculoskeletal system via tendons and thus lead to movements. If there is an intention of movement, an impulse is transmitted from the brain through the central nervous system to the motoneurons, stimulating the skeletal muscles to contract. The resulting electrical potential can be measured as an electromyographic (EMG) signal prior to the occurrence of the movement itself [17].

EMG signals can be acquired using surface electrodes or intramuscular needle electrodes [18]. The latter yield a more precise signal than surface electrodes [19] but require an invasive procedure [20]. Therefore, surface EMG electrodes are more feasible, especially in an industrial context. They detect electrical signals by measuring the skin's surface tension (voltage) directly above the targeted muscle [18]. Consequently, the measured voltage is directly dependent on the position of the electrodes and, due to the different physiology of individuals (e.g., fat content and sweating of the skin), varies from subject to subject [10, 21].

The raw EMG signal cannot directly be used as an input to control machines or other technologies [22] as it is contaminated by various noise signals and artifacts [23]. Hence, the first step involves filtering the signal [24, 25]. The control features can then be extracted using methods such as root mean square (RMS) [22, 24, 26], wavelength (WL), mean absolute value (MAV) [13], or zero-crossing rate (ZC) [17]. The EMG signal can further be normalized by referencing each individual's maximum voluntary contraction (MVC) to compare measurements across individuals [17].

2.2   State of the Art

In order to achieve seamless interaction, the exoskeleton has to be able to adapt to the support context. An exoskeleton's adaptation is generally triggered by an event, a measured value, or based on models [27]. For myoelectric control, the adaptation can be measurement-based by the EMG signal passing a defined threshold or a proportional adjustment relative to the measured value. It can also be model-based, using EMG signals or its features as an input variable for muscle models or in a set of measured data for statistical models such as classification, regression (function), reinforcement learning, or neuro-fuzzy approaches.

The following figure (Fig. 1) overviews the procedure for implementing myoelectric control and existing myoelectric control strategies in exoskeletons, according to [13]. The different approaches will be discussed in more detail in the following.

In threshold-based control, the pre-processed EMG value is used as a trigger to adjust support. Here, the amplitude level or an extracted statistical feature is compared to a predefined value [13]. By combining this trigger with a predefined finite-state machine, support can be adjusted to the task. Threshold-based myoelectric control has the lowest computational cost but can only initiate sequential tasks, supporting only a limited number of motions at a time [13]. Once triggered, the exoskeleton will perform a predefined movement routine. This routine supports the task with a given speed and direction of movement, which cannot be changed at runtime [28].

In order to enable a continuous adaptation, the support can directly be linked to the EMG signal or its features using a proportional transfer function. This adaptation aims to artificially strengthen the muscle through the exoskeleton by coupling the support torque to the level of muscle activation [21]. It exploits the direct correlation between muscle activity and the amount of contraction [29]. By adjusting the gain of the measured EMG signal to alter the torque of support, it can be adapted to the user and the task [21].

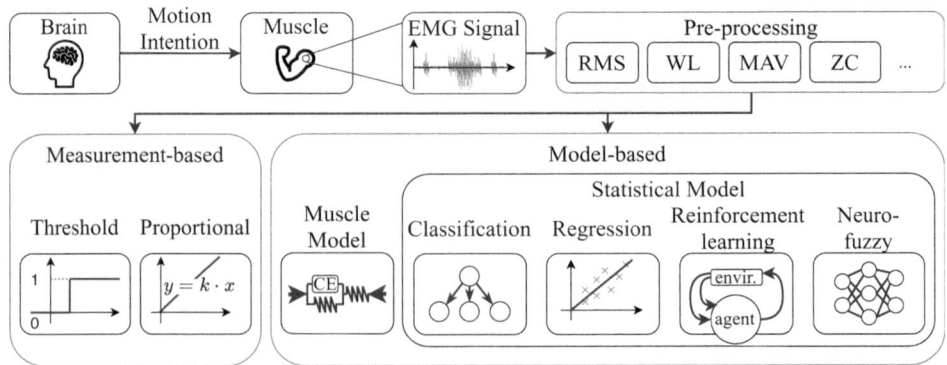

Fig. 1 General approach and overview of myoelectric control strategies in exoskeletons [13]

It enables continuous control for simple motion tasks (e.g., flexion, extension), can be easily implemented, and requires few computational resources. However, it cannot support complex motion tasks involving multiple joints [13].

Muscle models provide another opportunity to adapt the support to the measured EMG signal continuously. These models describe the functioning of the muscle based on the joint kinematics and the measured muscle activity. Compared to proportional fitting, these models enable a more accurate transfer function of muscle activity to the expected muscle force by modeling muscle function. The most common application is the Hill-type muscle model [13]. It understands the muscle–tendon complex as parallel and serial mechanical elements such as springs. The force generated by the muscle is calculated based on the current muscle length (geometric relationship) and its activation (derived from the EMG signal) [30]. Other parameters, such as the spring stiffness of the tendon, vary from user to user and need to be tailored to the user. Accordingly, optimization algorithms are used to simplify calibration and adapt the parameters to the individual user [28]. Muscle model-based control is generally effective and versatile in application, which supports its application in industrial contexts. However, the control has a high computational cost, a regular need for calibration, and is sensitive to unmodeled or external disturbances [13]. In addition, the Huxley model considers the stiffness distribution at the microscopic level. However, it is rarely used for time-critical applications due to the high computational cost of solving nonlinear partial differential equations [31].

Statistical models offer the possibility of adaptation based on trained algorithms and a corresponding data set. In myoelectric control, these can be further divided into classification algorithms, regression, reinforcement learning, and fuzzy logic.

Classification methods assign the extracted EMG features to the corresponding body movements. This way, movements that need to be supported can be identified by the occurrence of a specific activation pattern across an array of muscles and be supported accordingly. Commonly, support-vector machines (SVM), artificial neural networks [17], linear discriminant analysis (LDA), and K-nearest neighbors (KNN) are used [13]. Classification based on the available EMG features allows reliable assignment of movement patterns. However, these patterns are fixed and discrete and cannot be used for continuous motion control [25].

Regression describes forming a transfer function from a desired input variable to an existing output variable using an available data set. This procedure can be used to continuously estimate joint angles or joint moments based on EMG features. For this purpose, different methods like linear regression, artificial neural networks, or Kalman filters can be used [13]. These methods successfully match the EMG data to the corresponding joint angles and moments [13]. However, they require a broad data basis.

In reinforcement learning, optimal policies for control are independently learned by interacting with the environment. The derived control strategy is optimized by maximizing the reward function [13]. For myoelectric control, the kinematic states of the joint and the EMG signal can be used to follow a given trajectory [32]. The deviation of the motion from the given trajectory can be chosen as a reward function. By training within an unknown environment, the resulting control can interact with more complicated and uncertain environments [13]. However, the approach requires a high computational effort [33], and direct mapping to a multi-degree-of-freedom exoskeleton is limited [21].

Fuzzy logic enables the literal description of the desired action: if<premise>, then <action>. The approach of the neuro-fuzzy modifier combines a neural network with fuzzy logic [20]. During application, the fuzzy logic is fine-tuned iteratively using the neural network [20, 22]. In myoelectric control, the premises could be provided by the state of motion for each EMG signal. The motor actuation signal and gain can specify the desired actions [22]. The neural network trains the fuzzy logic with each control cycle, making it adaptive concerning different environments. However, the complexity of the logic increases with increasing degrees of freedom, which makes implementing and training the fuzzy logic more difficult [13].

2.3   Application to Exoskeletons

Seamless integration and intuitive control of an exoskeleton enable the system to become a natural extension of the human body [10]. In addition to covering the entire range of human motion, the system should also be capable of supporting the human if needed without interfering with movements and tasks. The existing myoelectric control strategies lay a foundation for enhanced human–robot interaction and improved adaptability with respect to the support context compared to conventional controllers [13].

Nevertheless, according to [22], there are general difficulties in implementing a control system based on EMG signals. The reproducibility of measurements is not given even with comparable movements and the same user. The use of muscles for a movement and thus also their muscle activity varies from user to user and especially between task executions, even for the same user. In addition, various muscles are used for a particular movement, with one muscle participating in different movements. The role of the muscle during movement changes depending on the angle of the joint.

Therefore, repeated calibrations for changing users and tasks are necessary. This is accompanied by a high setup time, including the placement of electrodes [12] and the need to capture many muscle groups, especially in complex joints (e.g., the shoulder joint) [13]. Further, depending on the joint, these muscles cannot be measured with non-invasive methods due to a lack of access (muscles not directly below the skin) [10].

According to [34], both myoelectric and dynamic model-based control can significantly contribute to the reduction of muscle activity when using an exoskeleton. Myoelectric control, however, plays to its strengths, especially by being adaptive when the dynamics of the task change. Due to the inherent dependence on the human motor command, this control benefits from the human's adaptability. The support can be adapted seamlessly to different speeds and handled weights, leading to a more symbiotic and intuitive function-ing. On the other hand, intrinsic mechanical control relies on an external adaptation (e.g., manually by the user) to respond to external changes, such as a variation in the dynamics of the task. If no adaptation is performed, the support will inadequately provide a force that is too high or too low.

Consequently, EMG-based control is particularly advantageous in dynamic motion sequences with broad variations. Here, the myoelectric-based system's adaptation is situation-dependent, adapted to the user and the respective task. In the case of consistent activities, dynamic models can provide sufficient support. Furthermore, integrating EMG sensors can provide additional information about the user's physiological state. For exam-ple, fatigue can be detected by an increasing envelope of muscle activity while muscle force remains constant [17]. This data can support the further development of exoskele-tons or be used as an objective assessment tool of support, for example. Therefore, sensor fusions are being pursued to gain information and overcome individual sensor classes' limitations. This process combines information from different sensor types to enable improved and more intelligent control [24]. Especially concerning the collection and fusion of physiological information, EMG-based technologies represent a central value in developing future technologies [35]. Accordingly, integrating EMG sensors is gaining importance as an extension of available sensors for more information and, thus, improved adaptability of the exoskeleton. In the following, these findings will be transferred to the characteristics of the work context.

3   Applicability in Industrial Contexts

3.1   Prerequisites

In the industrial context, adaptation focuses on the support context, i.e., the exoskeleton, the human, and the task within the corresponding work environment (Fig. 2A). For the exoskeleton's adaptation to the user, the level of support, the application time of force, and the movement dynamics are at least as relevant as the physical interface's morphology. The exoskeleton's support should be adapted to the user with their physical constitution (e.g., body size and stature), physiological changes during work (e.g., fatigue process), qualifications, and the associated range of tasks. It should adapt to the work environment and the requirements of the primary manual tasks (e.g., dynamics, posture, movement, work frequency, weights, variance) and the work equipment used (e.g., tools) [6, 36, 37].

Considering these general prerequisites, the following elaborates on how myoelectric control can contribute to successfully implementing exoskeletons in industrial contexts and, more specifically, for the exoskeleton Lucy [38]. The exoskeleton has been designed to support industrial tasks at and above head level. It uses pneumatic actuators to support upper arm elevation, which adjusts pressure depending on the angle between the upper

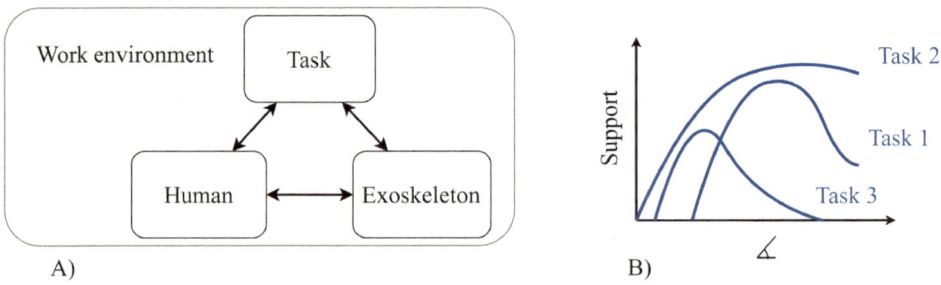

Fig. 2 Visual description of the support context's interrelationships (**A**) and set of exemplary force curves displaying the support (e.g., torque, force) over joint angle for different tasks (**B**)

arm and the upper body. The allocation of the required pressure, i.e., the level of support force, depends on the measured angle defined by force curves (Fig. 2B).

3.2   Proposed Implementation

Today, the user can manually adjust the adaption to the task. Therefore, the user can pick from a selection of predefined force curves (Fig. 2B). The force curves are defined for specific tasks and differ in the progression of the applied force along the arm angle. For example, in the repetitive task of overhead screwing, the user receives support only for arm movements at and above head level (Fig. 2B, Task 1). However, the movement of sorting boxes during logistics tasks requires support already at a lower arm position (Fig. 2B; Task 2 and Task 3). The ongoing research on adaptation with respect to the task enables the exoskeleton to select automatically force curves initiated by triggers [27]. For example, classification algorithms can adapt the force curve for the detected task. As described above, movement patterns can be classified very reliably using the measured muscle activities. However, the effort for setup and calibration does not compare to using already implemented sensors, such as angular encoders. These sensors allow the estimation of the position and movements, providing an equally reliable basis for classification algorithms. Long-term development should instead increase the focus on the handled objects and tools. Consequently, the force curves are extended to include the characteristics of the items. Forces and inertia caused by the movement of the objects can be detected by accelerometers along the arm segments and described by dynamic mechanical models. As a result, these mechanical models generate situation- and object-specific force curves, which can be optimized and adapted to the user in the next step.

The current version of Lucy provides a control interface to adjust the level of support to the user's needs. This manual adaptation scales the support force curve in its height. Advancing adaptivity towards the user should include an automatic and individualized adaptation of the force curve at runtime. For this purpose, data from the interaction between the user and exoskeleton, such as measured interaction forces and physiological parameters (e.g., muscle activities and findings on fatigue), need to be used. It is proposed to use continuous myoelectric control to adjust the control parameters to the human during use. Therefore myoelectric adaptation is intended to fine-tune the control with a focus on the individual user. On the other hand, muscle models are elaborated and highly accurate but out of scope because of their need for repeated calibration. The scaling in the level of support can already be done via the implementation of proportional myoelectric control. This adjustment allows a low-computation approach while continuously tracking the person's physiological state. Increasing muscle fatigue within an activity can be detected and taken into account with proportional myoelectric control. Regression models might be more suitable for complex motion sequences and a more accurate representation. However, the required model accuracy is directly related to the degree of desired adaptation to the user. The latter is taken into account by weighting the measured myoelectric input variables within the control. With a lower degree of adaptation to the user and a correspondingly low weighting, the inaccuracies in detecting the EMG signals also have a lower influence. Accordingly, the preparations for the EMG measurement can be reduced in the spirit of industrial applicability. As a result, the force curves will consider the dynamics of the items to be handled and the user's needs. Moreover, the processed data can be further used to improve the working environment in general.

At the moment, the data collected by the exoskeleton is primarily used for the implemented control. EMG can contribute to this data's meaningful use in future developments. Exoskeletons can be optimized by combining movement data with information on work processes and procedures, workloads, and workflow. Furthermore, it is possible to assess the user's posture at runtime ergonomically by fusing the measured muscle activity with current movement data and providing feedback to the user. The exoskeleton can also be considered a part of a connected manufacturing environment. There, information about the physiological state of the human being could facilitate interaction with the working environment.

4   Conclusion

Myoelectric controls have been analyzed to improve the interaction of exoskeletons with their users in the industrial work context. State-of-the-art control strategies have been presented, and possible applications for optimizing the interaction of the industrial exoskeleton Lucy have been proposed. The resulting strategy enables the force to be adapted at runtime optimized both to the dynamics of the handled item and to the user's needs via myoelectric control. The increased focus on the activity and the modeling of the physical impact through the handled item's dynamics allows for a good description of the necessary support using dynamic models. For this application, dynamic sensors, such as accelerometers, need to be implemented, while EMG sensors cannot account for their effort in setup and calibration. This contrasts with the individual adaptation of the support to the user. To be able to record the physiological state of the person, EMG sensors are of crucial importance. Through EMG-based control, fine-tuning the force curve to the user's physiological needs at runtime become possible. Therefore, the proposed control collectively enhances the exoskeleton's adaptation to the support context, essential to seamless and intuitive interaction. Accordingly, future research concerns integrating EMG sensors and implementing the control strategy as a proof-of-concept.

References

1. Mark, B.G., Rauch, E., Matt, D.T.: Worker assistance systems in manufacturing: a review of the state of the art and future directions. J. Manuf. Syst. **59**, 228–250 (2021)
2. Yao, Z., Linnenberg, C., Argubi-Wollesen, A., et al.: Biomimetic design of an ultra-compact and light-weight soft muscle glove. Prod. Eng. Res. Devel. **11**, 731–743 (2017)
3. de Looze, M.P., Bosch, T., Krause, F., et al.: Exoskeletons for industrial application and their potential effects on physical work load. Ergonomics **59**, 671–681 (2016)
4. European Agency for Safety, Health at Work, Kok, J., et al.: Work-Related Musculoskeletal Disorders: Prevalence, Costs and Demographics in the EU. Publications Office (2020)
5. Sgarbossa, F., Grosse, E.H., Neumann, W.P., et al.: Human factors in production and logistics systems of the future. Annu. Rev. Control. **49**, 295–305 (2020)
6. Hoffmann, N., Ralfs, L., Weidner, R.: Leitmerkmale und Vorgehen einer Implementierung von Exoskeletten. Zeitschrift für wirtschaftlichen Fabrikbetrieb **116**, 525–528 (2021)
7. Merkel, L., Berger, C., Schultz, C., et al.: IEEE International Conference on Industrial Engineering & Engineering Management: 10–13 Dec. IEEE, Piscataway, NJ (2017)
8. Weidner, R., Hoffmann, N.: Technische Unterstützungssysteme-Menschen gewollt (2020)
9. Di Pasquale, V., de Simone, V., Salvatore, M., et al.: Smart operators: how Industry 4.0 is affecting the worker's performance in manufacturing contexts. Procedia Comput. Sci. **180**, 958–967 (2021)
10. Cavallaro, E.E., Rosen, J., Perry, J.C., et al.: Real-time myoprocessors for a neural controlled powered exoskeleton arm. IEEE Trans. Biomed. Eng. **53**, 2387–2396 (2006)
11. Anam, K., Al-Jumaily, A.A.: Active exoskeleton control systems: state of the art. Procedia Eng. **41**, 988–994 (2012)
12. Gull, M.A., Bai, S., Bak, T.: Review on design of upper limb exoskeletons. Robotics. **9** (2020)
13. Fu, J., Choudhury, R., Hosseini, S.M., et al.: Myoelectric control systems for upper limb wearable robotic exoskeletons and exosuits-a systematic review. Sens. (Basel) **22** (2022)
14. Perry, J.C., Rosen, J., Burns, S.: Upper-limb powered exoskeleton design. IEEE/ASME Trans. Mechatron. **12**, 408–417 (2007)
15. Gantenbein, J., Dittli, J., Meyer, J.T., et al.: Intention detection strategies for robotic upper-limb orthoses: a scoping review considering usability, daily life application, and user evaluation. Front. Neurorobot. **16**, 815693 (2022)
16. Schmidt, R.F.: Neuro-und Sinnesphysiologie. Springer, Berlin Heidelberg (1998)
17. Abas, N., Bukhari, W.M., Abas, M.A. et al.: Electromyography assessment of forearm muscles: towards the control of exoskeleton hand. In: 2018 5th International Conference on Control, Decision and Information Technologies (CoDIT). IEEE, pp. 822–828 (2018)
18. Tiboni, M., Borboni, A., Vérité, F., et al.: Sensors and actuation technologies in exoskeletons: a review. Sens. (Basel) **22** (2022)

19. Gunasekara, J., Gopura, R., Jayawardane, T., et al.: Control methodologies for upper limb exoskeleton robots. In: 2012 IEEE/SICE International Symposium on System Integration (SII). IEEE, pp. 19–24 (2012)

20. Gopura, R.A.R.C., Kiguchi, K., Li, Y.: SUEFUL-7: a 7DOF upper-limb exoskeleton robot with muscle-model-oriented EMG-based control. In: 2009 IEEE/RSJ International Conference on Intelligent Robots and Systems. IEEE, pp. 1126–1131 (2009)

21. Lenzi, T., de Rossi, S.M.M., Vitiello, N., et al.: Intention-based EMG control for powered exoskeletons. IEEE Trans. Biomed. Eng. **59**, 2180–2190 (2012)

22. Kiguchi, K., Hayashi, Y.: An EMG-based control for an upper-limb power-assist exoskeleton robot. IEEE Trans. Syst. Man Cybern. B Cybern. **42**, 1064–1071 (2012)

23. de Luca, C.J., Gilmore, L.D., Kuznetsov, M., et al.: Filtering the surface EMG signal: movement artifact and baseline noise contamination. J. Biomech. **43**, 1573–1579 (2010)

24. Tang, Z., Yu, H., Yang, H., et al.: Effect of velocity and acceleration in joint angle estimation for an EMG-based upper-limb exoskeleton control. Comput. Biol. Med. **141**, 105156 (2022)

25. Lian, P., Ma, Y., Zheng, L., et al.: A three-step hill neuromusculoskeletal model parameter identification method based on exoskeleton robot. J. Intell. Robot. Syst. **104** (2022)

26. Nowak, M., Vujaklija, I., Sturma, A., et al.: Simultaneous and proportional real-time myocontrol of up to three degrees of freedom of the wrist and hand. IEEE Trans. Biomed. (2022)

27. Ott, O., Ralfs, L., Weidner, R.: Framework for qualifying exoskeletons as adaptive support technology. Front. Robot. AI. **9** (2022)

28. Buongiorno, D., Barsotti, M., Barone, F., et al.: A linear approach to optimize an EMG-driven neuromusculoskeletal model for movement intention detection in myo-control: a case study on shoulder and elbow joints. Front. Neurorobot. **12**, 74 (2018)

29. Burns, M.K., Pei, D., Vinjamuri, R.: Myoelectric control of a soft hand exoskeleton using kinematic synergies. IEEE Trans. Biomed. Circuits Syst. **13**, 1351–1361 (2019)

30. Liu, H., Tao, J., Lyu, P., et al.: Human-robot cooperative control based on sEMG for the upper limb exoskeleton robot. Robot. Auton. Syst. **125**, 103350 (2020)

31. Xu, X., Deng, H., Zhang, Y., et al.: Continuous grasping force estimation with surface EMG based on huxley-type musculoskeletal model. IEEE. Trans. Neural. Rehabil. (2022)

32. Hamaya, M., Matsubara, T., Noda, T., et al.: Learning assistive strategies for exoskeleton robots from user-robot physical interaction. Pattern Recogn. Lett. **99**, 67–76 (2017)

33. Xiao, F.: Proportional myoelectric and compensating control of a cable-conduit mechanism-driven upper limb exoskeleton. ISA Trans. **89**, 245–255 (2019)

34. Roche, E. (ed.): 2020 8th IEEE RAS/EMBS International Conference for Biomedical Robotics and Biomechatronics (BioRob). IEEE, Piscataway, NJ (2020)

35. Vélez-Guerrero, M.A., Callejas-Cuervo, M., Mazzoleni, S.: Artificial intelligence-based wearable robotic exoskeletons for upper limb rehabilitation: a review. Sens. (Basel) (2021)

36. Ralfs, L., Hoffmann, N., Weidner, R.: Method and test course for the evaluation of industrial exoskeletons. Appl. Sci. **11**, 9614 (2021)

37. Hoffmann, N., Prokop, G., Weidner, R.: Methodologies for evaluating exoskeletons with industrial applications. Ergonomics **65**, 276–295 (2022)

38. Otten, B.M., Weidner, R., Argubi-Wollesen, A.: Evaluation of a novel active exoskeleton for tasks at or above head level. IEEE Robot. Autom. Lett. **3**, 2408–2415 (2018)

Approach for Modeling and Monitoring Hazards Towards Safety Digital Twins for Robotic Applications

Ibrahim Al Naser, Aquib Rashid, Mohamad Bdiwi, and Steffen Ihlenfeldt

Abstract

Human–robot collaboration (HRC) is an advancing field which integrates the skills of humans and robots for realizing more flexible and efficient production lines. However, one of the main challenges of HRC is ensuring human safety. This paper presents a new approach for 1. modeling the most common hazards mentioned in ISO 12100, 2. developing a multilayer architecture for monitoring these hazards in a 3D environment as a digital shadow at present, and as a basis for safety-related digital twins in the future. All the safety-related data can be collected and analyzed over time. Consequently, possible hazard situations could be detected before they occur. The proposed method considers the safety regulations and standards and makes the user able to change the safety-related functions and sensors intuitively for dynamic processes and agile environments. The approach has been successfully implemented and tested in an industrial use case.

Keywords

Human–robot collaboration · Hazards modeling · Risk assessment · Hazards monitoring

I. Al Naser (✉) · A. Rashid · M. Bdiwi · S. Ihlenfeldt
Fraunhofer Institute for Machine Tools and Forming Technology IWU, Reichenhainer str.88, 09126 Chemnitz, Germany
e-mail: ibrahim.al.naser@iwu.fraunhofer.de

1   Introduction

HRC has the potential to improve efficiency and productivity in various industries like manufacturing, construction, healthcare, and transportation. However, the integration of robots into human environments poses certain hazards and safety concerns. To ensure the safety of human operators, hazard modeling and monitoring for human–robot collaboration is crucial. This involves the use of computer simulations and modeling techniques to predict the system behavior under different scenarios, such as system failure or abnormal operation based on standards such as 15,066. These simulations can be used to identify potential hazards and estimate the likelihood of a particular occurring hazard. This information is of high importance to designing safety mechanisms and protocols to mitigate the potential impact of the hazards. On the other hand, hazard monitoring for HRC involves the continuous observation of the system to detect potential hazards and safety issues in real time. This can be achieved through the use of sensors, cameras, and other monitoring devices that are integrated into the HRC system. By identifying potential hazards early on, authorities can take proactive measures to reduce the impact of the hazards and protect human lives and property. By connecting both monitoring and modeling of hazards, the gap between the life cycles of the machine can be filled, which will enhance overall safety and increases the HRC flexibility. Risks associated with human–robot collaboration (HRC) applications, as well as other types of machines, can occur at various stages of the machine life cycle, including the planning, operation, and maintenance phases. In the planning phase, the risk assessment process is to identify and mitigate the hazards described in ISO Standards 12,100 [1]. This requires engineering efforts, time, and expertise to complete it. Therefore, research conducted in the past regarding risk assessment has focused on developing methods enabling more efficient and adaptable processes. Inam et al. [2] apply the HAZOP method with HRC, based on a set of questions to conduct a risk assessment, which was simple, systemic, and flexible. However, it needs an expert to provide the questions and it is utilizing the information in the planning phase without any connection with the operation phase. Huck et al. [1, 3] show an interesting result to detect and simulate the collision of humans with collaborative robots in HRC. Nevertheless, the work focused on only cobots and one hazard which is a mechanical collision. This cannot be applied to other scenarios as most HRC scenarios include different hazards. Moreover, formal verification methods used by Askarpour et al. [4] evaluate the risk assessment for HRC based on the SAFER-HRC method. A flexible method is developed to mitigate and verify hazards in HRC but it presents a complex process and focuses on the planning phase and cannot communicate with the operation phase. Kroys A et al., investigate the possibility to model mechanical hazards in a 3D environment [5], which presents more flexibility to the process through virtual reality along with HRC. The work has considered mechanical hazards, which are based on the collision of objects in virtual reality. These considerations make the work specific to a use case and cannot be applied to HRC in general. Hornung et al. [6] conducted an intriguing survey on the tools available in the market for risk assessment in HRC. The survey evaluated the tools based on their flexibility and compliance with HRC and revealed that many of them do not adequately support risk assessment for HRC applications. The authors noted that the users need more expertise in safety to conduct risk assessments for HRC and emphasized the need for standardization and tools to simplify this process. On the other side, in the operation phase, many methods and algorithms have been developed to ensure human safety in HRC. This phase has many requirements such as real-time, flexibility, compliance with standards and norms, and safety, which are needed to be fulfilled. The interesting work of Rashid et al. [7] is to observe the minimum distance between the human and industrial robot in the operation phase accurately with high frame rate sensors based on a switching mechanism between two sensors. The results show an improvement in the productivity of the work and enhanced safety overall. However, considering just the mechanical hazards from the robot as a dynamic object is not enough to cover all the hazards in the HRC. Chin [8–10] uses computer vision to improve the safety of humans in HRC using different sensors such as RGB, Depth, and Thermal. Their algorithms detect the human position and safety parameter or the hazard which leads to a flexible solution but most of the time is for cobot application because of the delay from the algorithm. Despite the mitigation of the hazards in planning but still, accidents occur in HRC because of the unawareness of the worker or improper mitigations. In conclusion, improving the safety of humans in HRC depends on mainly two key aspects, which consist of filling the gap between the planning and operation phases and developing a reliable and efficient model for different types of hazards. In this context, the paper presents an approach for modeling and monitoring different types of hazards.

2　Proposed Approach

HRC can be used in different industrial applications which required various sets of different risks. This paper presents a new architecture for modeling the hazards in two layers: geometrical and behavioral. The first layer models the basic geometry of the hazards, while the second one describes their behavior. This approach will be presented in this section in more detail.

2.1　Geometry Layer (GL)

The common hazards in the field of human–robot collaboration are covered in this approach. As the user starts in the risk assessment process to detect the hazards, an additional step could be added to model the hazards at GL. By using simple 3D shapes such as Cuboid, Spheres, and Plans, the visualization part could be generalized to all hazards.

Table 1 Modeling parameter for GL with an example

Parameters	Description		Example
Position	x, y, z		25,125,21
Dimensions	Width, height, depth		5,1.5,2
Type	Static, dynamic		Dynamic
Attached	Attached to robot		False
Shape	Plane, sphere, cube		Cuboid

As a final result of this layer, a table to model the hazards at GL in a generic form is shown in Table 1.

2.2   Behavioral Layer (BL)

Each hazard has its behavior, and how it affects the human body depends on a variety of factors, including material, distance, particular body parts, etc. The structure of the BL and GL is connected in one direction, known as "Digital Shadow". This is currently being used for monitoring purposes, but it has the potential to be expanded to include two-way communication, known as "Digital Twin". The potential damage caused by these risks can differ depending on the amount of time a human spends in contact or exposure to the hazard. Thus, hazards can be classified into two categories: critical time and noncritical time hazards. For non-critical time hazards, time is considered the main parameter, while for critical time hazards, the distance between the human and the hazard is the primary factor to be considered. The classification of the hazards relative to their categories is presented in Fig. 1. For example, damage caused by exposure to radiation, thermal, or noise hazards can vary depending on the duration of exposure. This highlights the importance of understanding and modeling the hazards in these applications to identify and mitigate potential risks. Moreover, certain hazards are highly time-sensitive and require immediate response, within milliseconds. These hazards can be critical and can cause severe damage if not addressed promptly, highlighting the need for real-time monitoring and immediate response mechanisms like electrical, mechanical, or even thermal hazards. This brings the need to be proactive to such hazards by considering distance parameters as the real-time term can vary from one application to another. This highlights the importance of being proactive in addressing such hazards by utilizing distance parameters, as real-time requirements can vary depending on the application. In the context of BL, the safety distance refers to the distance between humans and hazards, and it serves as a trigger for the time accumulator, which is shown in Eq. 1. Additionally, the safety distance is used to create flags for critical time hazards, such as electrical, mechanical, or other hazards. For mechanical hazards related to robots, the safety distance is calculated according to

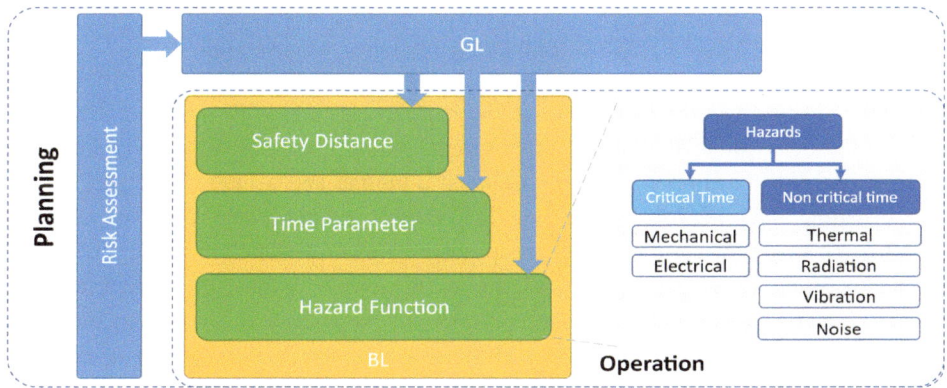

Fig. 1 Multilayer architecture and hazards classification

ISO Standards 13,855 and is derived from the risk assessment. Furthermore, Eq. 1 consists of different parameters in which S represents the calculated safety distance from the robot perspective based on ISO Standards 13,855, C_d and R_s are to consider the shape calculation cuboid or sphere respectively using enablers E_c and E_r. The time accumulator is utilized for non-critical time hazards and generates a flag based on the permissible quantities identified in the risk assessment and hazard function.

$$S_d = H_d - S * E_s - C_d * E_c - R_s * E_r \tag{1}$$

$$f_1 = \begin{cases} 1, S_d \leq D \\ 0, S_d > D \end{cases} \tag{2}$$

$$W_s = \begin{cases} t = t + 1, S_d \leq D \\ 0, \qquad\quad S_d > D \end{cases} \tag{3}$$

$$f_2 = \begin{cases} 1, W_s > t_s \\ 0, W_s \leq t_s \end{cases} \tag{4}$$

Additionally, the hazard function represents the fundamental behavior of the hazard and it is based on time or distance. The accumulated time is triggered when the allowed safety distance is violated. Then, time and distance are compared to the behavior function to predict the risk and alert users through flags f_1 and f_2 Eqs. 2, 3, and 4. Furthermore, because of the diversity in materials and how they respond, especially to thermal hazards, modeling all hazards is challenging. Therefore, the BL takes into account several hazards based on the common application of HRC. For critical time hazards e.g., Electrical and Mechanical, Eq. 1 is utilized to monitor the safety distance. In the following, the considered non-critical time hazards and their modeling are presented.

Thermal hazards. Limited research exists on generic modeling for thermal hazards. The ISO 17223 standards clearly define thermal hazards and outline a method for determining the behavior of the hazard by observing safety distances. Equations 3 and 4 are used to monitor the time of collision between a human and a hazard and provide accumulated time. ISO 17223 also specifies the types, conditions, and materials of thermal hazards. To model this behavior in this approach, a suitable function and optimized constants for fitting this curve are required. To investigate which function is suitable for such an effect, a comparison between different functions has been investigated. In addition, an interesting work from Hockett et al. [11] to model the thermal effect will be compared. Figure 2 illustrates the comparison of different functions with values has been digitalized from the original graph. Table 2 shows the root square minimum error of all functions and indicates that the Hocket-Sherby is the best fitting for the thermal effect.

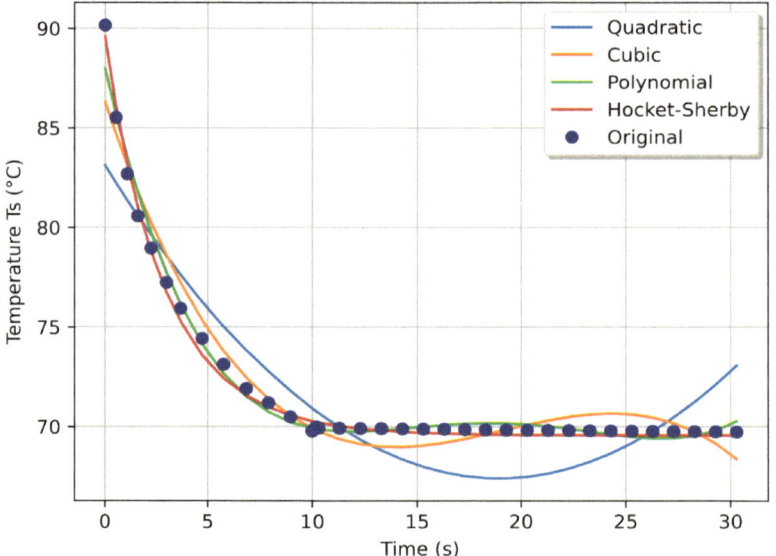

Fig. 2 Comparison between fitting functions for thermal hazards with ceramic material

Table 2 Thermal fitting functions with their accuracy and standard deviation (STD)

Function	RMS
Quadratic	0.9726215
Cubic	0.9935867
Polynomial 4th	0.0757873
Hocket-Sherby	0.0104166

Radiation hazards. There are two main types, ionizing, and non-ionizing. Ionizing radiation is high energy and can ionize atoms, causing DNA damage and increasing cancer risks such as X-rays and radioactive materials. Non-ionizing radiation has lower energy and cannot ionize atoms, but can still cause harm like radiofrequency (RF) radiation and electromagnetic fields (EMFs). Standards and regulations have been developed to limit radiation exposure levels to ensure safety. The focus of this work is on non-ionizing radiation types. Specific Absorption Rate (SAR) measures the rate of energy absorbed by the body when exposed to a radio frequency and electromagnetic field. SAR is expressed in W/kg and used to ensure RF energy levels are below safe limits based on DIN EN 12,198–1:2008–11. Furthermore, threshold values are measured in specific conditions for example time and distance. Radiation Standards for electrical, magnetic, and electromagnetic fields were measured with a distance of 0.25 m and 6 min for frequencies above 100 kHz based on DIN EN 12,198–1:2008–11. Radiation hazards could be monitored through Eqs. 3 and 4 with fixed time and distance as it is mentioned in DIN EN 12,198–1:2008–11.

Vibration hazards. Quantitative calculations of the vibration are divided into two parts: the vibration of the hand alone or the complete body. The calculation methods are described in EU 2002/44/EG in detail. The hazard is based on time and vibration value; both values can be taken from the user through risk assessment. Finally, from the standards 2002/44/EG, a threshold value is taken, and it will be classified based on that if it is harmful or not and its mitigation as is shown in Fig. 3. To monitor these hazards, Eqs. 3 and 4 could be used considering which part of the human is monitored along with the accumulated time.

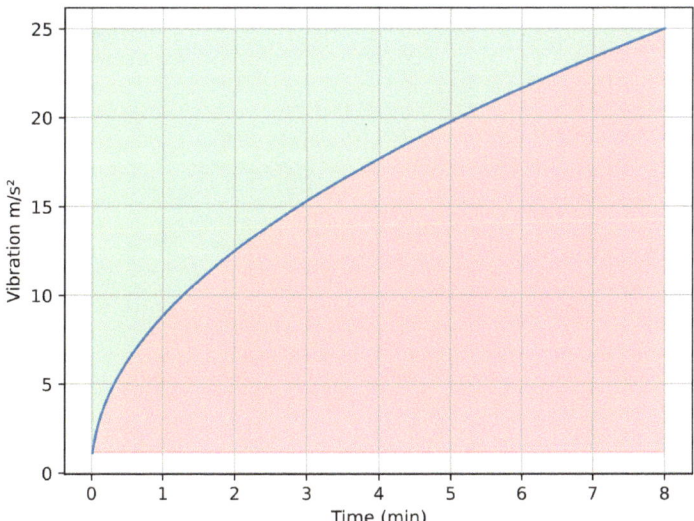

Fig. 3 Hazard function for vibration with time relation

Noise hazards. Modeling noise hazard is by finding a relation between the level of noise and distance. Furthermore, this will connect to the limit's values from the user to modeled function. Thus, a sound pressure level (SPL) can be taken as a measuring value with the distance. Sound attenuation describes how the SPL changes with increasing distance from the noise source. The sound attenuation formula is as follows:

$$SPL_2 = SPL_1 - 20Log\left(\frac{R_2}{R_1}\right) \tag{5}$$

where: SRL_2 Sound pressure level at the calculated point, SRL_1 Sound pressure level at reference point 2, R_1 Distance from the sound source to point 1, R_2 Distance from the sound source to point 2. Figure 4 shows the relation between the distance and SPL$_2$ which is set to 125 dB. As it is shown, the referenced points are measured at a distance of 1 m and then which is set to 125 *dB*. As it is shown, the referenced points are measured at a distance of 0.5 *m*. The noise levels increase as the object becomes closer to the sources. To have a full model of these hazards, exposure time needs to be calculated to be monitored during the operation phase. The information about the noise level based on the distance provides the information for the second layer of calculations based on Eq. 6, which is based on the exposure time and is expressed as follows:

$$T = \frac{480}{2^{(L-85)/3}} \tag{6}$$

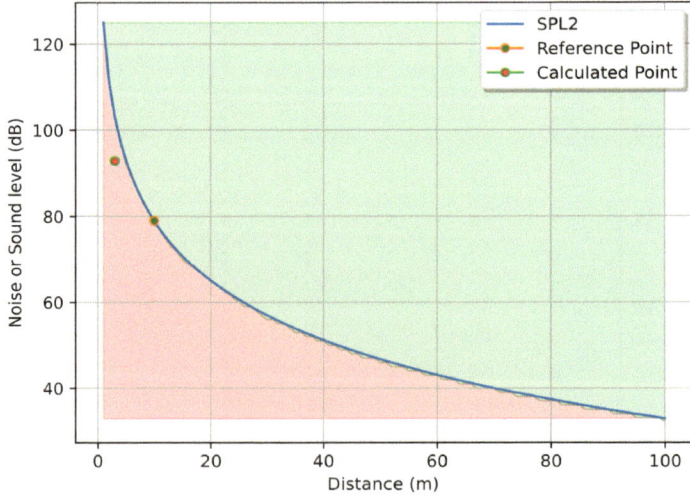

Fig. 4 85db noise level based on the distance from the source of the level

Finally, with the example from Fig. 4, the exposure time will be changed from 480 min which is a complete shift of 8h, and, with 0.5 m, it will become ~ 19.5 h because the SPL_2 increased to 98.86 *dB*. The time parameter is essential to monitor the noise hazards during the operation phase. Here, risk monitoring is done by using *3D* area modeling based on Eqs. 2, 3, and 4 to identify areas where noise is higher than others and keep the worker safe within the permissible limitations.

2.3   Hazards Monitoring

DynaRisk Framework serves as the basis for safety-related digital and encompasses various modules, with a focus on monitoring and modeling in this work. The modeling of hazards has already been thoroughly explained. The monitoring module is particularly important for ensuring safety and versatility in HRC applications, as illustrated in Fig. 6. A real-time hazard monitoring system that utilizes a combination of mathematical models and sensor data to assess potential hazards and alert operators of potential risks is shown in Fig. 6. The system operates by first utilizing information from the BL, which includes mathematical models of the hazards being monitored. In addition to the BL, the system also utilizes information from the GL, which includes geometry data such as position, size, and hazard type that provide information about the environment. This data provides a detailed understanding of the location and the movement of potential hazards in the physical environment. The algorithm then uses this information to set monitoring parameters in the "enablers" stage as shown in Fig. 5. This can include various types of sensors, such as lidar, radar, cameras, or other types of enabler's parameters, which activate and deactivate specific settings for the hazards. Once potential hazards are detected, the algorithm proceeds to the "check safety distance" stage, where it uses the information from the BL and GL to determine the minimum safe distance required to avoid the hazard. If the sensor data indicates that the safe distance is not being met, the algorithm proceeds to the "time accumulator" stage. This stage takes into account the amount of time that has passed since the hazard is detected and generates warning flags based on this time. The warning flags alert operators to take appropriate action to avoid the hazard. In case of critical-time hazards, the time accumulator is set to zero, and the flags are triggered directly, to ensure that the operator is alerted immediately and takes the necessary actions to avoid the hazard. This algorithm provides a robust and scientifically-based approach to real-time hazard monitoring, by utilizing mathematical models and sensor data to accurately predict and mitigate potential risks. It can be used as a basis for safety-related digital twins, which can provide a detailed understanding of the performance of physical systems in real time. Further, the method considers different types of hazards and sensors, which makes it more flexible to be implemented in different HRC scenarios. Safety-related digital twins (DynaRisk) collect all workplace hazard data to give the user a complete picture of all possible threats to workers in industrial environments.

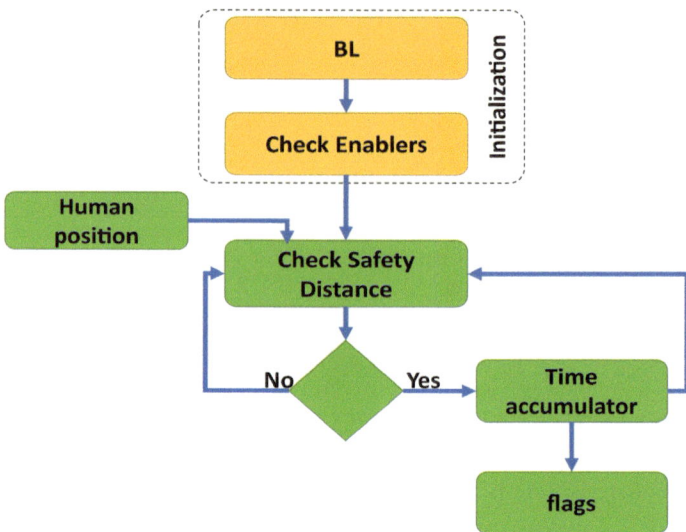

Fig. 5 Flow diagram shows the monitoring algorithm data flow

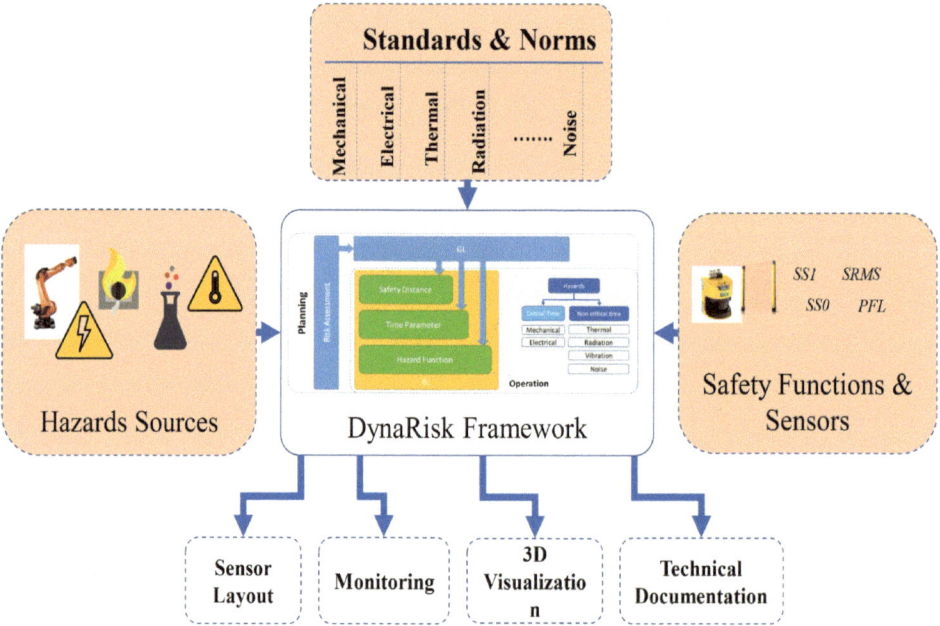

Fig. 6 Structure of DynaRisk framework and its modules

3 Results and Discussion

To examine the feasibility of the proposed approach in mitigating hazards, a practical scenario is developed and implemented within a simulation setting. The goal is to create a safe and efficient collaboration between human workers and robots in the workplace. The experiment involves the use of a KR-220 robot, a vehicle chassis as shown in Fig. 7. The main goal of the scenario is to assist human workers to perform heavy physical tasks in industry. To effectively model the scenario, a dynamic risk assessment is performed for both thermal and mechanical hazards, as shown in the modeled environment and physical world in Fig. 7. Based on the results of the risk assessment, three potential hazards are identified: one thermal hazard originating from the robot controller cabinet, and two mechanical hazards from the robot and the chassis. In addition, the human position within the cell is captured through the use of a Lidar and camera sensor system. The safety flags are activated based on the established distance during the planning phase. The robot moves with its first joint in a range of $0°-180°$ due to restrictions from the surrounding environment (walls). A human subject moves in the cell close to potential hazards to test the accuracy of the safety distance and triggering process. The blue line represents the dynamic mechanical hazard of the distance between the human and the robot, and the blue area represents the duration of the triggered safety flag. The human then moves near a thermal hazard, which is depicted in a green line and area.

Lastly, the human moves around the chassis of a car, which is represented by the red line and area. Fig. 8 displays the fluctuations in distance and safety flags during a single cycle of the scenario. The robot decelerates in response to triggered safety flags and comes to a halt once it reaches a designated distance. Meanwhile, the car chassis must be repositioned with each new cycle and the human remains in close proximity. By maintaining a consistent safety distance, incidents involving other objects within the cell are minimized, as the robot remains focused on its tasks. The performance of the monitoring algorithm is evaluated by measuring the average time needed to execute the

(a)   (b)

Fig. 7 Demo scenario: **a** front view cell. **b** side view cell

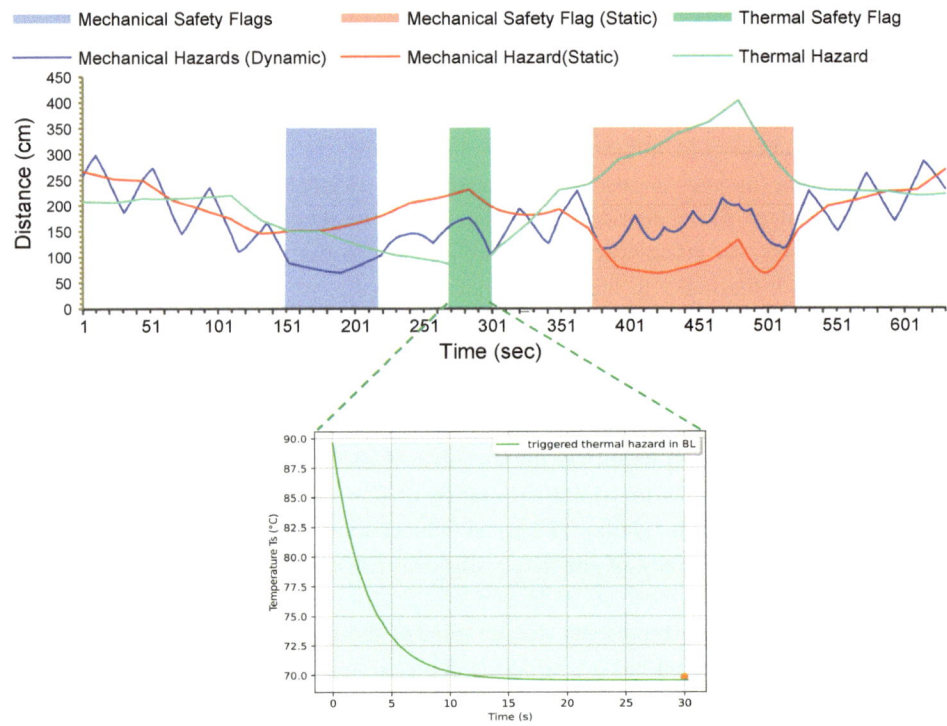

Fig. 8 Monitoring of 3 hazards with triggered safety flags and thermal behavioral layer function

monitoring section. This time can vary as each hazard has its own hazard function, which can be computationally intensive.

4   Conclusion

This paper introduces a new method for modeling and monitoring safety risks in human–robot collaboration (HRC). The proposed hazard model is comprised of two layers, the geometrical and behavioral layers, and considers safety distance and time as crucial parameters for hazard modeling. The model incorporates simulations to account for different types of hazards including radiation, thermal, noise, vibration, electrical, and mechanical hazards. The paper also provides evidence of the proposed solution's efficacy through an industrial scenario and evaluates the algorithm's performance based on the average time required to monitor the thermal and two types (static and dynamic) of mechanical hazards. This new approach provides real-time control for HRC safety.

References

1. Huck, T.P., Ledermann, C., Kröger, T.: Virtual adversarial humans finding hazards in robot workplaces. In: Proceedings-IEEE International Conference on Robotics and Automation (2021). https://doi.org/10.1109/ICRA48506.2021.9561668
2. Inam R., et al.: Risk assessment for human-robot collaboration in an automated warehouse scenario. In: IEEE International Conference on Emerging Technologies and Factory Automation, ETFA, pp. 743–751 (2018). https://doi.org/10.1109/ETFA.2018.8502466
3. Huck, T.P., Ledermann, C., Klose, S., Dai, F., Matthias, B., Byner, C.: Development of a simulation-based risk assessment tool for HRC applications. In: ISR Europe 2022; 54th International Symposium on Robotics, pp. 1–8 (2022)
4. Askarpour, M., Mandrioli, D., Rossi, M., Vicentini, F.: SAFER-HRC: safety analysis through formal verification in human-robot collaboration. In: Lecture Notes in Computer Science (including subseries Lecture Notes in Artificial Intelligence and Lecture Notes in Bioinformatics), vol. 9922 LNCS, pp. 283–295 (2016). https://doi.org/10.1007/978-3-319-45477-1_22/COVER
5. Kroys, A., Maertins, R., Ortmeier, F.: An approach to semi-automatically determine mechanical hazards in VR models. In: IEEE/ASME International Conference on Advanced Intelligent Mechatronics, AIM (2012). https://doi.org/10.1109/AIM.2012.6265981
6. Hornung, L., Wurll, C., Hein, B.: Evaluation of software solutions for risk assessment focusing on human-robot collaboration, pp. 3–14 (2023). https://doi.org/10.1007/978-3-031-22216-0_1
7. Rashid, A., Bdiwi, M., Hardt, W., Putz, M., Ihlenfeldt, S.: Efficient local and global sensing for human robot collaboration with heavy-duty robots. In: IEEE International Symposium on Robotic and Sensors Environments, ROSE 2021-Proceedings (2021). https://doi.org/10.1109/ROSE52750.2021.9611766
8. Chin, R.T., Dyer, C.R.: Model-based recognition in robot vision. ACM Comput. Surv. (CSUR). **18**(1) (1986). https://doi.org/10.1145/6462.6464
9. Hoshino, S., Niimura, K.: Robot vision system for human detection and action recognition. J. Adv. Comput. Intell. Intell. Inform. **24**(3) (2020). https://doi.org/10.20965/jaciii.2020.p0346
10. Al Naser, I., Dahmen, J., Bdiwi, M., Ihlenfeldt, S.: Fusion of depth, color, and thermal images towards digital twins and safe human interaction with a robot in an industrial environment. In: RO-MAN 2022-31st IEEE International Conference on Robot and Human Interactive Communication: Social, Asocial, and Antisocial Robots, pp. 532–537 (2022). https://doi.org/10.1109/RO-MAN53752.2022.9900548
11. Hockett, J.E., Sherby, O.D.: Large strain deformation of polycrystalline metals at low homologous temperatures. J. Mech. Phys. Solids. **23**(2) (1975). https://doi.org/10.1016/0022-5096(75)90018-6

Optimization Problems in Production and Planning: Approaches and Limitations in View of Possible Quantum Superiority

Maximilian Zwingel, Christopher May, Sebastian Reitelshöfer, and Wolfgang Mauerer

Abstract

Modern production and process planning is characterized by complex and diffuse inter-relationships of parameters, properties and control values. New materials, innovative production technologies, differing degrees of automatability and application dependency form a multidimensional problem space for optimization, which cannot be efficiently solved by today's technologies. Approximations in form of genetic algorithms, different heuristics and simplifications exist, but lack applicability due to high runtime and estimation errors. Quantum computers, quantum annealers and hybrid algorithms show potential to offer added value and better performance over established approaches for optimization, planning and production control, but are often incomprehensible for production engineers. Based on an analysis of industrial problems in different domains and a definition of relevant problem cases, the potential of quantum systems for optimization in production and planning is explored. An approach to close the gap between classical and quantum optimization from an engineering standpoint is

M. Zwingel · C. May (✉) · S. Reitelshöfer
Institute for Factory Automation and Production Systems (FAPS), Friedrich-Alexander-Universität Erlangen-Nürnberg, Egerlandstraße 7–9, 91058 Erlangen, Germany
e-mail: christopher.may@faps.fau.de

M. Zwingel
e-mail: maximilian.zwingel@siemens.com

W. Mauerer
Technical University of Applied Sciences Regensburg, Galgenbergstraße 32, 93058 Regensburg, Germany

Siemens AG, Corporate Technology, Otto-Hahn-Ring 6, 81054 München, Germany

S. Ihlenfeldt et al. (eds.), *Annals of Scientific Society for Assembly, Handling and Industrial Robotics 2023*, https://doi.org/10.1007/978-3-031-74010-7_23

made by describing the transformation process for a real-world problem and discussing performance indicators of model implementations.

Keywords

Production planning • Optimization • Quantum algorithms

1   Introduction

Globalized markets with international competition and shorter lifecycles of product variants put pressure on modern production companies to optimize production and process costs. Reactivity in operations and in the planning of processes and production programs is a significant lever for reducing the overall cost. Due to increasing digitization of production, interconnected machines and digital interfaces for all partners in the supply chain, countless pieces of information are available for decision-making. Owing to the sheer volume and the complex interrelationships, information can long since not be interpreted and processed by humans. Instead, the modeling of individual problems using formal methods is required to optimize relevant parameters [1].

Depending on the specific use case, the corresponding number of set parameters, the dynamic input and output variables and the employed modelling formalism, both expressiveness and computation time of the chosen optimization approach vary vastly [2]. For industry-relevant optimization problems, performance and capacity of common off-the-shelf (mass-market) computing hardware is most often insufficient. Various heuristics and approximation approaches exist on the level of modelling scenarios, and with regard to the mathematical methods employed to perform the actual optimization. Yet, they come with different advantages and drawbacks [3]. During the last years, quantum computers and quantum annealers (despite still being at an early technological stage from an application perspective) have reached sufficient maturity to explore possible approaches to problems inspired by industrial use-cases. In particular, quantum optimization methods can be mapped to multi-order optimization problems in production as well as process planning and control. In this paper, we gently introduce general principles of possible quantum approaches, as they differ from known and widely used techniques in many important aspects, and assess their potentials by highlighting advantages, disadvantages and predicted computational capabilities in contrast to classical approaches for industry-relevant optimization problems. In Chap. 2, we describe classical optimization problems in the field of planning and production, and exemplify how to mathematically model such problems. Chapter 3 reviews various established classical approaches to solve optimization problems in production and manufacturing using approximation techniques or heuristics. Chapter 4 then classifies and reviews possible quantum approaches before Chap. 5 shows how to cast an exemplary optimization problem based on a real-world application into the mathematical formalism that underlies two seminal approaches to quantum optimization.

2 Standard Optimization Problems in Planning and Production

Optimization problems in planning and production can be divided into different problem classes depending on their time horizon and the sector in view. In the area of planning, available resources are distributed according to applicable constraints and demands. Constraints can be process-related (e.g., assembly sequence), time-related (e.g., shelf life), organizational (e.g., personnel availability) or technological (e.g., setup sequences, maintenance downtimes). Optimization problems that have received consideration in the literature stem from a wide range of aspects:

- Resource allocation [4]
- Scheduling [5]
- Capacity planning [5]
- Supply chain optimization [6]
- Layout optimization [7]
- Quality control [8]
- Maintenance planning [9]

In production optimization, more specific variants of the above general problem classes must be considered; among many other examples, the following topics have been considered in the literature:

- Optimization of cellular manufacturing systems [10]
- Flexible shop-job scheduling problem [11]
- Lot size optimization for production [12]
- Lot size and preventive maintenance [13]
- Robot path optimization [15]
- Energy optimization [14]

Solving all the aforementioned problems with numerical methods requires, of course, to formulate them as mathematical models. Most of the underlying decision problems are NP-complete, as they can be solved in polynomial time (increasing in the size of the input, for instance constrains or variables) on a non-deterministic Turing machine. An elaborate classification of the complexity of the associated optimization variants is known [15], and many of them can be efficiently approximated at the expense of solution quality. Some can be solved by linear programming (resource allocation and quality control) or be formulated as convex optimization problems (robot path and energy optimization) under certain conditions, which allow for efficient numerical solutions or approximation. Often, seemingly small details in formulation and boundary conditions can lead to substantial differences in computational complexity.

3   Heuristics for Solving Optimization Problems

Genetic algorithms are well-known approaches [16] modelled on the processes arising in natural selection and genetics. The approach represents a schedule as a series of numbers (chromosomes) that correspond to a sequence of tasks, and uses genetic operations such as selection, cross-over, and mutation to generate new solutions [17]. In each iteration that applies these operations, the algorithm evaluates the quality of each schedule by calculating the objective function and selects the best schedules to use for generating the next generation. Over time, the algorithm converges to a near-optimal solution as the best-performing schedules are recombined and improved through genetic operations. The class of genetic algorithms is particularly useful in planning problems where the optimal solution is difficult to find using conventional algorithms, such as problems with many variants and a large number of variables.

Ant Colony Optimization (ACO) is an alternative optimization algorithm based on the foraging behavior of ants. In planning problems, ACO can be used to find the optimal solution for planning tasks by modeling the problem as a search for the shortest path through a graph. The ants, represented as agents in the algorithm, construct solutions by dropping virtual pheromones on the edges of the graph that represent paths between tasks. The algorithm uses this information to update the probabilities with which the ants choose particular paths, leading to exploration of the solution space and convergence toward the optimal solution. The algorithm continues until a satisfactory solution is found or another stopping criterion, such as a maximum runtime, is met [18].

More approaches to solve optimization problems that have been considered in the literature include:

- Particle Swarm Optimization [19]
- Bees Algorithm [20]
- Water-Flow-like-Algorithms [21]
- Multiagent models [22]
- Multivariant Bayesian control methods [23]
- Petri Nets [24]

All of these algorithms trade a decreased possibility of reaching the optimal solution (or deterministically reaching a non-optimal solution) for a runtime improvement over competing deterministic algorithms. However, for large problems with a significant number of variables and parameters, execution on sufficiently powerful and cost-appropriate hardware the runtime may still be in the order of hours or even days, which can be too costly for many scenarios of practical interest.

Recently, quantum algorithms (pure, hybrid or annealing approaches) have been devised [25–27] that are known to exhibit possible computational advantages over classical approaches for certain optimization problems, given widely accepted complexity-theoretic assumptions. While actual quantum advantage on real machines—that suffer from noise and imperfections perturbing computational operations as well as the information represented in quantum bits, feature very limited availability of resources, and are therefore referred to as noisy, intermediate-scale quantum (NISQ) machines—has not been observed so far except for specially crafted, artificial problems [28] they are nonetheless seen as potential future solutions to optimization problems of practical interest [29]. Given that possible advances in hardware or improved algorithmic insights make it hard to predict when actual quantum advantages will be available, it seems reasonable to consider how problems can be cast such that they allow us to process them on future quantum machines, and identify quantum computational primitives that could find deployment in production logistics and manufacturing tasks.

4   Potentials of Quantum Inspired Algorithms in Optimization of Planning and Production

In the past years, many approaches to quantum computing based on entirely different physical principles—from superconducting circuits via trapped ions and neutral atoms to optical techniques—have matured from foundational physical experiments to commercially available systems that are accessible to non-experts via remote cloud access, for instance via commercial offerings by Amazon, IBM, and other vendors. Many quantum algorithms have been devised [30]. Unfortunately, while it is possible to prove computational speedups over the best possible (or known) classical approaches for some of them, these advantages do not yet manifest on currently available NISQ systems. Other algorithms are explicitly adapted to the limitations of NISQ systems, and can be executed by them with reasonable quality; unfortunately, it is still unclear if and under what circumstances they can outperform any of their classical alternatives.

There are many physical implementations of quantum computers that utilize different quantum degrees of freedom to represent and transform information, and even their basic approach to computation differs (applying gates, applying measurements in certain orders, or transforming systems in appropriate ways). Yet, they can exploit non-classical features like entanglement, superposition, and interference that are widely believed to provide improved computational power over classical approaches [31]. Most quantum algorithms that can be applied to scheduling problems operate by encoding the constraints and objectives of the scheduling problem into a specific mathematical objective function, whose variables can be encoded into quantum states. The variables are binary, and formulas can contain up to quadratic contributions in the variables. Additionally, it is not possible to specify explicit constraints that apply to the objective function–these need to be implicitly included in the formula. The resulting optimization problem (choose values for the variables that minimize or maximize the objective function) is then called a quadratic unconstrained binary optimization (QUBO) problem—we provide an example for this class of problems below.

QUBO problems are amenable to solving by quantum annealers, such as those developed by D-Wave Systems [32, 33]. These devices are special-purpose quantum optimizers designed to find an assignment of variables that corresponds to a global minimum of a QUBO formula. Intuitively (and roughly speaking), the method relies on the physical principle that sufficiently slow ("adiabatic") transformations of physical systems that start in an energetic ground state remain in the ground state; by choosing an initial system whose ground state can be determined and prepared, and by slowly transforming it into a system that represents the problem of interest, a ground-state solution to the latter can be inferred [34] (note that efficiently finding minimum energy states of physical systems is a classically intractable problem as well). Exact performance guarantees for annealing are not available except in special cases, and determining advantages over classical heuristics is, especially under the influence of noise and imperfections, an open research topic; nonetheless, also quantum annealing is believed to outperform classical optimization approaches in the long run.

Another example of a quantum algorithm that can be used to solve QUBO problems (albeit it also extends to more general tasks) is the quantum approximate optimization algorithm (QAOA). It has been applied to problems such as job shop scheduling, resource-constrained project scheduling, and task scheduling [35]. It is a hybrid quantum–classical algorithm, meaning that it combines quantum computing with classical computing to find an approximate solution to an optimization problem. The algorithm uses a certain problem-specific sequence of parameterized quantum gates to evolve the initial state of a system towards a state that encodes the optimal solution. After sampling the outcome (which is given by a statistical distribution, and makes multiple computational runs and measurements necessary to estimate the probability distribution), a classical optimization procedure is used to update gate parameters such that the next computational run delivers results closer to a desired optimum. This combination of quantum and classical computations is iterated until convergence to a sufficiently good result. While exact performance guarantees for QAOA are not yet known except for very special cases [36], it has been established that classically simulating the algorithm is impossible (again, based on reasonable complexity-theoretic assumptions), and the strong belief prevails that QAOA will lead to performance advantage on sufficiently reliable and scalable quantum hardware. Other quantum algorithms, such as Grover's algorithm, have also been proposed for scheduling problems, albeit these rely on perfect quantum hardware, and only deliver a quadratic speedup for exponential search spaces [37].

5   Casting a Real World Problem for Quantum Algorithms

To illustrate the steps required to cast optimization tasks in form of a QUBO problem, let us commence with a specific example, a job shop scheduling problem that is based on a real-world scenario.

Suppose a company in the toy industry picks and places approximately 1.000 individual single orders (J) on three flexible assembly lines (M) in one shift (28.800 s, T). Each order consists of a number (varying between 1–5) of individual articles to be picked. Due to the cost structure and lead time promises to customers, the orders should be processed as fast as possible across all three lines (goal function).

Additionally, the following constraints are given:

- Every job must start and run exactly once.
- Only one job can be running on each line at any given time.
- The order of operations within a job is irrelevant to its makespan.

To formulate the QUBO, the use case has to be cast in binary variables. For the specific use case, this can be achieved by using the equation detailed in [38]:

$$\sum_{j=1}^{J}\sum_{m=1}^{M}\sum_{t=0}^{T}(t + d_{j,m}) \times x_{j,m,t} \tag{1}$$

$$+p_1 \times \sum_{j=1}^{J}\sum_{m=1}^{M}\left(\sum_{t=0}^{T} x_{j,m,t} - 1\right)^2 \tag{2}$$

$$+p_2 \times \sum_{j=1}^{J}\sum_{(m_1,m_2)\in P_j}\sum_{t_1+d_{j,m_1}>t_2} x_{j,m_1t_1} \times x_{j,m_2t_2} \tag{3}$$

$$+p_3 \times \sum_{m=1}^{M}\sum_{j_1\neq j_2}\sum_{t_1\leq t_2<t_1+d_{j_1,m}} x_{j_1,m,t_1} \times x_{j_2,m,t_2} \tag{4}$$

$$+p_4 \times \sum_{j=1}^{J}\left(\sum_{t=0}^{T}\left(t \times x_{j,\sigma_M^j,t}\right) + d_{j,\sigma_M^j} + \hat{s}_j - \sum_{j=1}^{J}\sum_{m=1}^{M}\sum_{t=0}^{T}(t + d_{j,m}) \times x_{i,m,t}\right)^2 \tag{5}$$

$$\hat{s}_j \in \mathbb{Z}, x_{j,m,t} \in \{0,1\}, \forall j = 1, ..., J, \forall m = 1, ..., M, \forall t = 1, ..., T$$

Line (2) of the equation describes the completion time of all operations. The additional terms represent the different constraints (one start (3), precedence (4), no overlap (5) and max make-span (6) constraint), each weighted by penalty coefficients (p1–p4). Note that while manually determining QUBO representations may be a somewhat challenging task, automated means of appropriately transforming various constrained optimization problem formulations that domain experts may be better accustomed to have become available, for instance quark [39].

To solve the formulated optimization problem using quantum algorithms, further transformation steps are required. The QUBO must be mapped to the physical quantum computer or annealer, for which automated approaches are available on quantum annealers [40] and gate-based quantum computers [41]. Typically, this process induces significant overhead in the number of required qubits.

This process of mapping the optimization problem to the physical quantum systems is where the gap between today's optimization in planning and production and quantum potentials persists. For instance, current-generation quantum annealers feature several thousand qubits, yet only scale to problems with 10s or hundreds of variables. It is needless to say that such small-scale problems can at the moment also be solved using performant hardware and deterministic algorithms in acceptable time.

Hybrid approaches to optimization, like the D-Wave Leap Hybrid solver, divide problems into sub-parts, and combine part solutions obtained by the quantum computer. That way, they can scale to substantially larger problem instances, albeit introducing additional overhead–both in splitting and recombining the problem, and in the time required for communication. Their approximated solutions are similar in quality to heuristics in standard approaches [42, 43], although a close and nuanced look is required to characterize performance and achievable qualities. It has been shown that hybrid solvers can come close to a satisfiable solution in industry relevant optimization scenarios [44].

Supported by the continuing momentum of developments and progress in the field of hybrid approaches, one can assume that the solution of optimization problems on a larger scale may become possible in the future through quantum systems.

6   Conclusion and Outlook

Classical, hybrid and quantum approaches each have their strengths and weaknesses when it comes to solving optimization problems in production and planning. Deterministic approaches and heuristics are well-established and often provide exact or near-exact solutions to problems. These methods are often used in production and planning where there is little room for error.

Quantum approaches, such as quantum annealing and QAOA, offer the potential for significant speedups in solving certain optimization problems, however the technology is still in its early stages. Nevertheless, an increasing number of examples, like [45], show the applicability of quantum technologies to real problems in production and planning.

In order to solve more problems with larger numbers of variables and high complexity efficiently, quickly and optimally in the future, we expect the use of quantum technologies to become a viable option. Even if practical applications have not yet materialized, we believe that it is worthwhile to start investigating the use of novel computational approaches to problems in our domain.

This work was supported by the German Federal Ministry of Education and Research in the funding program "quantum technologies—from basic research to market" under contract number 13N16096.

References

1. Abass, S.A., Elsayed, M.S.: Modeling and solving production planning problem under uncertainty: a case study. AJOR (2012). https://doi.org/10.5923/j.ajor.20120205.03
2. Abramson, M, Asaki, T., Dennis, J., Magallanez, S., Magallanez, R., Sottile, M.: Solving Computationally Expensive Optimization Problems with CPU Time-Correlated Functions (2008)
3. Fathi, M., Khakifirooz, M., Pardalos, P.M.: Optimization in large scale problems. In: Industry 4. 0 and Society 5. 0 Applications. Springer Optimization and Its Applications Ser, vol. 152. Springer, Cham (2020)
4. Hegazy, T.: Optimization of resource allocation and leveling using genetic algorithms. J. Constr. Eng. Manage. (1999). https://doi.org/10.1061/(ASCE)0733-9364(1999)125:3(167)
5. Carvalho, A.N., Scavarda, L.F., Oliveira, F.: An optimisation approach for capacity planning: modelling insights and empirical findings from a tactical perspective. Prod. (2017). https://doi.org/10.1590/0103-6513.001417
6. Zhou, Z., Cheng, S., Hua, B.: Supply chain optimization of continuous process industries with sustainability considerations. Comput. Chem. Eng. (2000). https://doi.org/10.1016/S0098-1354(00)00496-8
7. Stelle Chemim, L., Christine Sotsek, N., Kleina, M.: Layout optimization methods and tools: a systematic literature review. GEPROS (2021). https://doi.org/10.15675/gepros.v16i4.2806
8. Al-Sultan, K.S., Rahim, M.A.: Optimization in quality control. Springer, US, Boston, MA (1997)
9. Pandey, D., Kulkarni, M.S., Vrat, P.: A methodology for joint optimization for maintenance planning, process quality and production scheduling. Comput. Ind. Eng. (2011). https://doi.org/10.1016/j.cie.2011.06.023
10. Abdelmola, A.I., Taboun, S.M., Merchawi, S.: Productivity optimization of cellular manufacturing systems. Comput. Ind. Eng. (1998). https://doi.org/10.1016/S0360-8352(98)00119-3
11. Pezzella, F., Morganti, G., Ciaschetti, G.: A genetic algorithm for the flexible job-shop scheduling problem. Comput. Oper. Res. (2008). https://doi.org/10.1016/j.cor.2007.02.014
12. Adacher, L., Cassandras, C.G.: Lot size optimization in manufacturing systems: the surrogate method. Int. J. Prod. Econ. (2014). https://doi.org/10.1016/j.ijpe.2013.07.026
13. Qu, L., Liao, J., Gao, K., Yang, L.: Joint optimization of production lot sizing and preventive maintenance threshold based on nonlinear degradation. Appl. Sci. (2022). https://doi.org/10.3390/app12178638
14. Opoku, R., Obeng, G.Y., Osei, L.K., Kizito, J.P.: Optimization of industrial energy consumption for sustainability using time-series regression and gradient descent algorithm based on historical electricity consumption data. Sustain. Anal. Model. (2022). https://doi.org/10.1016/j.samod.2022.100004
15. Crescenzi, P., Kann, V.: A compendium of NP optimization problems. Citeseer
16. Brindle, A.: Genetic algorithms for function optimization (1980)
17. Gen, M., Cheng, R.: Genetic algorithms and engineering optimization, vol. 7. John Wiley & Sons (1999)
18. Dorigo, M., Birattari, M., Stutzle, T.: Ant colony optimization. IEEE Comput. Intell. Mag. (2006). https://doi.org/10.1109/MCI.2006.329691

19. Hassan, R., Cohanim, B., Weck, O. de, Venter, G.: A comparison of particle swarm optimization and the genetic algorithm. In: 46th AIAA/ASME/ASCE/AHS/ASC Structures, Structural Dynamics and Materials Conference. 46th AIAA/ASME/ASCE/AHS/ASC Structures, Structural Dynamics and Materials Conference, Austin, Texas. American Institute of Aeronautics and Astronautics, Reston, Viriginia (2005). https://doi.org/10.2514/6.2005-1897

20. Pham, D., Ghanbarzadeh, A., Koç, E., Otri, S., Rahim, S., Zaidi, M.: The Bees Algorithm Technical Note. Manufacturing Engineering Centre, Cardiff University, UK, pp. 1–57 (2005)

21. Wu, G.-H., Cheng, C.-Y., Yang, H.-I., Chena, C.-T.: An improved water flow-like algorithm for order acceptance and scheduling with identical parallel machines. Appl. Soft Comput. (2018). https://doi.org/10.1016/j.asoc.2017.10.015

22. Melnichuk, A.V., Sivakova, T.V., Sudakov, V.A.: Solving optimization problems using multiagent models. KIAM Prepr. (2019). https://doi.org/10.20948/prepr-2019-100

23. Makis, V.: Multivariate Bayesian process control for a finite production run. Eur. J. Oper. Res. **194**, 795–806 (2009)

24. Balduzzi, F., Giua, A., Menga, G.: First-order hybrid Petri nets: a model for optimization and control. IEEE Trans. Robot. Automat. (2000). https://doi.org/10.1109/70.864231

25. Chuang, I.L., Vandersypen, L.M.K., Zhou, X., Leung, D.W., Lloyd, S.: Experimental realization of a quantum algorithm. Nature (1998). https://doi.org/10.1038/30181

26. Grant, E.K., Humble, T.S.: Adiabatic quantum computing and quantum annealing. In: Foster, B. (ed.) Oxford research encyclopedia of physics. Oxford research encyclopedias. Oxford University Press, New York, NY (2018)

27. Weigold, M., Barzen, J., Leymann, F., Vietz, D.: Patterns for hybrid quantum algorithms. In: Proceedings of the 15th Symposium and Summer School on Service-Oriented Computing (SummerSOC 2021), pp. 34–51. Springer International Publishing (2021). https://doi.org/10.1007/978-3-030-87568-8/_2

28. Madsen, L.S., Laudenbach, F., Askarani, M.F., Rortais, F., Vincent, T., Bulmer, J.F.F., Miatto, F.M., Neuhaus, L., Helt, L.G., Collins, M.J., Lita, A.E., Gerrits, T., Nam, S.W., Vaidya, V.D., Menotti, M., Dhand, I., Vernon, Z., Quesada, N., Lavoie, J.: Quantum computational advantage with a programmable photonic processor. Nature (2022). https://doi.org/10.1038/s41586-022-04725-x

29. Bayerstadler, A., Becquin, G., Binder, J., Botter, T., Ehm, H., Ehmer, T., Erdmann, M., Gaus, N., Harbach, P., Hess, M., Klepsch, J., Leib, M., Luber, S., Luckow, A., Mansky, M., Mauerer, W., Neukart, F., Niedermeier, C., Palackal, L., Pfeiffer, R., Polenz, C., Sepulveda, J., Sievers, T., Standen, B., Streif, M., Strohm, T., Utschig-Utschig, C., Volz, D., Weiss, H., Winter, F.: Industry quantum computing applications. EPJ Quantum Technol. (2021). https://doi.org/10.1140/epjqt/s40507-021-00114-x

30. Bharti, K., Cervera-Lierta, A., Kyaw, T.H., Haug, T., Alperin-Lea, S., Anand, A., Degroote, M., Heimonen, H., Kottmann, J.S., Menke, T., Mok, W.-K., Sim, S., Kwek, L.-C., Aspuru-Guzik, A.: Noisy intermediate-scale quantum algorithms. Rev. Mod. Phys. (2022). https://doi.org/10.1103/RevModPhys.94.015004

31. Forcer, T.M., Hey, A., Ross, D.A., Smith, P.: Superposition, entanglement and quantum computation. QIC (2002). https://doi.org/10.26421/QIC2.2-1

32. Pastorello, D., Blanzieri, E.: Quantum annealing learning search for solving QUBO problems. **10** (2018). https://arxiv.org/pdf/1810.09342

33. Scherer, A., Guggemos, T., Grundner-Culemann, S., Pomplun, N., Prüfer, S., Spörl, A.: OnCall Operator Scheduling for Satellites with Grover's Algorithm. In: Paszynski (ed.) Computational Science-ICCS 2021, vol. 12747. Lecture Notes in Computer Science, pp. 17–29. Springer International Publishing, [S.l.] (2021)

34. Cruz-Santos, W., Venegas-Andraca, S.E., Lanzagorta, M.: A QUBO formulation of minimum multicut problem instances in trees for D-wave quantum annealers. Sci. Rep. (2019). https://doi.org/10.1038/s41598-019-53585-5

35. Hadfield, S., Wang, Z., O'Gorman, B., Rieffel, E., Venturelli, D., Biswas, R.: From the quantum approximate optimization algorithm to a quantum alternating operator ansatz. Algorithms (2019). https://doi.org/10.3390/a12020034

36. Zhou, L., Wang, S.-T., Choi, S., Pichler, H., Lukin, M.D.: Quantum approximate optimization algorithm: performance, mechanism, and implementation on near-term devices. Phys. Rev. X (2020). https://doi.org/10.1103/PhysRevX.10.021067

37. Paszynski (ed.).: Computational Science-ICCS 2021. Lecture Notes in Computer Science. Springer International Publishing, [S.l.] (2021)

38. Zhang, J., Lo Bianco, G., Beck, J.C.: Solving job-shop scheduling problems with QUBO-based specialized hardware. ICAPS (2022). https://doi.org/10.1609/icaps.v32i1.19826

39. Finzgar, J.R., Ross, P., Holscher, L., Klepsch, J., Luckow, A.: QUARK: a framework for quantum computing application benchmarking. In: 2022 IEEE International Conference on Quantum Computing and Engineering (QCE). 2022 IEEE International Conference on Quantum Computing and Engineering (QCE), Broomfield, CO, USA, 9/18/2022−9/23/2022, pp. 226–237. IEEE (uuuu-uuuu). https://doi.org/10.1109/QCE53715.2022.00042

40. D-Wave Systems Inc.: D-wave ocean software documentation. https://docs.ocean.dwavesys.com/en/stable/

41. Qiskit: Open-source quantum development. https://qiskit.org/

42. D-Wave Systems Inc.: Hybrid solvers. https://docs.ocean.dwavesys.com/en/latest/overview/hybrid.html#leap-s-hybrid-solvers. Accessed 15 Feb 2023

43. Raymond, J., Stevanovic, R., Bernoudy, W., Boothby, K., McGeoch, C.C., Berkley, A.J., Farré, P., Pasvolsky, J., King, A.D.: Hybrid quantum annealing for larger-than-QPU lattice-structured problems. ACM Trans. Quantum Comput. (2023). https://doi.org/10.1145/3579368

44. Carugno, C., Ferrari Dacrema, M., Cremonesi, P.: Evaluating the job shop scheduling problem on a D-wave quantum annealer. Sci. Rep. (2022). https://doi.org/10.1038/s41598-022-10169-0

45. Schworm, P., Wu, X., Glatt, M., Aurich, J.C.: Solving flexible job shop scheduling problems in manufacturing with quantum annealing. Prod. Eng. Res. Devel. (2023). https://doi.org/10.1007/s11740-022-01145-8

Development and Testing of an Elbow Exoskeleton Prototype with Pneumatic Actuation for Industrial Tasks

Samet Ersoysal, Benjamin Reimeir, and Robert Weidner

Abstract

Despite the increasing use of automation and digitalization in industrial workplaces, workers still have to handle heavy loads and have to perform strenuous, repetitive, long-term assembly tasks at head level or above, which may lead to degenerative musculoskeletal disorders. The growing trend towards wearable support systems has already resulted in a large number of exoskeletons being in research or commercially available. However, most of the support systems for industrial workplaces focus on the back and shoulders, but not the elbow joint. In a preliminary study, we presented a soft passive elbow exoskeleton, which was limited to static tasks, although dynamic support is necessary for most industrial tasks. Building upon that work, this paper presents an elbow exoskeleton that can be coupled to an existing shoulder exoskeleton. The developed prototype is designed to support the elbow flexion with a pneumatic actuator for

S. Ersoysal (✉) · B. Reimeir · R. Weidner
Chair of Production Technology, Universität Innsbruck, Institute of Mechatronic, Innsbruck, Austria
e-mail: Samet.Ersoysal@uibk.ac.at

B. Reimeir
e-mail: Benjamin.Reimeir@uibk.ac.at

R. Weidner
e-mail: Robert.Weidner@uibk.ac.at; Robert.Weidner@hsu-hh.de

B. Reimeir
Department of Sport Science, Universität Innsbruck, Innsbruck, Austria

R. Weidner
Laboratory of Manufacturing Technology, Helmut-Schmidt-University/University of the Federal Armed Forces Hamburg, Hamburg, Germany

© The Author(s) 2025
S. Ihlenfeldt et al. (eds.), *Annals of Scientific Society for Assembly, Handling and Industrial Robotics 2023*, https://doi.org/10.1007/978-3-031-74010-7_24

industrial applications. The functionality of the prototype was tested on three male participants in one static and two dynamic tasks. The laboratory tests have shown that the exoskeleton reduced the mean muscular activity of the brachioradialis and biceps brachii in all tasks. Based on the results, the developed exoskeleton may potentially support the elbow flexion in industrial tasks, but further testing is needed to evaluate its biomechanical effects on the user.

Keywords

Exoskeletons • Support systems • Elbow • Wearable technologies • Human–machine interaction

1   Introduction

With the implementation of industry 4.0 technologies and the increase in automation, the need for physical labour has been reduced. These technologies are making great progress and some of the physical tasks that humans perform are being gradually replaced. However, they cannot be applied to all industrial sectors because of technical difficulties [1]. A high level of flexibility in the systems is required because of the increasing personalization of products to customer requirements, which is why humans cannot be completely substituted due to their cognitive and physical abilities [2, 3]. Therefore, humans will continue to play a key role in the future of the industry [4]. Nevertheless, workers in production halls, logistics centres, and construction are exposed to high levels of physical and/or cognitive stress by performing demanding, repetitive, and long-term tasks which can be a risk factor for musculoskeletal disorders (MSD). Industrial companies are facing challenges from MSDs as well as demographic changes, which is why wearable support systems, known as exoskeletons, have gained traction. In various studies, the use of exoskeletons reduced muscle activity during industrial tasks, which may possibly prevent musculoskeletal disorders in the upper and lower extremities, and the back [5–8]. The high acquisition costs of exoskeletons, however, prevent them from being widely used in companies.

In [9] we presented a soft passive exoskeleton, which supports the flexion of the elbow joint. The developed prototype is limited to static tasks, whereas dynamic support is often needed in industrial workplaces, too. In addition, due to the soft structure, tensile forces imposed by the actuator cause shear forces on the skin, harming the soft tissue and elbow joint of the user. Based on these considerations, a novel active exoskeleton with rigid structures is presented, which can be coupled to an existing shoulder support system Lucy [7].

2   State of the Art

The main purpose of an exoskeleton is to support and stabilize vulnerable areas of the body, such as the back and joints of the extremities, and reduce the physical load. In addition, they can improve the endurance, performance, and work capacity of the user [2].

2.1   Classification Characteristics of Exoskeletons

Exoskeletons can be classified into various features. The main classifications are based on supported body regions and morphological structure (including actuation technology). In general, they are grouped into upper extremity systems (shoulder, elbow, wrist, and fingers), back systems (upper and lower back), lower extremity systems (ankle, knee, hip, and foot), and full body systems [1]. Regarding the morphological structure, these systems can be divided into rigid vs. soft textile-based as well as end-effector vs. anthropomorph-designed systems [10]. Moreover, exoskeletons are often distinguished in passive, active, or semi-active systems.

Passive exoskeletons have no external energy supply and support the body segments with mechanical components such as gas springs or spring elements. These absorb and store the energy generated by the user's movements and support them during a counter-movement in the opposite direction [11]. Despite their ease of design and development, they can only be adjusted manually by the mechanical preload in order to adjust the force curve or support level. Active exoskeletons have a power supply, actuators, and a control unit and are therefore more complex than passive systems. With these systems, the desired support level can be determined by sensors or adjusted by the user and realized with various actuation technologies. The used sensor technology is essential for a harmonic human–machine interaction, as it provides information about the user's pose or intention and the exoskeleton's state. In order to ensure the safety of the user, sensors can be used to detect the interaction of the exoskeleton with the environment [12]. In addition, actuation technology has a significant effect on the performance and dynamics of an exoskeleton. Different types of actuation technologies, such as pneumatic, hydraulic, and electric have their advantages and disadvantages and play a major role in the acceptance, usability, and portability of an exoskeleton [13].

According to a study, only half of the considered industrial exoskeletons addressed the upper extremity [14]. It is shown, that most of them support the shoulder and only a few of these systems target the elbow joint. Elbow exoskeletons are rarely developed as a sole system and are usually an extension of shoulder support systems, since the shoulder is involved in most industrial tasks. In addition, any further distal extension is challenging as more degrees of freedom have to be considered.

2.2   Analysis and Musculoskeletal Disorders of the Elbow Joint

The elbow joint is the connection between the humerus in the upper arm and the radius and ulna in the forearm. It consists of three partial joints and has two degrees of freedom (flexion/extension and pronation/supination), which allow a wide range of motion for the forearm. In the development of an exoskeleton, flexion and extension can be simplified as a hinge joint. The three flexor muscles in the elbow joint are the brachialis, the brachioradialis, and the biceps brachii. Extension of the elbow joint is made possible by the triceps brachii. In most cases, the active flexion of a healthy individual is in an angular range between $0°$ and $145°$. For daily activities, the range is between $30°$ and $130°$ [15].

Both the joints and the muscles are burdened with high stress in daily life and in industrial workplaces. As a result of their complexity, they are more likely to get injured, resulting in reduced mobility and long-term pain, such as arthritis, lateral/medial epicondylitis, or biceps tendon rupture [16]. The major causes of musculoskeletal disorders of the elbow joint are carrying and handling heavy objects, and working with power tools in manufacturing and assembly halls [14, 17]. These tasks, along with exposure to vibrations and repetitive motions, are the most widespread main causes leading to work-related MSD [18].

An analysis of existing industrial exoskeletons, as well as diseases of the elbow joint show that there is a need for exoskeletons to support the elbow. In particular, there is a need to support elbow flexion against gravity during dynamic activities and to stabilize the elbow joint during static activities.

3   Main System Requirements

In the development of exoskeletons for industrial applications, the requirements for the systems have been specified to some extent. However, due to the wide variety of application scenarios, these cannot be generally clarified [17]. In order to provide the proper support, the interaction, the task, the technology, and the environment must be considered. The first important step towards ergonomic design is a good model of the kinematics and dimensions of the human limbs [19]. Therefore, human anthropometry plays an essential role in the development of human–machine systems.

3.1   Requirements of the Morphological Structure

The developed elbow exoskeleton is an extension of the shoulder exoskeleton Lucy. Lucy is an active and rigid exoskeleton that uses pneumatic actuators to support anteversion or flexion of the shoulder joint in industrial tasks [7]. The elbow system should be provided with an easy coupling by the upper arm structure and must not interfere with the basic components, such as the kinematic structure and the control of the existing system. The developed system should support flexion and relieve the elbow joint and the elbow flexion muscles. Furthermore, it should accommodate the occurring loads and load peaks without restricting other movements such as extension, pro-, and supination.

Basically, it is important to consider the number of joint axes, the length of the kinematic structure, and the type of control for required functionality and optimal ergonomics. The weight and inertia of the exoskeleton should be as small as possible to improve comfort and mobility. Low mass is also an important aspect for mobility and also brings advantages when safety requirements are considered. Since long-term use of the system is possible, it should be adaptable to different anthropometries and have a skin-friendly and comfortable interface to prevent the user from experiencing fatigue and pain. In addition, the exoskeleton should cover a large range of motion of the arm so as not to hinder the human in the main and secondary activities. Natural human movement can be made possible with passive degrees of freedom, but this increases the complexity of the system. Efforts should be made to reduce the complexity of the system to a minimum, thus increasing reliability and reducing development and production costs [20].

3.2   Requirements of the System Control

In designing an active exoskeleton, the support level or the control of the system must be considered. As a prerequisite for the acceptance of the exoskeleton, an optimal interaction between sensors and actuators is necessary to recognize the situation and intention. The support should only be provided if needed based on the user's movement during the task. Furthermore, the support level should be adjustable during the task and it should be possible to turn the system off at any time. In addition, the user should have control over the system and be able to extend the forearm without applying great force.

4   Exoskeleton Design

In [9], we presented four different design strategies for exoskeletons that support the flexion of the elbow joint. Regarding to the modularity or the integrability of the system to the shoulder exoskeleton Lucy, a cable-driven support system seems to be beneficial. The concept of the functional prototype is a combination of a cable drive, a direct drive, and a system with a lever principle (see Fig. 1). Here, the cable is not attached to the forearm interface as in the soft and passive concept but is wound onto a winch mechanism at the elbow joint to convert a linear motion into a rotational motion. The cable is encased in a Bowden housing, with its endpoints attached to the actuator and the upper arm structure of the shoulder exoskeleton. The winch is made of additive manufactured elements connected to the upper and lower arm interfaces. The radius of the winch or the attachment point of

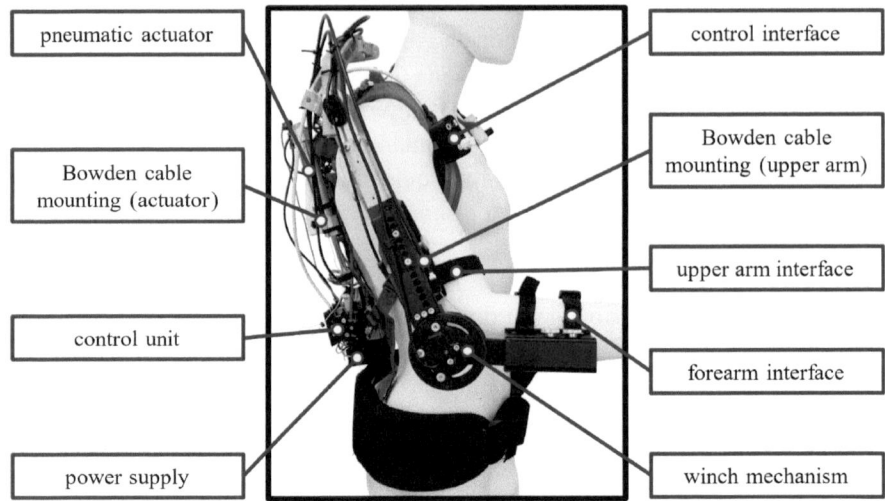

Fig. 1 Overview and central elements of the active elbow exoskeleton

the cable acts here as a lever arm, allowing torque to be generated. With this approach, the advantages of the concepts presented in [9], such as reduced weight at the distal end, redirection of the force to more stable body regions, and continuous torque throughout flexion, remain. Furthermore, the proximal forces or shear forces in the interface are redirected by the rigid structures to avoid discomfort and injuries. In this concept, the control unit and the actuator can be placed on the back structure of the shoulder support system where the existing elements are not affected.

The dimensioning of the actuator is based on the average weight of a CEP (courier, express, and parcel shipment) of 7.4 kg [21], which yields a necessary torque of 12.7 Nm. Using a simple lever and moment equation, the necessary force of the actuator can be determined. A lever arm with a length of 40 mm is chosen to avoid a voluminous structure at the distal end and thus not hinder the user in various activities. With the torque requirements and with the desired range of motion (flexion of 0–145°), a required force of 317 N over a path of 101 mm is obtained. According to the shoulder support system, the same drive technology is used, which leads to the selection of a pneumatic cylinder DSNU-S-25–100 with a theoretical force of 294.5 N and a torque of 11.8 Nm at 6 bar.

In order to control the support force, pressure sensors are used to detect the supply pressure and the system pressure in the pneumatic cylinder. The elbow angle can be measured by using a rotary potentiometer. The user can power the system and adjust the support level with a control interface attached to the shoulder strap of the shoulder exoskeleton.

To meet the requirements for comfort, design, position, and clamp of the interfaces were considered. Due to its flexible characteristics, the interface can be adjusted to the forearm of different users. In addition, its position is adjustable on the forearm structure to align the rotation axis of the exoskeleton and the elbow joint.

5   Evaluation of the Exoskeleton

The evaluation aims to demonstrate whether the exoskeleton supports the elbow flexion of the user in industrial tasks. For the evaluation of exoskeletons, different subjective and objective approaches exist. However, despite the high number of exoskeletons in research, there is no standardized method for evaluating exoskeletons. In various studies, different industrial tasks are performed in different postures with and without a support system to evaluate the effects of the exoskeleton on the user [22]. The most widely used method in the evaluation and assessment of exoskeletons is electromyographic analysis (EMG), which measures muscle activity [23]. Therefore, the preliminary test of the developed exoskeleton is an EMG analysis.

5.1   Study Design

Based on the common use cases in the industry which can be a risk factor for MSDs, one static and two dynamic application scenarios were considered (see Fig. 2). The tasks were performed with three healthy male participants (age: 26 ± 1 years; height: 184.3 ± 3.8 cm; weight: 75 ± 3 kg; right-handed). The exoskeleton was worn during the support (first trial) and no-support (second trial) condition to ensure a comparison of the elbow angle. Between the trials, a pause of at least five minutes was taken to allow the muscle groups to recover. In order to avoid falsifying the evaluation of the developed elbow system, the shoulder support was left off.

For the static task, a dumbbell weighing 5.75 kg was held in the right hand for 30 s to represent carrying heavy packages over long distances. To achieve maximum leverage, an elbow angle of approximately $90°$ was maintained by each participant. In the first dynamic task, the participants had to perform six repetitions of biceps curls with the same dumbbell in the right hand to replicate repetitive loading. In order to standardize movement speed and to indicate the start of the flexion/extension movement at the top and bottom position, an acoustic signal was provided at 50 beats per minute. In the second dynamic task, repetitive loading and unloading of packages from shelves were replicated. The participants had to lift a package weighing 10 kg from a shelf with both hands and place it on a higher shelf. The lifting task was performed three times by each participant. To ensure data comparability, the distance to the shelf had to be the same and the task had to be completed in a given timeframe.

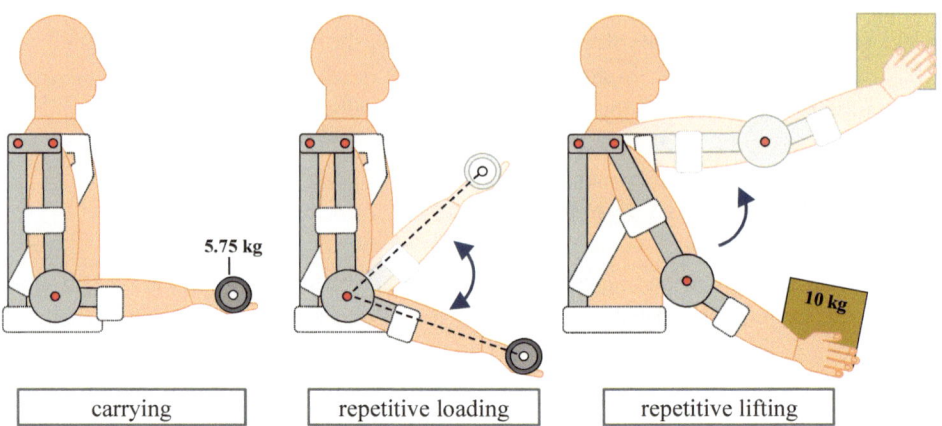

| carrying | repetitive loading | repetitive lifting |

Fig. 2 Schematic representation of the carrying, repetitive loading, and repetitive lifting tasks

First, the tasks and the operation of the exoskeleton were described to the participants. After that, surface electrodes were attached to the elbow flexors biceps brachii and brachioradialis, the antagonist/elbow extensor triceps brachii, and the shoulder muscles deltoideus anterior and deltoideus medialis. This was done to investigate the effects of the exoskeleton on other upper extremity muscle groups in addition to the elbow flexor muscles. At the beginning of each study, the maximum voluntary contraction (MVC) of the observed muscle groups was measured for each participant in order to normalize the EMG signal to a reference value.

It turned out that the cable attachment on the winch could not withstand the high forces of about 294.5 N at a maximum support level. Therefore, the evaluation had to be performed with 4 bar supply pressure and a support level of 50%, which corresponds to a torque of 3.9 Nm.

5.2 Results and Discussion

Figure 3 shows one participant's EMG signal during the repetitive loading task, where a lower biceps brachii muscle activity with support (50%) and without support (0%) is illustrated (reduction in mean muscular activity of -13.4%). However, a higher muscle activity on the antagonist triceps brachii (growth in mean muscular activity of +40.8%) can also be observed when the participant extends the forearm, whereas the shoulder muscles did not show any change in muscle activity. Similar behavior was observed with the EMG signal of the other participants. In Fig. 4, the mean muscle activities and the relative difference in all tasks is visualized. The figure shows that a reduction of the

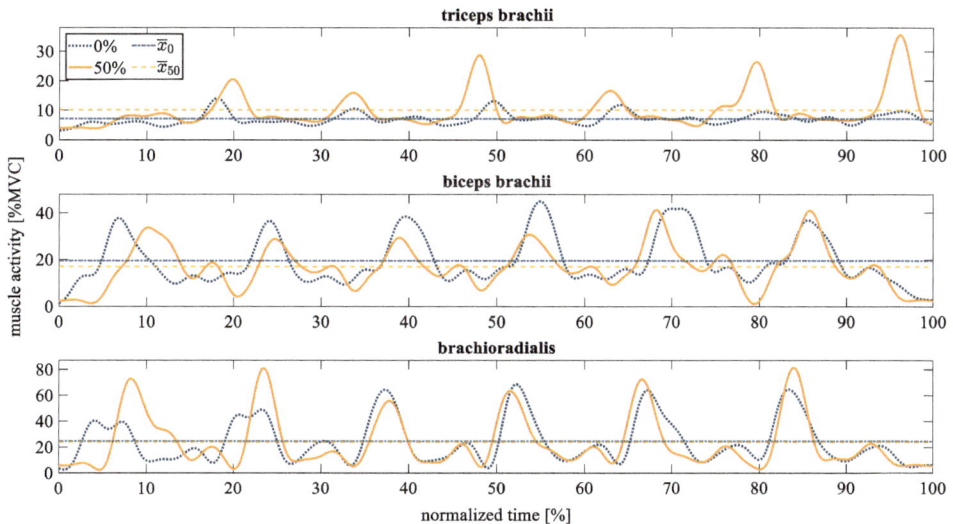

Fig. 3 Illustration of muscle activities during the repetitive loading task with six repetitions with support (50%) and without support (0%)

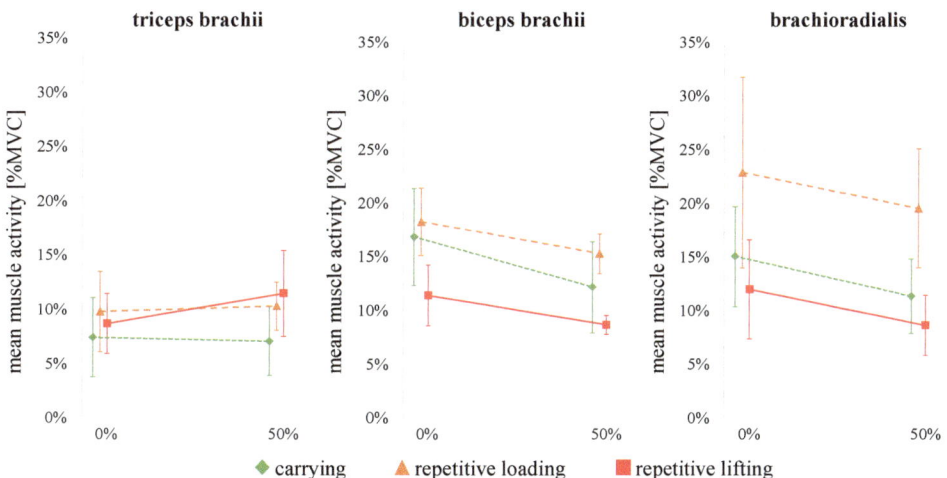

Fig. 4 Visualization of the mean values and the standard deviation of muscle activities in various tasks with support (50%) and without support (0%) for each participant

muscle activity of the elbow flexors by the exoskeleton occurs in all tasks, but an increase can also be seen in the elbow extensor.

Based on the results of electromyography, it can be said that the developed exoskeleton provides a support torque on the elbow, which can be seen in the muscle activities of the flexors. Although the exoskeleton supports the flexion of the elbow, the additional strain on the triceps caused by the extension may have negative effects on the user. This effect could be reduced or avoided by optimizing the control system and improving the situation detection. For the initial evaluation of the exoskeleton, electromyographic analysis was well suited but not sufficient. It should be considered that the sole determination of muscle activity can lead to misleading statements about the overall benefit of the exoskeleton. To evaluate an exoskeleton effectively, further methods, such as biomechanical and physiological measurements should be utilized [22]. In addition, the evaluation of the prototype should be done with a wide variety of participants with different anthropometric characteristics. Finally, the studies should be conducted in field studies under real working conditions rather than in optimal laboratory environments, since they only represent a fraction of the real workloads that occur.

6   Conclusion

Despite the elbow joint's proneness to injury and its stress in many industrial activities, few elbow support systems exist. In comparison to the soft passive exoskeleton presented in [9], dynamic activities with a higher support level could be achieved with the presented prototype. In addition, support torque across the entire flexion of the elbow could be provided with the winch concept. Furthermore, the rigid structure solved the problem of proximal forces caused by the cable's tensile force. Nevertheless, rigid systems face the challenge of aligning the rotation axis of the exoskeleton and the elbow joint, which in turn can lead to displacements and shear forces between the interface and the skin. As a result, a rigid exoskeleton does not provide the same range of motion as a soft exoskeleton. Additional passive degrees of freedom would be needed to allow natural human movement. However, this would significantly increase the complexity and weight of the exoskeleton. Due to the concept of the winch mechanism, the weight at the distal end was significantly reduced compared to the other active mechanisms, such as direct or indirect drive presented in [9].

References

1. Weidner, R., Linnenberg, C., Hoffmann, N., et al.: Exoskelette für den industriellen kontext: systematisches review und klassifikation. In: GfA (ed) Digitale Arbeit, digitaler Wandel, digitaler Mensch?: 66. Kongress der Gesellschaft für Arbeitswissenschaft, TU Berlin, Fachgebiet Mensch-Maschine-Systeme/HU Berlin, Professur Ingenieurpsychologie, 16–18. März 2020, Berlin. GfA-Press, Dortmund (2020)

2. Weidner, R., Redlich, T., Wulfsberg, J.P.: Technische Unterstützungssysteme. Springer, Berlin Heidelberg, Berlin, Heidelberg (2015)
3. Huysamen, K., Bosch, T., de Looze, M., et al.: Evaluation of a passive exoskeleton for static upper limb activities. Appl. Ergon. **70**, 148–155 (2018)
4. Hölzel, C., Knott, V., Schmidtler, J., et al.: Unterstützung des Menschen in der Arbeitswelt der Zukunft. In: Technische Unterstützungssysteme, die die Menschen wirklich wollen
5. Huysamen, K., de Looze, M., Bosch, T., et al.: Assessment of an active industrial exoskeleton to aid dynamic lifting and lowering manual handling tasks. Appl. Ergon. **68**, 125–131 (2018)
6. van Engelhoven, L., Poon, N., Kazerooni, H., et al.: Evaluation of an adjustable support shoulder exoskeleton on static and dynamic overhead tasks. Proc. Hum. Factors Ergon. Soc. Annu. Meet. **62**, 804–808 (2018)
7. Otten, B.M., Weidner, R., Argubi-Wollesen, A.: Evaluation of a novel active exoskeleton for tasks at or above head level. IEEE Robot. Autom. Lett. **3**, 2408–2415 (2018)
8. Gillette, J.C., Stephenson, M.L.: Electromyographic assessment of a shoulder support exoskeleton during on-site job tasks. IISE Trans. Occup. Ergon. Hum. Factors **7**, 302–310 (2019)
9. Ersoysal, S., Hoffmann, N., Ralfs, L., et al.: Towards a modular elbow exoskeleton: concepts for design and system control. In: Schüppstuhl, T., Tracht, K., Raatz, A. (eds.) Annals of Scientific Society for Assembly, Handling and Industrial Robotics 2021, pp. 141–152. Springer International Publishing, Cham (2022)
10. Weidner, R., Argubi-Wollesen, A., Otten, B.M. et al.: Individuelle und aufgabenabhängige Unterstützung bei physisch beanspruchenden Tätigkeiten durch anziehbare Systeme. In: Müller, R., Franke, J., Henrich, D. et al. (eds.) Handbuch Mensch-Roboter-Kollaboration. Hanser, München (2019)
11. Schick, R.: Einsatz von Exoskeletten in der Arbeitswelt. Zbl Arbeitsmed **68**, 266–269 (2018)
12. Tiboni, M., Borboni, A., Vérité, F., et al.: Sensors and actuation technologies in exoskeletons: a review. Sens. (Basel. **22** (2022)
13. Gopura, R.A.R.C., Kiguchi, K., Bandara, D.S.V.: A brief review on upper extremity robotic exoskeleton systems. In: 2011 6th International Conference on Industrial and Information Systems, pp 346–351. IEEE (2011)
14. Voilque, A., Masood, J., Fauroux, J., et al.: Industrial exoskeleton technology: classification, structural analysis, and structural complexity indicator. In: 2019 Wearable Robotics Association Conference (WearRAcon), pp 13–20. IEEE (2019)
15. Kapandji, A.I.: Funktionelle Anatomie der Gelenke: Schematisierte und kommentierte Zeichnungen zur menschlichen Biomechanik, 6th edn. Thieme, Stuttgart (2016)
16. Nickel, S.: Das Ellenbogengelenk. Anatomie, Funktion & Erkrankungen (2022)
17. Bogue, R.: Exoskeletons-a review of industrial applications. IR **45**, 585–590 (2018)
18. Parent-Thirion, A., Biletta, I., Cabrita, J., et al.: 6th European working conditions survey: 2017 uppdate. Publications Office of the European Union, Luxembourg (2017)
19. Schiele, A.: Fundamentals of ergonomic exoskeleton robots. [s.n.], [S.l.] (2008)
20. Tsagarakis, N.G., Caldwell, D.G.: Development and control of a 'soft-actuated' exoskeleton for use in physiotherapy and training. Auton. Robot.. Robot. **15**, 21–33 (2003)
21. Esser, K., Kurte, J.: Kompendium 2018-Zahlen-Daten-Fakten der KEP-Branche: Transportaufkommen und durchschnittliches Gewicht
22. Argubi-Wollesen, A., Weidner, R.: Biomechanical analysis: adapting to users' physiological preconditions and demands. In: Karafillidis, A., Weidner, R. (eds.) Developing Support Technologies, vol. 23, pp. 47–61. Springer International Publishing, Cham (2018)
23. Grazi, L., Chen, B., Lanotte, F., et al.: Towards methodology and metrics for assessing lumbar exoskeletons in industrial applications. In: 2019 II Workshop on Metrology for Industry 4.0 and IoT (MetroInd4.0&IoT), pp. 400–404. IEEE (2019)

Camera Based Calibration of a Flexible, Reconfigurable Robotic System in Construction

Benjamin Kaiser, Timo König and Alexander Verl

Abstract

The use of robotics for prefabrication in timber construction is becoming increasingly important. Novel, modular, reconfigurable, and transportable manufacturing systems allow rapid adaptation to project-specific requirements. However, continuous reconfiguration of the system is a major challenge. Accurate calibration is a critical requirement for robotics. Typically, this requires expert knowledge and specialized measurement hardware. This issue arises in specific applications like timber prefabrication, where the availability of robotics experts is limited, and frequent reconfigurations necessitate calibration. To this end, this paper presents a camera-based calibration method for the described use case. The proposed method focuses on automatic calibration without the need for expensive measurement equipment or highly skilled personnel. The proposed method uses AruCo markers to determine the relative pose between the platforms. The automatic calibration process is performed by detecting the AruCo markers in the field of view of the camera of each robot and using them to estimate the pose of the platforms. The experimental results from a simulation evaluate the achievable accuracy and demonstrate the feasibility of the proposed method.

Keywords

Camera-based calibration · Reconfigurable manufacturing system · Timber prefabrication

B. Kaiser (✉) · T. König · A. Verl
Institute for Control Engineering of Machine Tools and Manufacturing Units (ISW), University of Stuttgart, Seidenstraße 36, 70174 Stuttgart, Germany
e-mail: benjamin.kaiser@isw.uni-stuttgart.de

S. Ihlenfeldt et al. (eds.), *Annals of Scientific Society for Assembly, Handling and Industrial Robotics 2023*, https://doi.org/10.1007/978-3-031-74010-7_25

1   Introduction

The construction industry faces major challenges due to a shortage of skilled labor and the prevailing low level of automation on the one hand, and the increasing demand for affordable housing and the reduction of emissions and resource consumption on the other [13]. Industrial robots can increase productivity and are increasingly present in timber prefabrication. A novel modular, reconfigurable robotic system from the Cluster of Excellence "Integrative Computational Design and Construction for Architecture" (IntCDC)[1] adapts to project-specific requirements [14]. The system is transportable and enables prefabrication close to the construction site, overcoming the transportation constraints of components and encouraging the use of local construction resources. The new approach facilitates the flexible configuration of robot cells from the system modules. However, precise calibration of the robots in relation to their environment is a fundamental requirement for effective and efficient task performance. With the IntCDC prefabrication system, the need for calibration after each reconfiguration presents a challenge. This involves determining the relative positions and orientations of the platforms, which cannot be placed precisely due to their size and weight. Unlike conventional robot cells, which only require calibration at the initial commissioning or after modifications, the considered system requires regular calibration due to its dynamic setup at different locations. In addition, the systems use case is project-specific timber construction on-site or at local, mostly small timber construction companies, where the availability of robotics experts and specialized measurement equipment such as laser scanners is limited. Thus, the IntCDC system needs an automated calibration process that does not require expert knowledge of robotics or metrology. Hence, this application requires a calibration procedure that is as accurate and automatic as possible.

The modular manufacturing system shown in Fig. 1 currently contains 40 and 20 ft long platforms equipped with industrial robots and have linear axes throughout their length. Additionally, a heavy-duty automated guided vehicle (AGV) connects the cells and increases the workspace. In this work, we use the cell layout shown in Fig. 1 consisting of one 40 ft (**FP1**) and one 20 ft (**FP2**) fabrication platform. We organize the remainder of this paper as follows. In Sect. 2 we review and discuss related work in the field of robot calibration. In Sect. 3, we present our method for the calibration of the IntCDC prefabrication system. This is followed by Sect. 4, where we evaluate the performance and achievable accuracy based on a simulation. We conclude and discuss our work in Sect. 5.

2   Related Work

Calibration is necessary to enable accurate and safe operation of industrial robots. This includes the absolute calibration of the robot arm itself and the calibration of the robot to

---

[1] https://www.intcdc.uni-stuttgart.de/.

Fig. 1 IntCDC prefabrication system

the environment. The absolute calibration of the robot arm is necessary to compensate for manufacturing tolerances and hence deviations from the idealized kinematics. Calibration with the environment aims to define a common coordinate system within a robot cell and is especially important when several robots interact within the same environment. This requires a common coordinate system that must be defined and calibrated. There are several methods described in related work about robot calibration. The use of laser trackers for robot calibration is common practice. The methods in [7, 9, 15] use this type of measurement device. Due to their high accuracy, they allow precise calibration. However, expert knowledge is required to successfully employ laser trackers. Also, these devices are expensive and are not widely available, especially in the use case considered here. Other methods use mechanical systems such as telescoping ballbars [10] or geometric measurement systems like drawstring displacement sensors [5]. The results show that both methods are suitable for calibration and the hardware used is significantly cheaper compared to the laser trackers. However, these methods also lead to a lower calibration accuracy. Due to the mechanical design, the method from [10] is particularly suitable for smaller industrial robots and small workspaces. Filion et al. use a portable photogrammetry system [4] and find that their method leads to a similar accuracy as laser trackers. These methods are time intensive and require expert knowledge and practical experience. Although cheaper than laser trackers, the systems used are still relatively expensive. Additionally, they are hard to automate. Many known methods use cameras as a measurement system. They are the most widely used systems, and as a result, there are numerous examples of their application in robot calibration. In [3] a camera mounted on the robot's hand and static chessboard targets are used. Balanji et al. changed the setup in [1] and used a fixed camera while moving a target with several ArUco markers. Other camera-based methods use 3D cameras to calibrate using a chessboard target [8] or a calibration sphere [12]. They find that their method can be conducted with minimal supervision and is suitable for automation. However, the achieved accuracy is relatively low compared to other methods.

3   Methodology

Considering our specific use case with the prefabrication system from IntCDC we choose to use a camera-based calibration approach based on the use of fiducial markers. The calibration problem at hand can be solved by using cameras and markers. We choose this approach for several reasons: firstly, related work has successfully used camera-based approaches with fiducial markers in robotics applications, and secondly, this approach is suitable for automation and doesn't require metrology experts and specialized measurement technology. However, the accuracy achieved with this method must be compared to typical tolerances in timber construction, which are much lower (1–5 mm) than in production technology.

3.1   Calibration Procedure

To solve the calibration problem, we attach a unique set of markers parallel to the linear axis on each container. By attaching these markers to the platforms, the cameras, that are in the robot's hand (eye-in-hand) can determine the relative position and orientation of the platforms. Figure 2 shows a robot platform with attached markers. We assume that a model like shown in Fig. 3 that represents the desired arrangement of the containers and the positions of their markers exists. This model specifies the expected position of the containers and their markers. In practice, we need to ensure that the containers and their markers are approximately where the model expects them to be. In this paper we choose AruCo markers as marker system because of their effectiveness demonstrated in various robotics applications as well as their position and angular stability [6].

   ArUco markers are small, square fiducial markers that can be easily recognized by cameras. The calibration process consists of moving to the markers and using the cameras to capture images of the ArUco markers. The position and orientation of the markers can be estimated using computer vision techniques, and the relative transformations between the platforms can be determined based on the estimated marker positions and orientations. By

Fig. 2 Robot with attached markers

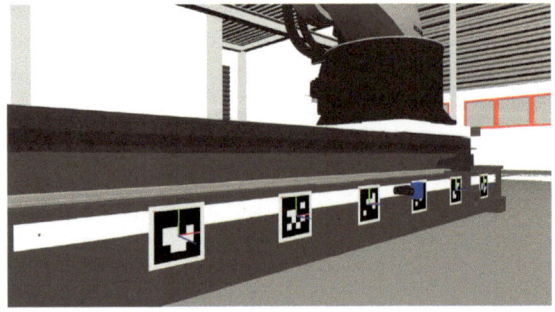

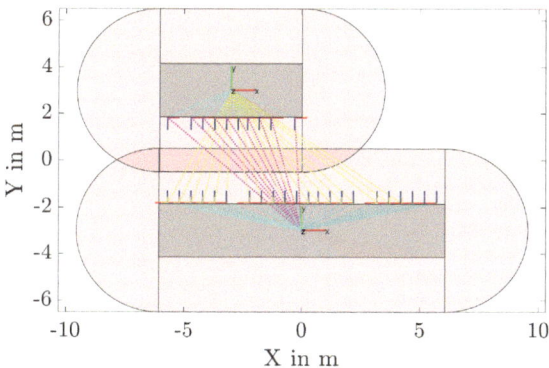

Fig. 3 Geometrical representation of the calibration scene

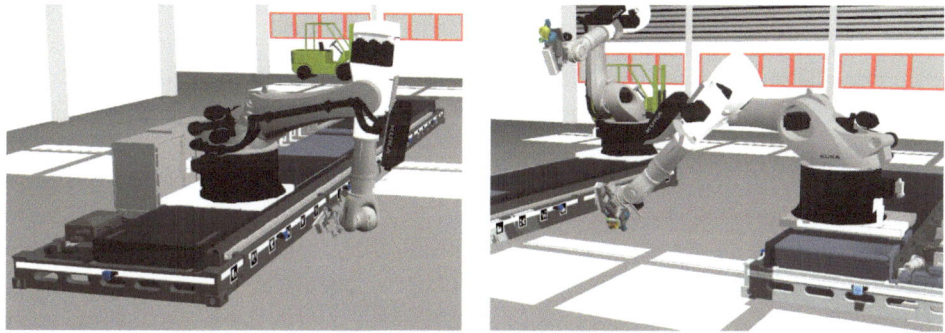

Fig. 4 Simulation model of the reconfigurable, robotic system

repeating this process multiple times, the relative transformations can be refined and the calibration can be improved.

The calibration procedure as exemplified in Figs. 3 and 4 is as follows. First, each robot detects the poses of its own set of markers P_{b_i} by searching for them within the expected area as defined by the model. The robots proceed to estimate the poses P_{t_i} of the markers on the other robot's platform. However, only the markers that fall within a specific minimum distance are considered detectable. This detectability is determined based on the robot's workspace and model, which dictate the extent to which the robot can reach the markers. Each marker is approached and detected several times from different poses. We compute the relative positions and orientations of the robot platforms in 3D space from the estimated positions of the markers. The detected marker poses undergo filtering, with RANSAC being employed to effectively eliminate outliers. From that we get the estimated transformations T_{b_1,m_1} and T_{b_2,m_2} between the robots $i = 1, 2$ base and their markers as well as T_{b_1,m_2} and T_{b_2,m_1} between the base and the target platforms markers. From that we derive the observed transformations $T_{b_1,b_2} = T_{b_1,m_2} T_{b_2,m_2}^{-1}$ and $T_{b_2,b_1} = T_{b_2,m_1} T_{b_1,m_1}^{-1}$ between the robot bases.

Fig. 5 Sampled poses on
parametric surface

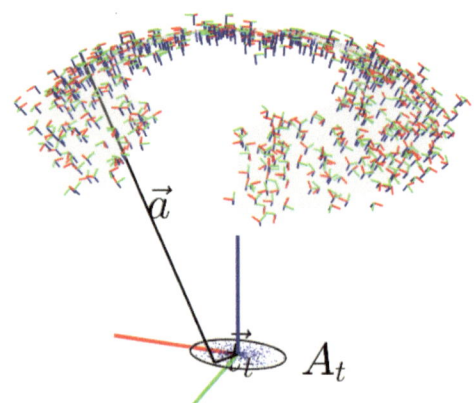

With $T_{b_1,b_2} = T_{b_2,b_1}^{-1}$ we get the set of all observed homogeneous transformations between platform **FP1** and platform **FP2** that we then use to estimate the transformation between both robot platforms. We do this by calculating the average translation and rotation of the observed homogeneous transformations.

To accurately calibrate the robot-world coordinate system, each marker must be observed several times from different poses. For that, as shown in Fig. 5, we randomly select points $\vec{p} = S(r, \phi, \theta)$ on the surface of a sphere S to position the camera for observing the markers. With $\vec{v} = -\vec{p} \times \vec{b}$ and $\vec{b} = [0, 0, 1]^T$ we use the skew symmetric cross-product matrix $[\vec{v}]_x$ to calculate the rotation $R = I + [v]_x + [v]_x^2 + \frac{1-c}{s^2}$, $s = \|v\|, c = a \cdot b$ that points the camera towards the target. This allows for the capture of marker images from multiple angles and positions, aiding in achieving precise calibration.

Accurate estimation of the marker position in the robot base coordinate system requires hand-eye calibration. The hand-eye calibration problem refers to the determination of the relative pose between a camera and a robot frame. This relative pose is critical in robotic systems that require accurate alignment of the camera with the end effector. In a typical hand-eye calibration, the end effector is moved to a series of known positions and orientations, and the corresponding marker poses are estimated in the camera image. The relative transformation between the camera and the robot frame is then estimated from these recorded positions by solving the hand-eye transformation problem $AX = XB$. The accuracy of the hand-eye transformation depends not only on the camera but also on the number and quality of the selected poses. We select the sampled poses according to the methods from [11] and use the dual-quaternion-based algorithm from [2] for calibration.

4   Validation

In this section, we validate the method described in Sect. 3. For this purpose, we use a simulation model of the reconfigurable manufacturing system, which corresponds to the real setup in our laboratory. The use of the simulation model, in this case, allows us to evaluate the achieved accuracy of the hand-eye calibration as well as the estimation of the transformation between the robot platforms since these transformations can be obtained from the kinematic model.

4.1   Kinematic Error Model

Industrial robots are not absolutely accurate. Deviations during the production of the robot bodies as well as dynamic and static influences such as gravity or deformations of the elements affect the accuracy. Deviations in the link lengths can lead to inaccuracies in the forward kinematics, which describes the mapping from joint space to cartesian space. These inaccuracies can result in deviations between the desired or measured and actual end-effector position and orientation. These deviations affect the accuracy of the hand-eye calibration and hence also the camera-based calibration by causing inaccuracies when estimating the marker poses. To include this effect in the validation, we employ a kinematic error model $e(J) = f(j_1, j_2, j_3, j_4, j_5, j_6)$ to reflect the influence of these errors on the accuracy of the camera-based calibration. To represent the error model, we introduce a random deviation to the robot's bodies lengths into the kinematic model. We sample the workspace of the robot in the joint space and compare this deviating kinematic model $f_e(J)$ with the nominal kinematic model according to the robot's datasheet. This results in the configuration-dependent error model $e(J)$ of the forward kinematics. We use this error to distort the simulation model. Table 1 shows the modified Denavit-Hartenberg parameters (mdh) of the nominal and the deviating kinematic chain. Figure 6 shows the resulting error model used in this paper. The resulting errors have a magnitude between 0.5 and 2 mm. This error model assumes that the

Table 1 Modified Denavit-Hartenberg parameters of the KUKA KR420-R3330

	Nominal model				Error model			
	θ	a	d	α	θ	a	d	α
j_1	0	0	1.045	0	0	0	1.04327	0
j_2	0	0.5	0	90	0	0.49864	0	90
j_3	0	1.3	0	180	0	1.30175	−0.00171	180
j_4	0	0.055	1.525	90	0	0.0534	1.52727	90
j_5	0	0	0	−90	0	0	0	−90
j_6	0	0	0.29	90	0	0	0.28949	90

Fig. 6 Error model in
workspace

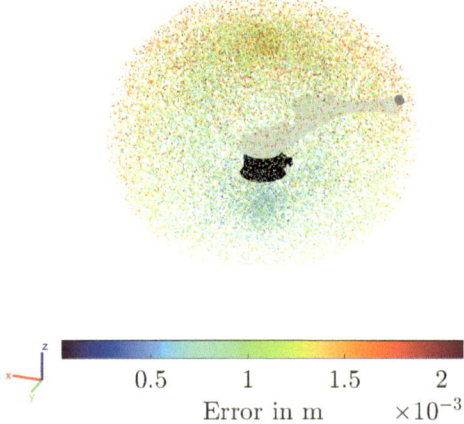

0.5   1   1.5   2

Error in m $\times 10^{-3}$

Table 2 Results of the hand-eye calibration

	y (m)	y (m)	z (m)	q_w	q_x	q_y	q_z	e_{pos} (m)	$e_{angular}$ (rad)
T_{mdl}	0.37749	−0.10537	0.18691	0.35355	−0.61237	0.61237	−0.35355	0	0
T_1	0.37729	−0.10549	0.18684	0.35340	−0.61236	0.61246	−0.35355	0.00024	0.00035
$T_{1,e}$	0.37746	−0.10557	0.18685	0.35342	−0.61236	0.61247	−0.35351	0.00021	0.00034
T_2	0.37736	−0.10547	0.18687	0.35334	−0.61238	0.61246	−0.35356	0.00017	0.00045
$T_{2,e}$	0.37736	−0.10543	0.18689	0.35336	−0.61238	0.61246	−0.35356	0.00014	0.00041

deviations in the link lengths are the only sources of error and does not consider dynamic errors or errors from gravitation or deformation.

To evaluate and compare the precision of the hand-eye and robot-robot transformations, we use the distance between the estimated and the nominal transformations obtained from the kinematic model. This distance is calculated with $|T_2, T_1|_2 = \sqrt{(T_{2,x} - T_{1,x})^2 + (T_{2,y} - T_{1,y})^2 + (T_{2,z} - T_{1,z})^2}$ and the angular difference between two quaternions is determined by $\theta = cos^{-1}\left(2\langle q_1, q_2\rangle^2 - 1\right)$.

4.2   Hand-Eye Calibration

In the first step, we perform a hand-eye calibration to get the transformation from the robot flange to the camera frame. We use a DICT5X5 Charuco board with 12×9 markers and a marker size of 6 cm. The camera has a resolution of 2048 by 2048 px. In total, 200 poses are automatically used to record samples. We performed this for both robots with and without added errors from the error model. For the Calibration we obtain the hand-eye transformations T_1, $T_{1,e}$, T_2 and $T_{2,e}$ using the algorithm from [2].

Table 2 presents the obtained transformations along with the position and angular distance in comparison to the nominal transformation T_{mdl} from the kinematic model. The results demonstrate that the hand-eye calibration is accurate to within 0.25 mm for the position and $0.4°$ for the orientation.

4.3   Robot Calibration

We proceeded to calibrate the transformation between the platforms, using markers of 18 cm in size. The platforms were equipped with 10 markers for **FP2** and 22 markers for **FP1**. The hand-eye transformations previously determined were used, with and without errors. Initially, each robot calibrated its own markers in relation to its base coordinate system. We recorded a total of 50 samples of different poses per marker, all pointing towards the marker.

Figure 7 shows the estimation error of the position and orientation of the self-observed markers of **FP2** for the case without (blue) and with (red) the error model.

Figure 8 shows the estimation error of the position and orientation of the self-observed markers of **FP1** for the case without (blue) and with (red) the error model. For both platforms, the errors are in the range of 1–2 mm or $0.28°$. It can be seen that the deviations are higher when the error model is used. The robots then detect the reachable markers of the other platform. Figure 9 shows the estimation error of the position and orientation of **FP1** markers observed by **FP2**. The figure shows that markers 22–28 could not be reached by the shorter

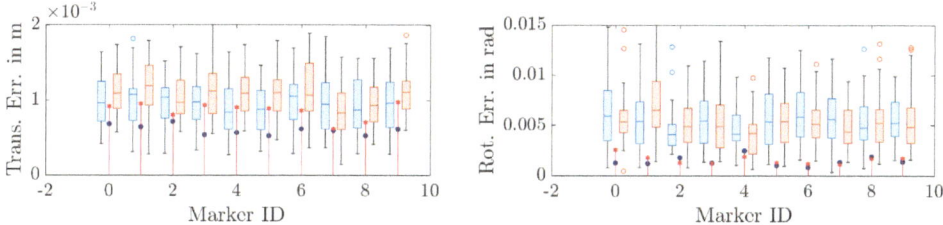

Fig. 7 Estimation error of the self-observed markers of **FP2**

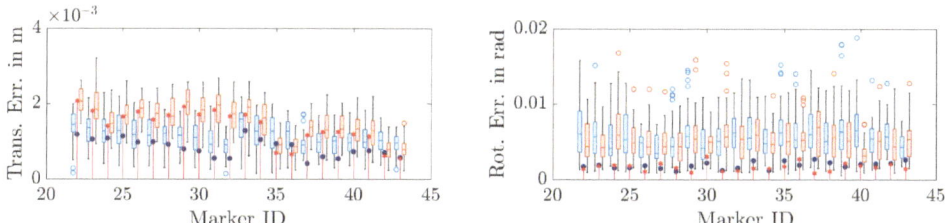

Fig. 8 Estimation error of the self-observed markers of **FP1**

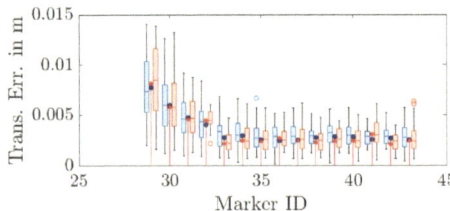 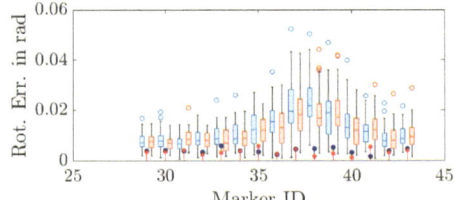

Fig. 9 Estimation error of **FP1** markers observed by **FP2**

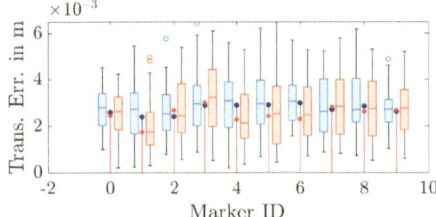 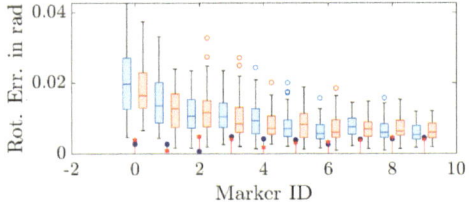

Fig. 10 Estimation error of **FP2** markers observed by **FP1**

Table 3 Results of the robot-robot calibration

	x (m)	y (m)	z (m)	q_w	q_x	q_y	q_z	e_{pos} (m)	$e_{angular}$ (rad)
T_{mdl}	3	−6	0	1	0	0	0	0	0
T_{12}	3.00347	−6.00383	−0.00138	0.99999	−0.00050	0.00015	−0.00025	0.00535	0.00116
$T_{12,e}$	3.00260	−6.00676	0.00056	0.99999	−0.00101	0.00012	−0.00026	0.00726	0.00210

FP2. Markers 29–32 are reachable within the specified tolerance of 1 m, but are still further away than desired, resulting in higher scatter and greater estimation errors.

Figure 10 shows the estimation error of the position and orientation of **FP2** markers observed by **FP1**. Here every marker is easily accessible with the exception of marker 0 which is located on the outer edge and can only be reached from one side of the long **FP1**, limiting the accessible positions for sampling.

The recorded poses are used to calibrate the transformation between the robot platforms. And lead to the results shown in Table 3.

The obtained accuracy, although relatively low, is still within the acceptable range, at least in the case where no error model is used. The achieved accuracy of about 5 mm, measured against the workspace of the cell with a distance of 6 m between the platforms, is sufficient for avoiding collisions and carrying out tasks like pick and place, nailing, or gluing. Nevertheless, the accuracy falls short of what is required for milling high-precision joints for wooden components.

5   Conclusion

In summary, the camera-based calibration approach using ArUco markers has demonstrated its suitability for the IntCDC system. This approach provides a straightforward and automated method to calibrate the system without requiring specialized measurement equipment or expertise in robotics. However, it should be noted that the achievable accuracy is lower than the typical tolerances for timber construction, and further research and evaluation are necessary. One potential method for enhancing calibration accuracy involves integrating supplementary sensors into the system, such as a 3D probe. Incorporating these sensors has the potential to enhance accuracy beyond the standard tolerances commonly present in timber construction. To accomplish this, the robots can be probed from various angles within the shared workspace, thereby building upon the outcomes of the camera-based calibration process. In future work, this method could be utilized to enhance the local navigation accuracy of the AGV between the robot platforms. Overall, implementing this method could improve the efficiency and efficacy of the IntCDC system, resulting in more sustainable and cost-effective timber construction practices.

Acknowledgements Supported by the Deutsche Forschungsgemeinschaft (DFG, German Research Foundation) under Germany's Excellence Strategy—EXC 2120/1—390831618.

References

1. Balanji, H.M., Turgut, A.E., Tunc, L.T.: A novel vision-based calibration framework for industrial robotic manipulators. Robot. Comput.-Integr. Manuf. **73**, 102,248 (2022). https://doi.org/10.1016/j.rcim.2021.102248
2. Daniilidis, K.: Hand-eye calibration using dual quaternions. Int. J. Robot. Res. **18**(3), 286–298 (1999). https://doi.org/10.1177/02783649922066213
3. Du, G., Zhang, P.: Online robot calibration based on vision measurement. Robot. Comput.-Integr. Manuf. **29**(6), 484–492 (2013). https://doi.org/10.1016/j.rcim.2013.05.003
4. Filion, A., Joubair, A., Tahan, A.S., Bonev, I.A.: Robot calibration using a portable photogrammetry system. Robot. Comput.-Integr. Manuf. **49**, 77–87 (2018). https://doi.org/10.1016/j.rcim.2017.05.004
5. Gan, Y., Duan, J., Dai, X.: A calibration method of robot kinematic parameters by drawstring displacement sensor. Int. J. Adv. Robot. Syst. **16**(5), 172988141988307 (2019). https://doi.org/10.1177/1729881419883072
6. Kalaitzakis, M., Cain, B., Carroll, S., Ambrosi, A., Whitehead, C., Vitzilaios, N.: Fiducial markers for pose estimation. J. Intell. Robot. Syst. **101**(4) (2021). https://doi.org/10.1007/s10846-020-01307-9
7. Lattanzi, L., Cristalli, C., Massa, D., Boria, S., Lépine, P., Pellicciari, M.: Geometrical calibration of a 6-axis robotic arm for high accuracy manufacturing task. Int. J. Adv. Manuf. Technol. **111**(7–8), 1813–1829 (2020). https://doi.org/10.1007/s00170-020-06179-9
8. Miseikis, J., Glette, K., Elle, O.J., Torresen, J.: Automatic calibration of a robot manipulator and multi 3d camera system, pp. 735–741 (2016). https://doi.org/10.1109/SII.2016.7844087

9. Nubiola, A., Bonev, I.A.: Absolute calibration of an ABB IRB 1600 robot using a laser tracker. Robot. Comput.-Integr. Manuf. **29**(1), 236–245 (2013). https://doi.org/10.1016/j.rcim.2012.06.004

10. Nubiola, A., Bonev, I.A.: Absolute robot calibration with a single telescoping ballbar. Precision Engineering **38**(3), 472–480 (2014). https://doi.org/10.1016/j.precisioneng.2014.01.001

11. Shi, F., Wang, J., Liu, Y.: An approach to improve online hand-eye calibration. In: Pattern Recognition and Image Analysis. Lecture Notes in Computer Science, vol. 3522, pp. 647–655. Springer, Berlin, Heidelberg (2005). https://doi.org/10.1007/11492429_78

12. Švaco, M., Šekoranja, B., Šuligoj, F., Jerbić, B.: Calibration of an industrial robot using a stereo vision system. Procedia Eng. **69**, 459–463 (2014). https://doi.org/10.1016/j.proeng.2014.03.012

13. Wagner, H.J., Alvarez, M., Groenewolt, A., Menges, A.: Towards digital automation flexibility in large-scale timber construction: integrative robotic prefabrication and co-design of the buga wood pavilion. Constr. Robot. **4**(3–4), 187–204 (2020). https://doi.org/10.1007/s41693-020-00038-5

14. Wagner, H.J., Alvarez, M., Kyjanek, O., Bhiri, Z., Buck, M., Menges, A.: Flexible and transportable robotic timber construction platform—TIM. Autom. Constr. **120**, 103,400 (2020). https://doi.org/10.1016/j.autcon.2020.103400

15. Xie, H., Pang, C.t., Li, W.l., Li, Y.h., Yin, Z.P.: Hand-eye calibration and its accuracy analysis in robotic grinding. In: 2015 IEEE International Conference on Automation Science and Engineering (CASE), pp. 862–867. IEEE (2015). https://doi.org/10.1109/CoASE.2015.7294189

Inspection in High-Mix and High-Throughput Handling with Skeptical and Incremental Learning

Robert Schimanek, Pinar Bilge and Franz Dietrich

Abstract

In the circular economy, remanufacturing success relies heavily on the accurate identification and classification of used products. Processes, which rely on worker experience, lack objective validation, leading to the potential for mislabeling and inaccurate damage assessment. This, in turn, results in additional manual evaluations and unproductive costs, which run counter to the principles of sustainability. To address these issues, machine learning and artificial intelligence have been applied with promising results. However, producing reliable large amounts of labeled data remains a challenge, as workers are susceptible to human error. This paper addresses process design in production. It proposes a new design to ensure that only valid labels enter the prediction models, reducing the potential for false labels in the dataset. Through this, the aim is to improve the accuracy and reliability of remanufacturing, ultimately reducing costs and mitigating the carbon footprint in the manufacturing, repair, and maintenance industries.

Keywords

Inspection • Machine learning • Incremental learning • Skeptical learning • Anomaly detection

R. Schimanek (✉) · P. Bilge · F. Dietrich
Technical University Berlin, Institute for Machine Tools and Factory Management, Chair for Handling and Assembly Technology Research, Pascalstraße 8-9, 10587 Berlin, Germany
e-mail: r.schimanek@tu-berlin.de

© The Author(s) 2025   305
S. Ihlenfeldt et al. (eds.), *Annals of Scientific Society for Assembly, Handling and Industrial Robotics 2023*, https://doi.org/10.1007/978-3-031-74010-7_26

1   Introduction

Raw material production and manufacturing contribute 50% to global greenhouse gas emissions and 90% to biodiversity loss and water stress [1]. One major aim of circular economy is the extension of product life through multiple life cycles, for instance via remanufacturing. Remanufacturing is already well established in the automotive sector. It involves dismantling, testing, repairing, and offering products as new, reducing demand for new products.

Accurate labeling is vital for optimal remanufacturing. It identifies remanufacturing need and action, optimizes value, and promotes sustainable consumption through transparent information of remanufactured products. However, the absence of objective validation of product labels can result in inaccurate labeling, leading to misidentification and improper damage assessment. This ultimately results in unproductive costs and undermines sustainability in circular economy.

In the pursuit of advancing circular economy, practitioners and researchers have employed the use of machine learning (ML) and artificial intelligence (AI) to enhance the reutilization of products [4, 6]. The development, training, and deployment of effective ML/AI models for product identification and defect detection can have a significant financial and environmental impact on industries that deal with manufacturing, repair, and maintenance. Some applications result in a reduction of carbon footprint and decreased expenditure for new product purchases [5]. In order to fully realize the benefits of these applications, it is crucial to develop a system that provides workers with an interactive interface displaying key product characteristics, thereby enabling informed decision-making in the implementation of these technologies.

For the multi-class classification of products in a production system, ML algorithms require substantial amounts of labeled data. However, producing accurate data can be challenging, as workers perform the labeling and are prone to human error. The presence of noise and mislabeled data can result in poor supervised ML outcomes [7]. The quality of the dataset, including the identification numbers of products, is crucial for delivering optimal results to support the inspection decision. Mislabeled products with incorrect identification numbers can have a significant impact on the training and results of the model.

Several strategies have been proposed to minimize false labels in datasets. These strategies encompass a range of techniques, such as data cleaning and outlier detection [12], learning annotations with skepticism [8], and interactive improvement of annotations [9, 10]. Additionally, there are methods for adjusting the confidence scores of predictions, and holistic confident learning approaches that focus on the uncertainty of individual class confidence [11]. Many of these strategies are applied post-data generation. In this paper, we take a new proactive approach by implementing principles in the production process to ensure that only accurate labels are incorporated into the models with the input data.

This paper addresses the issue of false input data and labels inspection in the production environment. Its aim is to reduce the mislabeling of returned products using (incremental) ML methods. First, the challenges in high-variant product inspection are analyzed in

chapter "Experimental Investigation of the Influence of Different Nozzle Exit Geometries on the Depositing of Strands in Fused Layer Modeling". Second, existing methods for the reduction of the reproduction of false labels are presented and discussed in Chap. 3. Third, a new process for in situ incremental learning for high-variant product inspection is presented in Chap. 4. Finally, Chap. 5 discusses the properties of the proposed methods and concludes on the achievable optimization.

2   Challenges in High-Mix Product Inspection

In this paper, we focus on the inspection of used products. The inspection process involves manual evaluation of the product features by employees. This includes identifying the product ID, classifying it based on its features, and recording the attribute data. The overall objective of this inspection process is to ensure that only high-quality products are selected for further processing, repair, maintenance, and resale.

The return of used automotive parts poses challenges in identifying them due to their variability from the exchange components. The dataset used in this paper includes over 500,000 inspections of used parts from 18 product types, such as starters, recorded at manual workbenches. The high-throughput inspection processes were conducted under regular industrial circumstances, where each inspection took, on average, 35 s. With more than 10,000 unique parts (classes) in this high-mix inspection case, multi-class classification models are required. The authors have highlighted the significant class imbalance present in the dataset, with some classes occurring over 10,000 times and others appearing less than ten times. Previous research has demonstrated the feasibility of using incremental learning methods to continuously learn from the first sample on a cleaned dataset significant class imbalance [5]. In this contribution, we present processes to implement incremental Learning (IL) in a production environment to minimize the need for off-site adjustments of the models during the product inspection process.

In ML, batch learning is a strategy where a model is trained with the current dataset and put into production when it achieves acceptable performance on a test set. This approach can require high computational effort and substantial memory for training, making it suitable when the dataset is large enough to fit in memory or when concept drift is unlikely.

IL is an alternative to batch learning in which the model is trained incrementally over time, with small datasets, as new data is generated. This approach enables the model to make predictions based on the accumulated data and reduces computational effort as the model only needs to be updated with new data rather than retrained from scratch. IL can be performed with limited resources and has a wide range of applications, including in autonomous systems where continuous learning from sensor values is required. IL has been extensively studied in the literature, demonstrating its effectiveness in real-world applications.

Effective deployment of ML models in production requires addressing the issue of incorrect inputs and class labels. Batch learning enables automated data cleaning through regular updates, while incremental learning demands careful consideration of in situ data quality.

To maintain high-quality training data, it is advisable to implement ongoing monitoring of incoming data in the process.

ML applications assume a static system but systems experience changes, known as concept drift, which affect the accuracy of predictions. Concept drift occurs due to the alteration of training data's statistical properties over time and can appear suddenly, periodically, gradually, or incrementally.

Figure 1a–c shows the business data input features from the inspection. Direct product-related and indirect delivery/customer features are included. Three input features' change over time is analyzed. A Modified Hockett-Sherby Law was used for regression, where x represents weeks and y unique values, with,

$$y(x) = -b - (b - a) \cdot e^{-c \cdot x^4}. \tag{1}$$

The unique entries of business data features 1 and 2 see a rapid rise at the beginning of the observation period, resulting in a 5-fold increase in feature 1 within 20 weeks. The increase then flattens and the trend is shown in Fig. 1a. The number of unique entries was 36 in the first week and reached 329 by the end of the observation period. The future trend is predicted through regression with a determination coefficient of 0.988, indicating a slow increase in the number of feature 1.

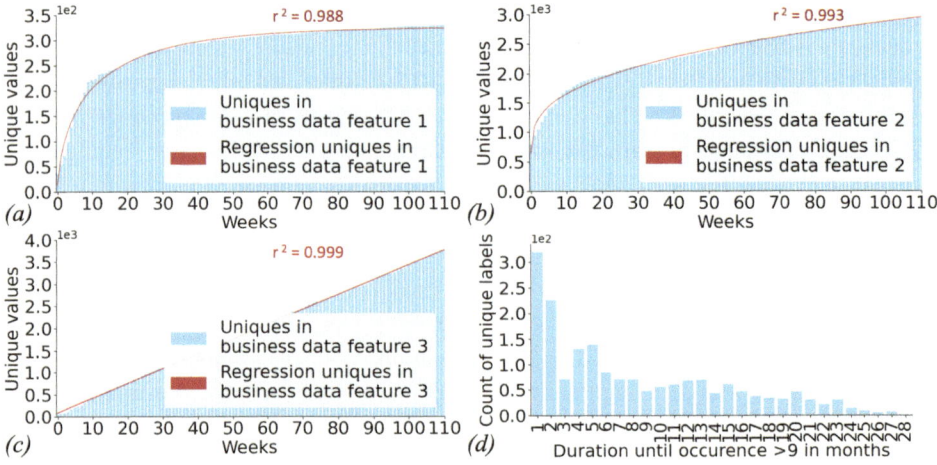

Fig. 1 Unique values in business data feature 1 (**a**), 2 (**b**), 3 (**c**) over time in weeks; Amount of samples until occurrence > 9 (**b**)

In the case of delivery data feature 2, a total of 2914 unique values were recorded in the dataset, with 94.6% of the values being valid. The reduction in the increase of Feature 2 is significantly smaller compared to Feature 1. Features 1 and 2 exhibit a continuous drift of input quantities. The regression analysis predicts a significant increase for Feature 1.

Business data feature 3 showed more than 3600 unique values recorded consistently. It has a proportional trend with the number of weeks, represented by a linear regression function. The trend in Fig. 1c shows a weekly increase of 25–50 new values, starting from 42 in the first week. The average time, until it becomes irrelevant in predicting labels, was calculated and shows over 99% of products retain relevance within 4 d. A small number took almost a month.

In previous research, we found that different learning thresholds, the number of samples per label, had varying impacts on batch and incremental learning [5]. Figure 1d shows in further analysis that with over 300 unique labels and a learning threshold of >9, recognition time was less than a month, which is low in consideration of a, e.g., weekly batch learning. This emphasizes the need for a holistic IL approach in high-variety product inspections, especially given the potential for faster recognition with tree-based ML models.

In addition to fast label additions, the uncleaned labels exhibit significant discrepancies. For example, in the case of the inspection of automotive parts for the remanufacturing process studied in this paper, approximately 8% of 435,956 labels were mislabeled which was identified by comparing them to an expert-created valid label master. The uncertainty in the labeling process is reflected in a study that distinguishes twice-labeled products without manual data cleaning. A comparison of 161,695 products labeled twice shows that the consecutively made labels only match 89%. This motivates stronger in situ intervention to protect IL methods from errors.

Product diversity presents a challenge to AI-enhanced inspection, with a growing number of product variants and evolving input data. Updating ML models to handle new labels and concept drift is necessary. Batch learning is impractical, consuming time and resources [5], while IL provides an efficient solution for updating the model with new data. However, errors in data collection can result in inaccurate IL predictions and costs. A data-centric, skeptical, and incremental approach is proposed for inspecting high-variation products.

3   Skeptical Learning Methods for Product Inspection

The accuracy of ML classification predictions depends on the quality of labeled data, especially in supervised learning. This section describes common techniques for improving data quality and validating product inspections. The methods' suitability for the proposed in situ process is also discussed.

3.1   Methods to Resolve Anomalies in AI-Enhanced Inspection

In ML modeling, data cleaning techniques for handling divergent data and labels are used. Some techniques address varying input variables but do not address mislabeled output labels in supervised learning. This section revisits methods for coping with mislabeled output labels and invalid input data.

Clustering is a common preprocessing step for ML model training. The approach starts as an unsupervised learning problem to form clusters for label prediction. Then, classification models are built based on these clusters. This approach may not be suitable for tasks with many classes, especially in applications requiring very high reliability such as automotive parts remanufacturing.

The implementation of **outlier and novelty detection** models through clustering or distances or angles can effectively detect anomalies in datasets and streams. The primary objective of anomaly detection is to automatically identify patterns in data that diverge from established norms. This can be achieved through the use of local outlier factor and local outlier probability methods [12]. Although we applied batch methods successfully in the inspection environment of the use case, it is important to note that batch-trained models required significant system memory, especially when dealing with a large volume of trained inliers. For instance, a model trained with 300,000 inliers required about 350 megabytes of RAM, which poses a challenge when handling millions of inliers in a year. In addition, these models had to be batch retrained and are thus not a valid option for a holistic IL approach to inspection.

Confident learning uses techniques such as pruning, counting, and ranking to handle label errors in ML models. It employs predicted probabilities and noisy labels to estimate the joint distribution and remove noisy data, producing a clean output [11]. However, these methods were designed for offline use and may require adaptation for IL.

Model confidence allows ML models to indicate their confidence in their result. Adjustment of the confidence score through temperature scaling or other confidence metrics allows for more elasticity of the deployed models in use [13]. These approaches depend on the data and modeling and are not well-suited for (incremental) trees but are more commonly used for artificial neural networks. Thresholds or detection of patterns in confidence can also indicate false predictions. Thus, confidence in label predictions is a crucial measure of agreement between human inspectors and AI, which is why it is included in our process.

3.2   Methods on Reduction of Mislabeling

There are various methods for reducing mislabeling. In the following, different concepts for the inspection of products are presented and discussed.

A basic idea for checking the labels is to compare the variables predicted during the inspection with the labels of the annotators. The prediction is not shown to the annotator, and if the AI prediction is highly reliable, it is used. However, if the annotator is very reliable, their label is preferred. This approach has drawbacks, as it requires performance measurement for the plant operator and may result in AI making decisions on product labeling, which is not permitted for one-to-one inspection. We thus apply a label hide element only in case of AI uncertainty or anomalies.

Skepitical learning described by Zeni et al. [8] consists of three components: A reference architecture, which includes the ML algorithm, user input, a knowledge component, and a supervised ML algorithm. The algorithm itself takes advantage of the crucial idea that this algorithm can be in one of three states, called train mode, refine mode, and regime mode. The final component is an algorithm for resolving conflicts that arise. In our process, we apply such a hidden conflict resolution.

With **interactive learning**, Zhang et al. propose asking the annotator for feedback whenever a conflict arises. Thus, in both train mode and refine mode, the user's input is no longer taken as ground truth [9]. These label processes are not designed for quality testing, where it is indispensable that the verification and confirmation of the label are done by human annotators. This paper takes up the previous research and supplements it to make it suitable for inspection. Teso et al. enhanced interactive learning by finding examples in the training data that conflict with each other and asking a human to correct them, which leads to better results [10]. Although this approach is sequential, it is unsuitable for our incremental learning approach which is intended to minimize mislabeling already in the production environment.

4   Design for Reduction of Mislabels in Inspection

This section presents a process design that reduces mislabeling and rework in product identification and damage assessment. In the proposed new design, we adopt skeptical learning principles to create an in situ fully incremental ML for inspection. The process design comprises feature aggregation, feature preparation, model evaluation, identification, manual labeling, and incremental learning in four phases. Figure 2 demonstrates the process flow of the new design.

The first phase, "feature aggregation", ensures that only valid features enter the ML models for label prediction. In the second phase, "feature preparation and model evaluation", anomaly detection determines if the inputs are novel parameters for predicting the desired label. The third phase requires "identification and manual labeling" in order to ensure human-validated labels for each product. In the final phase, the filtered samples are used for learning of the incremental models. The phases are explained in detail in the following paragraphs.

Feature aggregation: First, gathered features are checked for correctness. If the inputs are invalid, e.g., out of specification, then the process is not stable either for the inspection process, as these are required business data for subsequent processes like remanufacturing. The inspectors are required to redo inputs in such cases if the inputs are still not valid after the inputs have been repeated. The user is requested to enter the part into the database by an expert. The expert process is based on the procedure described by Schlüter et al. [4]. The expert can then decide whether the sensory is faulty or if the inputs are correct.

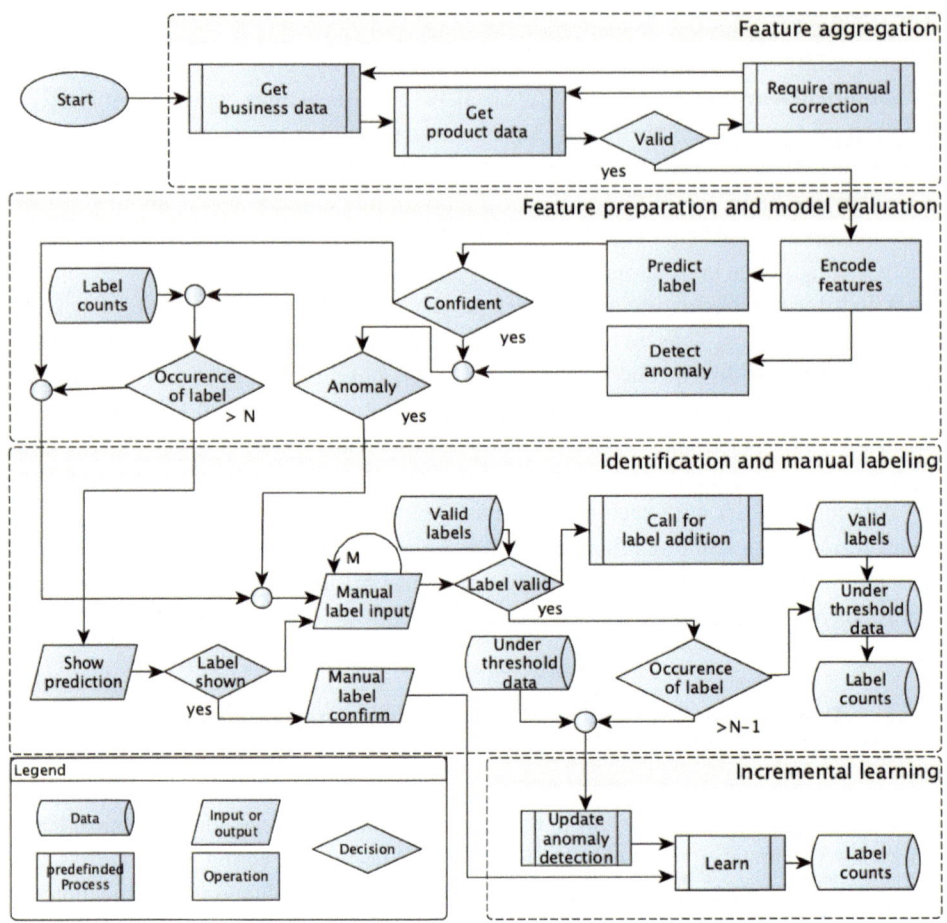

Fig. 2 Process flow for skeptical and class incremental learning

Feature preparation and model evaluation: Then, features are incrementally encoded and transferred to the IL model and anomaly detection. An IL model performs a prediction of the label. If the model is not confident with its predictions, e.g., over a fixed threshold, or an anomaly is detected, the prediction is not shown and manual label input is required. If the label occurs in the history more than N times, no anomaly and high confidence of the model are present. This prediction is shown to the worker. The proposed design ensures that only high-confidence predictions are shown to the worker for confirmation.

Identification and manual labeling: Human inspector is either called to confirm one label out of a suggestion list or manually enter a requested label. If the label is not in the suggestion list, the inspector can input a new label suggestion. If the inspector enters a number, it is checked on validity against a database of valid labels. These valid labels are created by experts in case a call is made. If the label is valid or confirmed after suggestion, the sample enters the training pool. If the label is invalid, the inspector is asked to re-enter the label. This process is repeated M times until an expert is called for identification and addition of the part information to a valid pool under threshold data. If the occurrence of label exceeds $N - 1$, the data and in case of additional under threshold data is allowed to enter the IL phase.

Incremental learning: The classification and anomaly detection models are incrementally updated for each entering sample. The anomaly detection is only incrementally updated for samples previously flagged as anomalies. Only fully completed measurements are considered for the class incremental learning. The class incremental model and the anomaly detection model can start to learn from the first sample. For a stable operation, it is advised to let the models enter productive operation after a successful warm-up phase.

5   Discussion and Conclusion

This paper proposes a new process design for product inspection that uses skeptical learning and incremental learning. The design is carefully crafted to maintain the validity of features and labels, for instance, by including a feature aggregation stage that allows only valid features into the ML models. Human validation of labels is also included in the identification and manual labeling phase to ensure the accuracy of the training process.

The paper highlights challenges in inspecting high-variation products for remanufacturing, with a need for incremental learning and ongoing data monitoring to maintain high-quality training data. Concept drift and product diversity require updating ML models with new data. in situ intervention is necessary to prevent errors in data collection and labeling that can result in inaccurate predictions and costs. The proposed skeptical and incremental approach can ensure accurate product identification and damage assessment.

The use of skeptical learning principles, specifically the adoption of anomaly detection and the requirement of manual labeling, ensures that only confident predictions are used in the training process. This reduces the risk of mislabeling and rework in product identification and damage assessment. The use of incremental learning also allows the models to continuously adapt and improve with more data, ensuring a high prediction accuracy in the inspection process.

The new in situ incremental learning process design enables the reduction of mislabeling and rework in product identification and damage assessment. The design adopts skeptical learning principles to create an in situ fully incremental ML model for inspection. The process design comprises of multiple phases including feature aggregation, feature preparation and model evaluation, identification and manual labeling, and incremental learning. Each phase ensures that only valid and high-confidence data are used to update the ML models, ensuring accurate predictions. The expert process is in place to validate inputs and labels. A manual labeling phase ensures human-validated labels for each product. The models are updated incrementally, ensuring that the models continue to learn and improve over time.

The proposed design is suitable for product inspection as it provides a robust and efficient solution to reduce the risk of mislabeling and rework. The incorporation of skeptical learning principles and incremental learning makes it an ideal choice for real-time product inspection applications, where accuracy is of the utmost importance. Additionally, the design is scalable, making it suitable for use in various industrial settings and for different types of products. In conclusion, the proposed process design is a suitable solution for product inspection due to its robustness, and efficiency.

The proposed inspection process does not require specialized expertise to develop, maintain, and improve, which is cost-effective. However, the input of experts is still necessary for valid features and labels, allowing for control of the AI-enhanced processes.

The proposed inspection process also has some limitations. The system's performance can be affected by drastic changes in the product design, packaging, or labeling from operation or getting products from another domain. Such changes can increase the risk of false predictions. In addition, the data is disregarded after IL. Relearning from historic data is not possible yet. The success lies in the quality of the feature and label validation algorithms and the expertise of the human annotators.

Further research and validation could address some of the limitations of the proposed design, including the quality of the in situ feature and label validation, and improve the system's overall performance as well as its elasticity to changes in the product spectrum or domain. With ongoing research, the proposed product inspection process has the potential to become a valuable tool for remanufacturers seeking to reduce cost and waste.

Acknowledgements The German Federal Ministry of Education and Research (BMBF) funded this research under the Resource Efficient Circular Economy—Innovative Product Cycles (ReziProK) program. This article is related to the collaborative project entitled "Sensory acquisition, automated identification and evaluation of old parts based on product data and information about previous deliveries" (EIBA, project ID 033R226). The TU Berlin team also acknowledges the support of Basel Alchick, Karin Betscher, Maurice Hoppe, Hanxiong Li, Sebastian Wenger, Zeming Yuan, Jiayi Zhu.

References

1. IRP: Global Resources Outlook 2019: Natural Resources for the Future We Want. Nairobi, Kenya (2019)
2. Chauhan, C., Parida, V., Dhir, A.: Linking circular economy and digitalisation technologies: a systematic literature review of past achievements and future promises. Technol. Forecast. Soc. Change (2022)
3. Noman, A., Umma A., Pranto T., Haque A.K.M.: Machine learning and artificial intelligence in circular economy: a bibliometric analysis and systematic literature review (2022)

4. Schlüter, M., Lickert, H., Schweitzer, K., Bilge, P., Briese, C., Dietrich, F., Krüger, J.: AI-enhanced identification, inspection and sorting for reverse logistics in remanufacturing. In: Proceedings of CIRP (2021)
5. Schlüter, M., Schimanek, R., Koch, P., Briese, C., Chavan, V., Bilge, P., Dietrich, F., Krüger, J.: Green incremental learning—energy efficient ramp-up for AI-enhanced part recognition in reverse logistics. In: Proceedings of CIRP (2023)
6. Kaiser, J., Lang, S., Wurster, M., Lanza, G.: A concept for autonomous quality control for core inspection in remanufacturing. In: Proceedings of CIRP (2022)
7. Nettleton, D.F., Orriols-Puig, A., Fornells, A.: A study of the effect of different types of noise on the precision of supervised learning techniques. Artif. Intell. Rev. **33**(4), 275–306 (2010). https://doi.org/10.1007/s10462-010-9156-z
8. Zeni, M., Zhang, W., Bignotti, E., Passerini, A., Giunchiglia, F.: Fixing mislabeling by human annotators leveraging conflict resolution and prior knowledge. Proc. ACM Interact. Mob. Wearable Ubiquit. Technol. **3**(1), 1–23 (2019). https://doi.org/10.1145/3314419
9. Zhang, W., Passerini, A., Giunchiglia, F.: Dealing with mislabeling via interactive machine learning. Künstl. Intell. **34**(2), 271–278 (2020). https://doi.org/10.1007/s13218-020-00630-5
10. Teso, S., Bontempelli, A., Giunchiglia, F., Passerini, A.: Interactive label cleaning with example-based explanations advances in neural information processing systems. In: Advances in Neural Information Processing Systems, Curran Associates, Inc., December 2021, vol. 34, pp. 12966–12977. https://doi.org/10.1007/s10462-010-9156-z
11. Northcutt, C., Jiang, L., Chuang, I.: Confident learning: estimating uncertainty in dataset labels. J. Artif. Intell. Res. **70**, 1373–1411 (2021). https://doi.org/10.1613/jair.1.12125
12. Breunig, M., Kriegel, H.-P., Raymond, T. Ng., Sander, J.: LOF: identifying density-based local outliers. In: International Conference on Management of Data, New York, NY, USA, Mai 2000, Bd. 29, S. 93–104. https://doi.org/10.1145/335191.335388
13. Guo, C., Pleiss, G., Sun, Y., Weinberger, K.Q.: On calibration of modern neural networks. In: 2017, Proceedings of the 34th International Conference on Machine Learning, vol. 70, pp. 1321–1330, 10, Sydney, NSW, Australia

Author Index